Engineering Textiles: Research Methodologies, Concepts, and Modern Applications

Engineering Textiles: Research Methodologies, Concepts, and Modern Applications

Editor

Jamshed A Khan

scitus
academics

Engineering Textiles: Research Methodologies, Concepts, and Modern Applications
Edited by **Jamshed A Khan**

ISBN: 978-1-68117-458-7
Library of Congress Control Number: 2016937634

SCITUS Academics LLC,
www.scitusacademics.com
Box No. 4766, 616 Corporate Way,
Suite 2, Valley Cottage,
NY 10989

Printed in United States of America on Acid Free Paper

Preface

Textile Engineering, also referred to as textile technology, studies various engineering and science principles. Textile technology can cover a wide range of subjects, from chemistry and physics in textile production to textiles for apparel and interior design. These principles are then applied to all types of textile fabrics and fiber-based yarns. The disciple studies extensively the chemical and physical principles, which are then used to study and analyze the behavior of polymers involved in the formation of textile fibers. In addition to textile engineering, the working mechanics of this industry include the design, manufacture, distribution and sale of apparel and clothing. Textile engineering is an innovation field, as it is a science area. It involves creative research on new techniques of production and improvisation. Textile manufacturers not only aim at developing and finding new techniques for producing new fabrics of better quality, but also at continuously improving existing textile fibers.

This title offers the textile scientific community an opportunity for critical, authoritative assessments of progress in textile engineering. The book covers a variety of important concepts and developments in the field, including: non-woven applications in textile engineering; textile waste treatment for use in emulsion rubbers, etc.

Table of Contents

CHAPTER 1

The sustainable future of the Scottish textiles sector: challenges and opportunities of introducing a circular economy model

Lynn Wilson

Zero Waste Scotland, Ground Floor, Moray House, Forthside Way, Stirling FK8, 1QZ, UK

ABSTRACT

Case description

Zero Waste Scotland introduced the concept of the circular economy to the Scottish textiles sector at events throughout 2013 to 2014. In April 2014, it commissioned research by independent consultants to examine the academic and industrial textile landscapes in Scotland, including developments in technical textiles and research into innovation in textile design and examples of circular economy models.

The research identified a number of initiatives, including projects producing an alternative to denim and one developing cavity wall insulation from processed natural fibres. It made recommendations to Zero Waste Scotland about shaping the future landscape of textile innovation in Scotland and also offered examples of the circular economy from Scandinavia that might be applicable.

Discussion and evaluation

The implementation of the circular economy into Scotland's textile sector, underpinned by the aforementioned initiatives, can learn too from its textile and fibre heritage. Zero Waste Scotland is implementing an action plan which will offer support to the textile industry exploring 'closed loop' manufacturing, as well as funds for fashion designers to explore concepts such as zero-waste pattern design, luxury apparel from alternative textiles such as recycled polyethylene terephthalate and natural fibres such as nettle. A master class skills programme, delivered by leading UK and international experts, will bring together industry, academia, and higher education professionals to engage in learning and information exchanges about the circular economy. This paper presents Zero Waste Scotland's role and the research findings.

KEYWORDS

Circular economy; Scottish textiles sector; Closed-loop manufacturing

1. BACKGROUND

Zero Waste Scotland (ZWS) was initially a programme of the Waste Resource Action Programme UK (WRAP UK). On 1 July 2014, ZWS was formally launched as an independent Scottish not-for-profit company.

On 11 February 2014, WRAP UK launched the Sustainable Clothing Action Plan, SCAP 2020 Commitment. Sixty-five signatories and supporters of the SCAP 2020 Commitment have pledged to reduce the carbon, water and waste footprints of clothing they supply or receive in the UK by 15% by 2020. Six of these companies are from Scotland.

ZWS supports SCAP 2020 and WRAP UK's Love Your Clothes campaign 2014 to 2016. In 2014, its key partners were John Lewis and Young Scot, the Scottish Youth information portal for 11 to 26 year olds. Both organisations supported the ZWS organised 'Spring Clean Your Wardrobe Weekend'.

ZWS' textiles portfolio and work programme for 2014 to 2016 is managed by its Circular Economy Team, which also covers Oil and Gas Decommissioning and Circular Economy Business Models. It identified various economic reports available from Scottish Enterprise (SE) and Scottish Development International (SDI) on the textiles industry in

Scotland[a]. However, there are few reports related to circular economy innovation in textiles in Scotland and evidence of closed-loop manufacturing.

In April 2014, ZWS commissioned independent research consultants to undertake an examination of the academic and industrial textile innovation landscapes in Scotland, including developments in technical textiles and research into innovation in textile design.

The aim of the commissioned research was to contribute to informing the direction and development of the 2014 to 2016 work programme delivered by ZWS' Circular Economy Team with the textiles sector in Scotland. The report, submitted to ZWS August 2014, examined the circular economy in relation to the textiles sector in Scotland.

The circular economy is, of course, more than just another name for 'reuse'. The Ellen McArthur Foundation says of it:

The circular economy is a generic term for an economy that is regenerative by design. Materials flows are of two types, biological materials, designed to reenter the biosphere, and technical materials, designed to circulate with minimal loss of quality, in turn entraining the shift towards an economy ultimately powered by renewable energy. [b]

The report submitted to ZWS found that there are pockets of innovation in the textiles sector in Scotland that are contributing to a circular economy; however, mainly, the driver is a commercial business goal rather than corporate social responsibility. This may reflect the relative weakness of the Scottish textile industry over the last decade and the lack of an overarching strategic vision for the industry as well as one that is drawing on, and support of, academic research.

Specifically, the report noted variations within the Scottish textiles industry's four sub-sectors: design, heritage/aesthetic, manufacturing, and technical textiles. It highlighted the following:

1 Examples of innovation or resource-efficient practice already happening in Scotland, such as (a) in its strong technical textiles - where opportunities may exist for closed-loop manufacturing and (b) computer-aided design.

2 Scotland should strengthen what it already does well and bolster its sustainability activities by linking to provenance, traceability, durability and quality.

3 But it also said that 'no true examples of circular economy business models in Scotland were identified'. To that end, it recommended that the search should be broadened to Europe.

4 It is hoped the creation of the Textiles Innovation Council will 'be instrumental in driving change towards a circular economy'[c].

The report, together with ZWS' analysis of its work with the textiles sector up to 2014, highlighted the challenges, opportunities and innovations that can make the Scottish textile industry more materially circular. For example, ZWS has been investigating innovative ways to protect and conserve resources and develop closed-loop manufacturing and high-value products and support services which contribute to a sustainable circular economy for the textiles sector.

In a broader context, this work is part of ZWS' support of the Scottish Government's Zero Waste Plan which has set ambitious targets to achieve 70% recycling and maximum 5% to landfill by 2025 for *all* Scotland's waste. 'Safeguarding Scotland's Resources', launched on 9 June 2013, is the Scottish Government's programme to reduce waste and create a more productive and circular economy[d] with one clear benefit: 'There is a potential saving of £2.9 billion through straightforward resource efficiency that this programme will help tap.' The Scottish Government has sealed its commitment by being the first government to join the Ellen MacArthur Foundation CE 100[e].

2. CASE DESCRIPTION

The research submitted to ZWS has informed its work programme for 2014 to 2016 (published internally August 2014) in terms of themes, policy, master classes, training events, campaigns and engagement with both industry and academia. The key objectives of the programme include the following:

To scope the potential for fibre processing in Scotland and early enquiry around establishing a sustainability standard for Scotland

To showcase and pilot further industry models of a circular economy (CE) and resource-efficient good practice

To seek greater engagement of industry *with* academia in terms of CE practice and theory

ZWS' textiles portfolio includes the following:

- The Scottish textiles sector
- Bulky textile waste - carpets and mattresses, public sector uniform procurement

- Commercial and consumer behaviour change

ZWS feels that the textiles sector has a significant part to play since it is the seventh most important contributor to the Scottish economy, with an annual turnover of £956 million. Exports of Scottish textile products are valued at £375 million. There are over 570 companies directly employing around 9,000 people across Scotland, with 55% of Scottish textile companies having fewer than 10 employees.

In areas such as the Scottish Borders, 50% of employment is within the textile industry. Recent investments from international global brands such as Chanel[f] demonstrate the high value and lucrative position of the Scottish textiles sector. Chanel announced at the beginning of 2014 that they would create 100 new jobs over the next 3 years at their plant in Hawick.

Global economic trends too have had a positive impact on the luxury export side of the Scottish textiles sector. In October 2013, the sector reported a level of growth 12% above the 2020 target and as a result, it has revised its targets to achieve between £1.2 and £1.5 billion in turnover growth by 2020. These revised figures recognise the significant gains to be made in international trade, and a 50% increase in exports is predicted by 2017[g]. Sixty-four percent of textile production in Scotland is for luxury export goods[b].

This increased productivity could contribute to the global waste and environmental impacts that textiles and apparel production create but it can be reduced through the design and manufacturing of products that follow circular economy guidelines or the creation of products with a low carbon impact. Scotland is already developing key products and customers in this market, such as hand-knitted apparel from Fair Isle and the Harris Tweed Authority, which reported an increase of 25% in sales in 2013.

The carbon and water impacts of the textiles supply chain and production are also significant, particularly within the luxury Scottish market, which relies heavily on importing all raw materials such as cashmere from Mongolia and China. Scotland imported £1.4 million of the highest-quality cashmere from Mongolia in 2012/2013, and the political importance of this was highlighted by a visit from a Mongolian government delegation to Scotland hosted by the Scottish Government in 2013[i].

The implementation of the circular economy within the textiles sector not only relies on the aforementioned initiatives and ZWS' work programmes but could also benefit substantively from reflecting on how Scotland's four sub-sectors that the ZWS-commissioned report identified, mentioned above, could be integrated into such an approach.

ZWS introduced the concept of the circular economy to the Scottish textiles sector through a series of events during 2013 to 2014 which were well received by both industry and academia.

These events were delivered in partnership with the Scottish Textiles and Leather Association (STLA) and the Scottish Textiles Academic Group, as well as an introductory event related to the SCAP, delivered in partnership with WRAP. Matters discussed included the concept of the circular economy and the challenges and opportunities ahead for textile design and manufacturing at a national and global supply-chain level, as well as assessing the SCAP Commitment targets of 15% reduction in carbon, water and waste by 2020.

The STLA events also gave the industry the opportunity to share their experiences and ideas as to how to make the industry more efficient. This included a case study delivered by the Scottish Leather Group presenting their 11 years' experience of working towards sustainability.

Feedback from these events identified the need for training and skills development in resource efficiency, sustainability and global design and manufacturing models such as closed loop. Training bodies such as Skills Development Scotland and the Scottish Textiles Academic Group, and other sector funding bodies such as Scottish Enterprise, have welcomed the opportunity ZWS is offering the textiles sector, via its master classes, to introduce new skills and global textile knowledge on the circular economy.

The ZWS-commissioned research also provided evidence of related change in other nations. Specifically, ZWS asked the researchers to show how Scotland could learn from, and adapt, closed-loop models throughout the design, manufacture and supply chain and to identify if there were any early adopters or textiles businesses currently working towards such models. The researchers were tasked with reviewing and referencing previous research by partner organisations such as WRAP UK and the Ellen MacArthur Foundation, and ongoing academic research that would help ZWS identify key interventions it could make to embed circular economy and resource-efficient business models in the textiles sector in Scotland.

The brief for the researchers was to identify three key areas of research:

1. To provide a summary of the academic and industrial textile innovation landscape in Scotland, including developments in technical textiles as well as research into innovation in textile design in Scotland

2. To present case studies that reflect the landscape of Scotland's textile industry in relation to innovation in resource-efficient and waste reduction practices that are, or could be, developed as models of good practice in the circular economy

3 To present recommendations for future areas of support in terms of resource efficiency and closed-loop textiles in Scotland

The research was required to conclude with recommendations for future engagement and demonstrate how ZWS can help shape the future landscape of textile innovation in Scotland in relation to the circular economy and closed-loop textile manufacturing. The research needed to align with, and reference evidence of, the global trends and examples of current closed-loop practice and circular economy textile models in other countries, including the UK, that could be applicable to Scotland.

The report's recommendations have informed ZWS's work programme 2014 to 2016 for the textiles sector on how to shape the future landscape of textile innovation in Scotland, and it offered relevant examples of the circular economy from Scandinavia and advised developing links with work being undertaken there. It identified nine companies as potential case studies. This paper presents four of those nine case studies and outlines developments with Scandinavia.

With the exception of one company, Kraft Architecture, whose objective is to recycle waste textiles into cavity wall insulation, no true examples of circular economy business models in Scotland were identified by the researchers. They then proposed that the project scope be broadened to Europe, arguing that the business environment is similar in landscape to Scotland and that any activities discovered might be transferable to the Scottish textile industry. They identified too ongoing projects that are exploring a number of initiatives, including producing an alternative to denim, an initiative first supported by WRAP UK and now being researched at Heriot-Watt University, Edinburgh.

The four case studies

Dawn Ellams - PhD Candidate, Heriot-Watt University

This research was sponsored by Lenzing AG to make 'denim' from TENCEL (Lenzing AG, Lenzing, Austria) using digital printing to impart the stone-washed look[j]. Given the high environmental impact of stonewashing denim, this could be a promising circular economy project if it is possible to commercialise it within Scotland. At present, the denim supply chains are located, typically, in countries such as Turkey, China or Pakistan. This work and research was further funded by WRAP through the Challenge Fund.

Jeans were created that have a cotton-like quality but only use one fifth of water, energy and chemicals needed to manufacture conventional jeans.

Manufacturing one pair of cotton denim jeans uses, on average, 42 litres of water and is energy intensive. Conventional denim production methods can also require up to 15 dyeing vats and an array of harmful chemicals [k].

The research identified several areas within the manufacturing process which offer opportunities for saving water and reducing carbon emissions.

Kraft Architecture - entrepreneur/inventor

Both virgin and waste wool and textiles can be converted into building insulation products, and there have been a number of attempts to set up businesses in Scotland supplying this as a product. Notable examples include Thermobond from Kraft Architecture (280 High Street, Glasgow G4 OQT, UK) and Earthwool® from Knauf Insulation (PO Box 10, Stafford Road, St Helens, Merseyside, WA10, UK). Generally sustainable insulation products fail to compete on price with mineral wool, which is also principally derived from waste materials, and fail to compete on performance with engineered insulation materials. The resulting market is very small and tends to be limited to applications like historic property restoration where technical and commercial factors are less important than heritage.

Kraft Architecture is a sole practitioner company, although this is augmented by part-time help as the need arises. The business focuses on low-carbon and low-energy housing and sustainable product development. The company started in 2009 and was awarded a SMART: SCOTLAND Feasibility Award in 2010 to develop the product from the concept.

One of the closed-loop products developed by Kraft Architecture is a natural fibre insulation that retails under the trade name 'Thermobond'. The insulation manufacturing process can work with a wide variety of waste streams including cellulose, cotton, wool, wood and polyester and so it can create insulation materials tailored for different applications. A research project undertaken by Glasgow Caledonian University measured the thermal conductivity of a range of waste products to verify their potential for use as insulation[l]. All the selected materials compare favourably with glass fibre on this performance metric. Keeping the proportion of wool at 60% or greater ensures that product flammability and resistance to smouldering complies with the British Standard BS5803-4:1985.

The manufacture of Thermobond requires 90% less energy compared to typical mineral wool insulation. In use, its durability and moisture control properties allow for a higher level of building fabric performance over the life of the building. It is also soft to touch and non-irritant, making handling and installation simple. Thermobond is treated with non-toxic, naturally occurring boron salts to heighten its resistance to vermin and insects over the life of the product.

Samples of Thermobond have been produced in sufficient quantities for use in demonstration installations. Kraft Architecture has secured investment of about £500,000 from the recently created Construction Scotland Innovation Centre[m] and further funding is being considered by ZWS. This will include suitable machinery to make insulation materials from recycled materials in its equipment set. Whilst this will not provide a volume manufacturing facility, it will provide a route to further product development, as well as sufficient material for the occasional demonstration installation.

Turnberry Rug Works-Johnstons of Elgin - a closed-loop approach

Turnberry Rug Works and Johnsons of Elgin are working collaboratively to achieve the up-cycling of cashmere.

Established in 1991, Turnberry Rug Works, based in Ayrshire, is a small-to-medium enterprise, well known for its rug-making expertise. Its products are aimed at the higher end of the market, being both handmade and often bespoke. Although most of its output uses merino wool, it can incorporate stones, leather, seashells, steel and glass, as well as silk, linen, worsted wools and felted yarns into the designs.

Johnstons of Elgin is a long-established knitwear manufacturer (since 1797) that specialises in cashmere products. The company is vertically integrated, with the incoming feedstock being cashmere sourced exclusively from Inner Mongolia. The company has both a knitting plant in Hawick and a weaving plant in Elgin.

Weaving and, to a lesser extent, knitting processes, produce waste that cannot easily be recycled. A common type of waste is selvedge, which is the semi-finished edge that is produced to stop the fabric from unravelling or fraying. The production of scarves produces long lengths of this material. Following a chance meeting, Johnsons of Elgin supplied samples of its scarf selvage to Turnberry Rug Works which, after some development work on the tufting machines, found it was able to use the selvage to manufacture rugs. Producing a rug using virgin cashmere would be a very expensive proposition; however, by using selvage, a luxurious product can be produced using material that would otherwise have been sold as waste. The rugs are currently in production and sold under the Ecochic brand.

Keela International

Keela International Ltd specialises in the manufacture of advanced clothing for outdoor wear. Customers include consumers, mountain rescue teams, police and military forces. This is a large but highly competitive market sector estimated to be worth about €5 billion in Europe alone. Because both the USA and European markets are saturated, effectively, innovation is important to help sustain market share. Manufacturer claims that can be used to promote sales include ‘free of toxins, environmentally friendly materials and sourcing, temperature control, reduced friction, water and wind resistant, moisture management'. The differentiating characteristics of Keela include expertise in fabric technology and garment construction.

Keela was awarded the Queen's Award for Enterprise: Innovation for its System Dual Protection fabric. This consists of two fabrics with very different characteristics, separated by an air gap. Originally developed for a military customer, it is now part of Keela's consumer product range. Independent research, conducted by Leeds University and Niederrhein University of Applied Sciences, substantiated the claims made for the material in terms of its ability to keep the wearer dry from external water and comfortable by regulating internal moisture.

Most modern outdoor clothing is made from materials that are difficult to dispose of, except in landfill; therefore, the most sustainable option is to maximise the life of the garment. This requires the customer to follow specific instructions on washing and reproofing and adds to the unfavourable environmental impact through the use phase of the textile.

As part of the company's drive towards sustainability, they have committed to SCAP and are the only apparel manufacturer member from Scotland.

International relations - Scandinavia

The researchers suggested to ZWS the need to explore case studies in Europe that might provide inspiration and examples for Scottish textile businesses to follow. Scandinavia was the prime suggestion, and a number of examples of the circular economy in textiles were identified.

The Research Council of Norway (NICE) funded a 3-year project (2010 to 2013) valued at €500,000, with a view to identifying and promoting the benefits of textiles made of Norwegian wool to local consumers. The budget was used to support a variety of promotional activities, including publishing a book, *Ren ull* (Pure Wool, Aschehoug). It is currently only printed in Norwegian, with a foreword in English by Prince Charles. Its approach is similar to the Campaign for Wool.

When the project ended, NICE wanted to continue its research and use the heritage and technology it had discovered to develop further consumer demand for Norwegian wool and create a label of provenance, similar to Harris Tweed, called the Nordic Swan Ecolabel. As part of this research, they chose to visit Shetland where fleece is gathered from over 700 crofters to produce Shetland wool.

Norway also developed a product which was given an international standard-approved Cradle to Cradle[n]. The Dutch airline, KLM, wanted to create Cradle to Cradle carpeting working with Desso. Norwegian wool was tested for chemicals, etc. and was approved. The research company brokered an introduction to NICE and ZWS and the potential to develop research collaborations around wool and revise the heritage of Scottish and Norwegian knitwear has been identified. In this way, each country can work towards a model of *provenance*, *traceability*, *durability* and *quality*, as per the four key values mentioned earlier that the ZWS-commissioned report recommends.

Another example is the 'Aran jumper', a well-crafted, distinctively patterned garment, synonymous with Scotland and Ireland, that is constructed so soundly, it can be disassembled, unravelled and reknitted, thus demonstrating an excellent closed-loop process. Families often deconstructed two small garments, knitting them into one larger garment, as their needs dictated.

The challenge today is to see how that heritage and evidence of closed loop can be harnessed to the circular economy to achieve a sustainable future for the textiles sector.

There is scope too to draw on the cultural craft heritage of Scotland; for example, there is a demand for hand-crafted products that draw on that heritage and once declining Scottish skills such as hand knitting, which have been invigorated by knitwear design companies such as Di Gilpin, who currently employs 90 freelance hand knitters throughout Scotland. Similarly, Eribe knitwear has 200 registered hand knitters who can each generate income of between £100 to £400 per month producing contemporary Scottish knitwear for export, mainly to China and Japan. Through the support from Scottish Enterprise, companies such as Eribe have doubled their turnover in 5 years, mainly through exporting.[o] These companies are challenged to source sustainable Scottish raw materials such as wool from Scotland (other than from Shetland) and a growing number of small sheep crofts. The textile sector has to be aware, however, that a revival in craft heritage matters has to consider existing industry frameworks. For example, since 1952, a sheep farmer with more than four sheep is required, legally, to sell wool only through the British Wool Marketing Board (BWMB). The penalty, if enforced, is still a 6-month jail term. In 1995, this legislation was relaxed to allow farmers to sell wool only for direct export. This was due to the low price, lack of demand and essentially oversupply, with BWMB not being able to clear the stocks.

3. DISCUSSION AND EVALUATION

The ZWS-commissioned research concluded that the Scottish textile industry is innovative in key sectors, particularly technical textiles and computer-aided design. However, in both cases, the overarching objective is either to increase business competitiveness or improve resource efficiency, rather than closed-loop sustainability. The report concluded that there is little innovation in business models or in new methods of making the sector more materially circular in ways that are commonly understood. It also concluded that Scotland has a strong technical textiles sector, and there may be specialist opportunities here for closed-loop recycling solutions.

It recommended that the Scottish textile industry should focus on strengthening what it already does well and bolster its sustainability activities by linking to *provenance*, *traceability*, *durability* and *quality*. Inspiration should come from the luxury goods sector and also, in a more modest way, from Scandinavian brands such as Filippa K, where longevity is an explicit brand strategy.

ZWS has integrated the report recommendations into its work plan to March 2016. It will focus on five key areas:

1. Industry and academia - skills training
2. New business model research
3. Fibre-processing facility
4. Funding to experiment with closed-loop technologies and systems
5. Textiles and apparel

3.1 Industry and academia - skills training

A series of master classes will be delivered from April to December 2015. The topics and content for the master classes are based on an analysis of the key characteristics of Scottish textiles design and manufacturing that will enable resource efficiency and skills for a circular economy such as design for disassembly and fibre processing and resource efficiency topics such as zero-waste pattern design and new technologies in dyeing and printing. One master class, Dyeing and Chemical Processing, will introduce global brand research such as dry dyeing by Stella McCartney for Adidas (Amsterdam, the Netherlands) and focus on resource efficiency such as a reduction of effluent from chemicals.

The majority of the 16 places for each course will be allocated to industry, with a quarter allocated to academia and education leaders who are developing curricula. Attendance is by invitation and application and candidates need to evidence how they will cascade the learning within their business or within departments and to colleagues.

Each facilitator will be travelling to Scotland from other parts of the UK or Europe (where possible, webinars will be used). To maximise the purpose of each visit, a student lecture series will also be part of the package accommodating up to 200 students at each lecture. Edinburgh University, Heriot-Watt University, Glasgow School of Art and Duncan of Jordanstone College of Art and Design in Dundee have provisionally agreed to host the these lectures.

Skills Development Scotland has confirmed that it is complementary to the Scottish Textiles Skills Action Plan[p] that has recently been funded by the Scottish Funding Council.

ZWS will source and invite experts from outside Scotland, and evaluation consultants will be appointed to assess the impact of the participants' learning experience and how they have cascaded their learning to other colleagues or implemented new skills in their business processes.

3.2 New business model research

ZWS has identified two textiles companies who would like to explore new business models, details of which will be published in 2015[q].

An example of a business model that was identified in the research and that has characteristics that could apply to Scotland is Waste to Wow! Established in 2013 and based in Italy, it is a start-up enterprise that has achieved the accolade of being winner of the 2014 European Social Innovation Competition, out of 1,254 entrants[r]. Their take on recycling is slightly different in that their waste stream is textiles from high-end fashion manufacturers that have failed quality control checks. Normally, these garments would be destroyed to make sure there is no leakage onto the grey market. From Waste to Wow! assures confidential and controlled disposal by using the material to make new, limited edition textiles that they sell under the brand name Quid (Progetto Quid, Verona, Italy). Thus, their feedstock tends to be of much higher quality than most similar recycling operations based on this model.

Scottish textile manufacturers who supply luxury global brand textiles must destroy their seconds and textile waste to avoid grey market leakage. An interior textiles company and some apparel textile companies have asked ZWS for advice as to how to dispose of luxury brand seconds currently going to landfill or being incinerated.

3.3 Fibre-processing facility

ZWS is interested in exploring a fibre-processing facility in Scotland that would help tackle the large fraction of lower grade bulk textiles which currently go to landfill, including carpets, mattresses and post-industrial waste from the textile industry. It also offers the opportunity to develop investment and jobs in fibre reprocessing in Scotland, adding value to materials which are currently worth little. As an example, if 60% of mattresses can be recycled (7,200 tonnes equates to 335,000 units), direct jobs created are estimated at 80 based on the ZWS business case of 8 jobs per 34,200 units.

A fibre-processing facility would enable Scotland to process post-industrial and post-consumer products from throughout the UK and gain some of the market share that other areas of the UK benefit from. Cotton and polyester are, respectively, around half and one quarter of the UK clothing fibre mix and hence are the priorities for fibre processing. Mixed fibres such as wool and nylon are characteristic of carpet construction, and mixed fibre extrusion is still a developing area. In the case of carpets, Anglo Carpets in England employs 35 people to process 1,200 tonnes. Carpet Cycle employs 50 people for 5,500 tonnes. Mid-point is one person per 75 tonnes processed.

The WRAP report, *UK textile product flow and market development opportunities*, has numerous recommendations regarding how the UK textiles market could be developed. ZWS' priority for this work, based on evidence in that WRAP report, is to investigate the potential for Scotland to reprocess cotton and polyester back to fibre or feedstock, at cost parity to virgin material. It is recognised that scale could limit the implementation of the reprocessing infrastructure and this may require a change of direction for this area of work.

The research areas for the fibre-processing facility are specific to the following:

- Identifying candidate technologies for fibre processing and identifying their needs for financial support to enable proof of design, trialling and scale-up
- A route map of the activities which might be needed to deliver a commercially viable service

Following the identification of suitable technologies and their applicability to Scotland, a financial and economic assessment will need to be conducted.

3.4 Funding to experiment with closed-loop technologies and systems

Funding will be offered by ZWS to three projects through an open call for industry and academic partnership projects. Project proposals should explore the feasibility of a new textile product or process that demonstrates closed-loop design or manufacturing. Awards will favour technical and performance textile companies.

Projects should be completed in January to December 2015. They will be monitored and evaluated externally to capture shared learning and secondary outcomes that may not be evident from the trialling of a product or process. Funding will be allocated for external monitoring and evaluation. Those projects which prove feasible maybe further supported by ZWS and taken to prototype/trial stage.

There will be *three* themes:

1 Closed-loop manufacturing - companies that already manufacture textiles and have post-industrial waste going to landfill that they think has the potential to be reprocessed

2 Piloting technologies/products for disassembly - companies that would like to trial technologies such as Wear2 microwave seam technology

3 Exploring new fibre opportunities - all proposals will be considered from recycled synthetic fibre to the introduction of large-scale natural fibre processing

3.5 Textiles and apparel

Scotland has a strong fashion and textile design community, with 600 textile and fashion graduates every year and graduates such as Lauren Smith from Edinburgh College of Art winning the prestigious Gold Award at London Graduate Fashion Week 2013. ZWS would like to encourage designers working in Scotland to consider supply chain sustainability and research textile design and garment construction that relates to circular economic and resource-efficient practices in design such as design for disassembly and sourcing and designing with recycled yarns such as recycled P.E.T., as well as alternative fibre and plant dying processes.

ZWS will provide funding for designers to develop prototypes working towards a truly sustainable circular economy product and/or using alternative fibres to imported yarn with a heavy carbon footprint. Mentors will be provided for each winner. The challenge will be finding suitably knowledgeable mentors in Scotland in this field.

4. CONCLUSIONS

Given the increasing demand for Scottish textile products, ZWS' work programme 2014 to 2016, and recent discussions with NICE, Norway, the Scottish Government's policy for the textiles industry should consider the following:

- Does Scotland need its own sustainability standard? Will this help or hinder business and how important is it to the provenance of Scotland's textile industry?
- How viable are pilots with the Scottish textiles industry to introduce new technology trials that will develop closed-loop processes?
- Can we source international investment in Scotland that not only provides job security but also invests in new technology and skills for a circular economy?

ZWS' support for the textiles sector in Scotland is already established. The key challenge now is to take forward the five aforementioned action points and evaluate and evidence the impact of embedding a circular economy in those areas. This paper has outlined our approach, why we have chosen it, and how we plan to implement it. The specific context is the potential for business growth in the textile sector in Scotland, within the broader context of ZWS' support for the Scottish Government's commitment to its zero-waste targets for 2025. Measuring the impact of a circular economy approach is a task for the long term and we know from industry examples, such as TENCEL® by Lenzing[s], that it can take up to 30 years to develop a product that revolutionises industry.

The production of TENCEL® is revolutionary. The production process is based on a solvent spinning process and represents the greatest accomplishment in cellulosic fibre technology. The unique closed loop production process makes TENCEL® the fibre of the future: eco-friendly and economical[t].

Options for development and investment in Scotland, therefore, not only rely on ZWS and Scottish Enterprise and other development agencies but also require collaborations and investment from the rest of the UK, including further and higher education, and global partners. ZWS' work programme from 2014 to 2016 will produce evidence and provide incentives to the textiles sector that show our commitment to underpinning a circular economy.

ZWS is intent on supporting a sustainable industry that will generate long-term growth and be supportive of the Scottish Government's zero-waste targets. That can only come about by facing up the challenges and opportunities that a circular economy model, as per our work programme for 2014 to 2016, is intent on discovering. ZWS will maintain alliances with all the key stakeholders mentioned in this paper, as well as forge stronger connections with the industry and academia. In our view, the theory and practice engendered by that connection can establish a body of knowledge on the circular economy that will determine the Scottish textile industry's short-, medium- and long-term goals and secure its future sustainability.

ENDNOTES

[a]Scottish Textiles Strategy 2011 to 2015

[b] http://www.ellenmacarthurfoundation.org/

[c]The Scottish textiles sector has been given £750,000 over 3 years to develop innovation and skills for industry. Heriot-Watt University, Edinburgh will administer the programme. Glasgow Clyde College will deliver an academy-style skills programme funded by a consortium of funders including Skills Development Scotland, and ZWS will be invited to a sub-group meeting when it is set up.

[d] http://www.gov.scot/Publications/2013/10/6262. Exploring the feasibility of circular economy business practices contributes to 'Safeguarding Scotland's Resources', Sections 6.1 Resource Efficient Scotland - Action 7 - Business engagement and an early adopter network and Action 12 - Remanufacturing and new business models, 6.2 Stimulating innovation and business opportunities and the provision of technical master classes for the textile design and manufacturing sector supports and 6.3 Sustainab`le Design Action, Objective - to stimulate innovation in the design and manufacture of products and packaging - to engender a shift towards design for longevity, reassembly or remanufacturing and useable by-products, designing out wasted energy, materials and pollution. In addition, by inviting teaching staff in education and academia to participate in the master classes with industry, it also supports Action 18 - Public engagement and education. ZWS works closely with Education Scotland, Young Scot and the Ellen McArthur Foundation on these activities

[e] http://www.ellenmacarthurfoundation.org/about/media - Richard Lochhead, Environment Secretary, statement on joining CE100 - 'Scotland's

economy will benefit from moving to a more circular model of production and consumption. Our Zero Waste Plan is already delivering important actions to make better use of resources, and we can accelerate progress if we join together with others on a global level'

[f] http://www.knittingindustry.com/chanel-to-create-100-new-jobs-at-barrie-knitwear-in-scotland

[g] http://news.scotland.gov.uk/News/Textiles-target-success-564.aspx

[h]Scottish Key Facts, compiled by the Economic Research Team, Scottish Enterprise, May 2014

[i] http://www.itv.com/news/border/story/2013-08-07/mongolian-delegation-in-hawick/

[j] http://www.hw.ac.uk/news-events/news/wooden-denim-to-cut-carbon-emissions-11090.htm - Dawn Ellams, Feb 2013

[k] http://www.hw.ac.uk/news-events/news/wooden-denim-to-cut-carbon-emissions-11090.htm

[l] http://www.cicstart.org/userfiles/file/AC-1-REPORT.PDF

[m] http://www.gcu.ac.uk/newsevents/news/article.php?id=73608

[n] http://www.c2ccertified.org/product_certification

[o]Scottish Enterprise case study

[p]The Scottish Textiles Skills Strategy & Action Plan has been developed to support the objectives of the industry strategy *A Strategy for the Textiles Industry in Scotland 2011-15* (http://www.scottish-enterprise.com/knowledge-hub/articles/publication/textiles-strategy-2011-2015) and sets out the key skills priorities that the textiles sector must address in order to fulfil its growth potential

[q]New business models are covered by a wider stream of the Circular Economy Team and therefore, were not part of the commissioned textiles research, although there are relevant documents within the report

[r] http://socialinnovationcompetition.eu/401/

[s]The Lenzing Group is a large publicly owned manufacturer of cellulosic fibres. It employs over 6,500 staff, making nearly 900,000 tons of fibres per annum, with sales of around €2 million per annum. Its three principal products, Viscose, Modal and Tencel are well known to the sustainable textile community. The feedstock is wood, sourced from sustainable forests. Fibre production by Lenzing achieves a wood utilisation rate of over 50%. Co-products include furfural, xylose and acetic acid with the remaining parts

used principally for energy production. In 2013, Lenzing launched a new fibre, EUROCEL, which is a 50/50 blend of Viscose and Tencel. The intended application is high-end non-wovens, mainly for hygienic applications as well as in the household and industrial sectors

[t] http://www.lenzing.com/en/fibers/tencel.html

CHAPTER 2

Self-Folding Textiles through Manipulation of Knit Stitch Architecture

Chelsea E. Knittel [1], Diana S. Nicholas [2], Reva M. Street [1], Caroline L. Schauer [1] and Genevieve Dion [2,*]

[1] Materials Science and Engineering, Drexel University, 3141 Chestnut Street, Philadelphia, PA 19104, USA

[2] College of Media Arts and Design, Drexel University, 3141 Chestnut Street, Philadelphia, PA 19104, USA

ABSTRACT

This research presents a preliminary study on finding predictable methods of controlling the self-folding behaviors of weft knit textiles for use in the development of smart textiles and garment devices, such as those with shape memory, auxetic behavior or transformation abilities. In this work, Shima Seiki SDS-One Apex computer-aided knitting technology, Shima Seiki industrial knitting machines, and the study of paper origami tessellation patterns were used as tools to understand and predict the self-folding abilities of weft knit textiles. A wide range of self-folding weft knit structures was produced, and relationships between the angles and ratios of the knit and purl stitch types were determined. Mechanical testing was used as a means to characterize differences produced by stitch patterns, and to further understand the relationships between angles and folding abilities. By defining a formulaic method for predicting the nature of the folds that occur due to stitch architecture patterns, we can better design self-folding fabrics for smart textile applications.

KEYWORDS

self-folding; computer aided knitting; origami; stitch ratios

1. INTRODUCTION

Knitting is the interlacing of yarns into loops resulting in fabrics that have inherent self-folding abilities. A greater understanding of these knit structures could facilitate the design of technical textiles for engineering applications and smart garments requiring compression and expansion properties. Study of origami paper folding techniques has gained popularity over the last decade as a means of designing these transformable structures in many types materials for a wide variety of applications. The unique characteristics of tessellated structures, which allow them to expand, compress and transform, have been used to develop nanoscale to macro scale products [1] from medical stents [2] to origami robots [3,4,5] and tunable antennas [6]. Recent interest in the use of self-folding–inspired structures for extraterrestrial applications has also grown, including research conducted through the National Aeronautics and Space Administration on folded solar arrays [7], where benefits lie in the ability of the structures to remain small while in transit, and then grow exponentially for use in outer space.

Origami structures have been explored in a variety of materials including paper, cardboard, plastic and metal sheeting, and textiles. Structures have also been created through the use of three-dimensional (3D) printing technology, where engineered mechanical instabilities produce folding behaviors as the object is compressed [8]. Using origami-like patterns and the self-folding behavior of knits, textiles exhibiting auxetic behavior have been developed [9,10,11,12,13]. However, this area of research would benefit from a deeper understanding of how and why the self-folding mechanisms are created. Discovering methods of controlling and understanding the self-folding mechanisms that occur with knit fabric relaxation would help expand the field of smart textiles, and benefit designs such as drug release [14] or compression bandages by affording smart textiles the ability to transform shape [9,15,16]. Additionally, creating these structures in fabric could capitalize on preexisting production methods with computer-programmed knitting machines that have successfully been used in the garment industry for decades [17].

The first step for successfully converting these paper-folding techniques into fabric form is to understand the mechanism through which folds in fabrics may be formed. While the paper-style folds may be mimicked through the use of processing techniques such as steaming, ironing and chemical processing (such as permanent press), utilization of the self-folding abilities of knit textile structures would allow for economical and repeatable production of a material without the need for additional manipulation [17]. Our research began through the study of knit stitch geometry arrangements, and the observation of paper-folding mechanisms in origami tessellation patterns. These remained an integral part of the pattern development throughout, as research was conducted to find a recognizable relationship to convert from one medium to another and to understand the effect of specific stitch architecture on folding behavior.

2. MATERIALS AND METHODS

2.1. Materials

Yarns used in this study included Supra-merino (7500 ypp, Silk City Fibers, Paterson, NJ, USA) Jaguar (85% Modal/15% Nylon, 8750 ypp, Silk City Fibers, Paterson, NJ, USA), high bulk polyester (150 denier) and spandex (0.12 mm diameter). Mechanical testing was carried out on samples made from the high bulk polyester yarn.

2.2. Computer-Aided Knitting

Industrial knitting technology is an important tool for the development of smart textiles and garment devices [17]. It provides advantages over other textile productions methods such as weaving by facilitating rapid prototyping without the need for large quantities of a material. Using industrial knitting machines, a wide variety of shapes and textures can be developed from various stitches, such as knit, purl, miss and tuck, as well as shaping through widening and narrowing of the fabric. All textiles produced in this body of work utilize only knit and purl stitches. While it should be noted that the purl stitch is actually the technical back of the knit stitch, here we designate each stitch as it is seen from one side of the fabric, the designated technical front. Knit stitches are those seen as small "V" shapes, and purl stitches are those appearing as small loops (Figure 1). All knit samples were produced on Shima Seiki Industrial knitting machinery, including the Shima Seiki SWG041N and the Shima Seiki SV122SG. Before prototypes of each textile were created, the Shima Seiki SDS-Apex

Design System was used to program each textile, and examine computer-simulated images of the knit architecture.

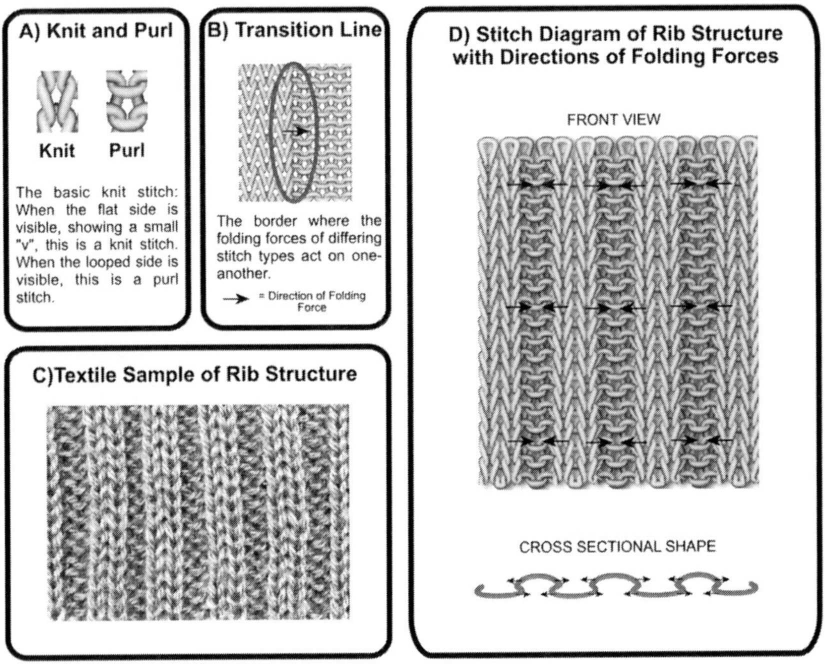

Figure 1. Self-folding in knits: (**A**) knit and purl stitches; (**B**) transition line; (**C**) textile sample of rib structure; (**D**) stitch diagram of rib structure with directions of folding forces.

While the computer simulations provided by this software accurately show a detailed image of the stitch architecture, they do nothing to predict the physical outcome of the textile that may be influenced by a number of factors such as yarn material, stretch and fabric relaxation [18]. Therefore, it is important to develop a better understanding of the movements and folds created by specific stitch arrangements in order to develop a predicative model for the design of self-folding textiles.

From traditional knit structures such as the 2 × 2 rib knit, we know that changes in the stitch pattern can produce self-folding in knitted textiles. In Figure 1A we see the basic two components of the knit textiles, the knit and purl stitch. When placed side by side, forming vertical stripes, this produces a knit structure commonly referred to as ribbing. Between each section there is a transition line (Figure 1B) where the folding forces of the two different stitches act upon each other, forcing one section of the fabric up while the other is pushed down. Figure 1C depicts a sample of an actual fabric that is created due to the directional folding forces (Figure 1D) of the rib structure. If these forces are unbalanced either in number or strength it can cause the fabric to fold on itself vertically. From this understanding, we predicted that we could manipulate these folding forces to create folds at predetermined angles and locations by changing the geometry of the stitch pattern.

2.3. Study of Origami Tessellation

Study of the folding mechanisms in knitted textiles began by recreating several distorting knit patterns described in the thesis work The Design of 3D Shape Knitted Preforms by Jennifer Underwood [19] in order to better understand and validate the folding and stretching movements afforded by their knit structures. Though this provided insight to the range of structures that could be created, the examples in the thesis did not explain the mechanics of the folding behavior. Further research led us to the work of Liu et al. [12] and Hu et al. [13], who developed weft-knitted negative Poisson's ratio fabrics. Upon observing these textiles in physical form, it was noted that they shared a number of similarities with origami tessellation patterns [20] (Figure 2). Tessellations are repeated geometric tiling patterns. Using these geometrical patterns, traditional paper folding techniques may be used to create paper structures with auxetic characteristics due to the folding mechanisms of these repeated patterns. Applied forces cause the folded structures to tessellate outwards, creating multidirectional expansion. Therefore, origami tessellations were studied as a means of understanding how to create self-folding structures through geometric principles, and to discover a relationship between paper folds, and self-folding fabrics.

Using this repeated zig-zag pattern, a fabric structure may be engineered that is similar to a common origami structure, featuring repeated zig-zag valleys and peaks [12]. What is different between these two mediums, however, is how and where these features are created. In the paper model, folds occur along the boundary lines of the geometric shapes. In a textile model, we see that strong points in the shape of the geometry, indicated by the grey circles in the diagram in Figure 2, cause the fabric to buckle. This creates a linear fold, running through the geometrical shapes instead of along the perimeters. Though the fold lines in the knit structure are shifted when compared to the paper folds, the resulting structures are similar.

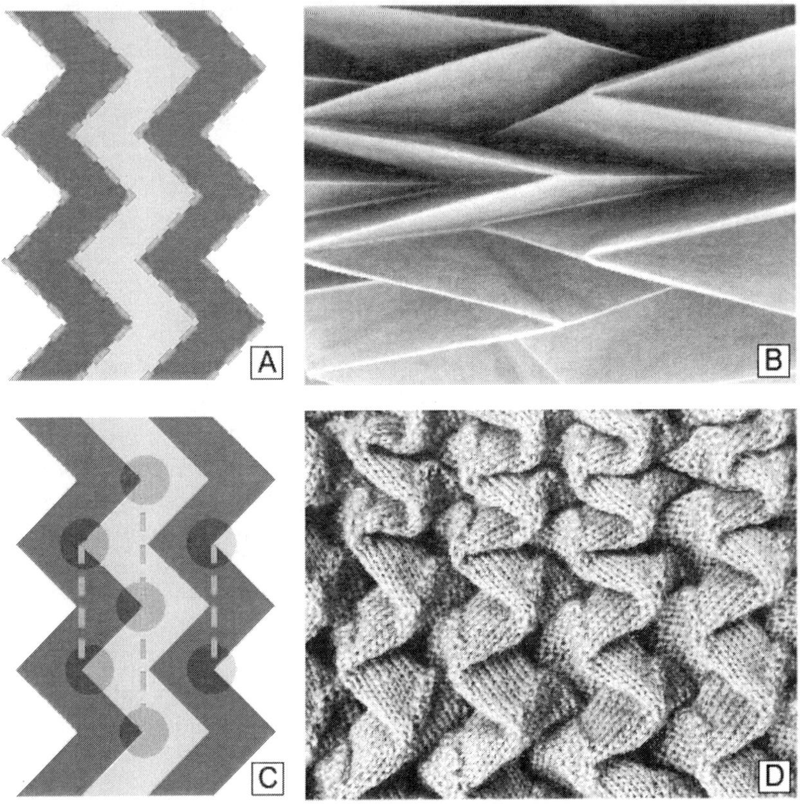

Figure 2. (**A**) Diagram of paper folding lines, represented by dotted lines; (**B**) paper sample; (**C**) diagram of textile folding lines, represented by dotted lines; (**D**) merino wool textile sample.

2.4. Mapping of Folding Patterns

After designing and knitting a wide variety of prototypes based on many different origami tessellation patterns, computer-simulated images were used to map the locations of folds on the physical models. Since the software is unable to show the folds that are present in the textiles, lines were drawn to map the self-folding actions occurring in each sample, and to indicate directionality of the fold. For each stitch architecture design, two knit samples were produced. The first sample was knit in a high bulk polyester yarn, while the second sample was knit with a combination of high bulk polyester and spandex (Figure 3). In some samples, folding mechanisms occurred along lines of stitch change. In others, it was an entire geometric segment of the sample that folded up or down.

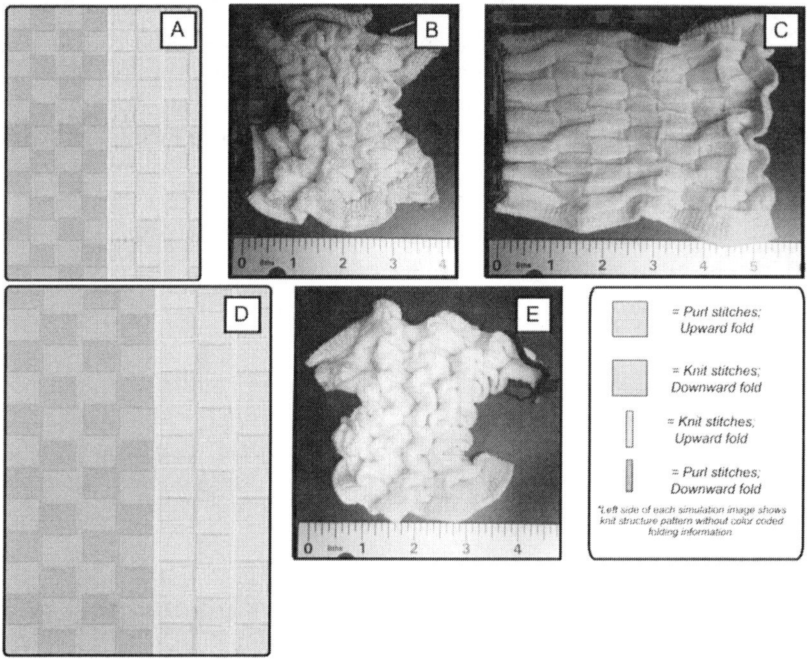

Figure 3. Mapping of folding patterns. (**A**) Computer simulation of original pattern; (**B**) original pattern knit with polyester yarn; (**C**) original pattern knit with polyester and spandex yarns; (**D**) computer simulation of readjusted pattern; (**E**) readjusted pattern knit with polyester and spandex yarns produces the same self-folding structure as the original sample. Shima Seiki SDS-One Apex Software.

2.5. Ratios in Knit Patterns

The design used in Figure 3 shows a key example of the importance of ratios in controlling self-folding behavior. This pattern was instrumental in determining methods of correcting and re-adjusting ratios in order to recreate effects using different materials. Figure 3A shows a simulation of the pattern, and Figure 3B shows the sample knit in a high bulk yarn. Strong self-folding abilities are clearly seen. The photograph in Figure 3C shows the first attempt at recreating this same pattern, with the addition of a fine spandex yarn. As seen, there is a clear discrepancy between the two fabrics. With the addition of spandex, the magnitude of the folding forces changed. Folding occurred normally between the squares of the checkerboard pattern (at the line along which the knit and purl stitch forces meet); however, the strong linear folds running down the length of the whole fabric became far less dominant. By calculating the ratio of shrinkage that occurred in the first spandex attempt, the pattern was scaled up proportionally to produce a fabric with the same size patterning dimensions as the original sample. This was achieved by measuring the dimensions of the two samples knit using the original computer program (Figure 3B,C). These dimensions were then used to calculate the proportional change necessary to produce the new program. It is important to note that due to yarn relaxation, stitch size change occurs differently in the horizontal and vertical directions, and must be accounted for separately. Therefore, proportional changes in the height and width must be accounted for separately.

The two computer simulations in Figure 3 are to scale, and show the difference between the size of the original pattern (Figure 3A) and the readjusted pattern (Figure 3D). While the computer program for the spandex sample is considerably larger, the fabric produced is to the same scale as the plain polyester version. While denser than the original, the fabric produced replicated the same folding effects as the plain polyester sample (Figure 3E).

2.6. Horizontal vs. Vertical Stitch Arrangements

In order to determine a formulaic method of designing self-folding textiles, it was important to observe and understand the effect of stitch arrangements. On the simplest level, we may note the difference between horizontal and vertical arrangements of knit and purl stitches. As explained earlier, when knit and purl stitches are arranged in a repeating pattern of vertical columns, a self-folding effect is exhibited where the knit stitches naturally overtake the purl and rise to the surface of the fabric. When knit and purl stitches are arranged in a repeating pattern of horizontal rows, an opposite effect is achieved and the purl stitches become dominant.

From the understanding of the behaviors of both vertical and horizontal rib knits and the effects of stitch height/width ratios, it was determined that a critical angle should exist that marks the change between a knit-dominant fold, and a purl-dominant fold. A study was conducted observing the effect of changing the angle of transition between the knit and purl fields of a textile. Knit swatches were created by dividing a square between knit and purl, with the line of transition placed at varying angles between 0 and 90 degrees, three of which are depicted in Figure 4. It is important to note that the angle range of 57–59 degrees is used, rather than one single angle. This is due to the nature of the grid-based programming. Only a limited number of angles are available that consist of a regular repeating pattern of stitches. While at 45 degrees our slope steps up one stitch at a time, and at 75 degrees our slope steps up four stitches and over one, the pattern repeats of angles in between become increasingly more complex. Since the pattern repeat around 58 degrees is quite long (occurring over a width of 21 stitches) depending on the size of the sample, different amounts of partial repeats may occur. This can change the degree of the angle slightly, but still produces a flat fabric.

Additionally, it should be noted that while each stitch is represented as a perfect square in the program, this is not the case in the actual fabric, and these dimensions will change depending on the material used as well as other parameters such as tension and takedown. Regardless of these changes, similar folding behavior is observed when knitting the same knit programs in a variety of materials, as well as on different types of knitting machines.

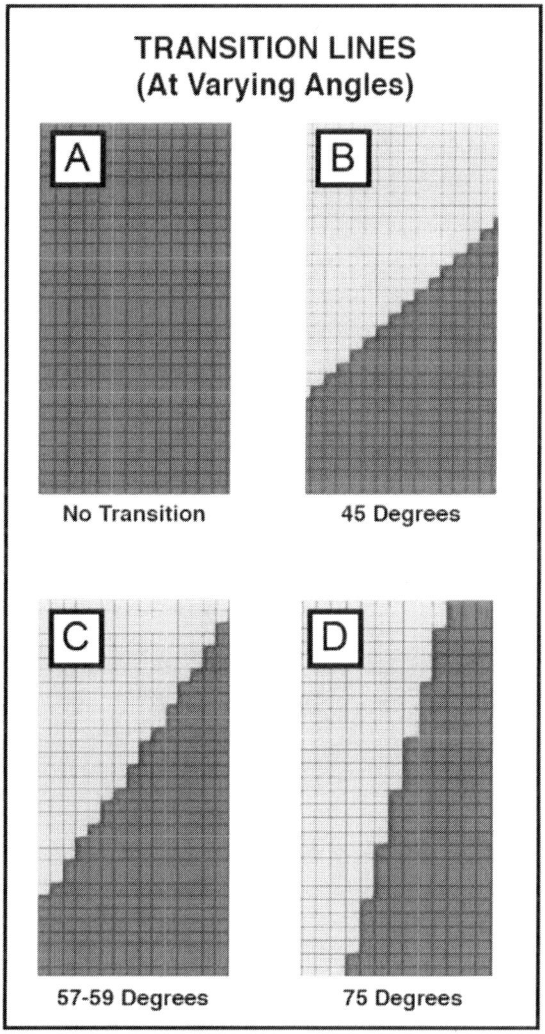

Figure 4. Transition line angles of knit samples: (**A**) No transition; (**B**) 45 degrees; (**C**) 57–59 degrees; (**D**) 75 degrees. Shima Seiki SDS-One Apex Software.

2.7. Sample Preparation and Mechanical Testing

Samples sets (N = 3) were knit on the Shima Seiki SWG041N industrial knitting machine for all possible transition angles between 0 and 90 degrees. While the angle of transition varied between sample sets, all other variables including stitch length, yarn tension and knitting speed remained constant.

Mechanical testing of sample segments was carried out on the two angles that we observed to have the strongest folding behavior, as well as the angle at which a flat fabric was produced. This was used to observe the effect of changes in the transition angle on the mechanical properties of the textile [21,22,23]. Each sample tested was 20 stitches in the course direction and 44 stitches in the wale direction. Each was secured on a sample card to facilitate improved grip into a Bose Electro Force 3200 Series III Axial Configuration tensile testing machine (Figure 5). After placement, the side of the sample card was cut before testing began.

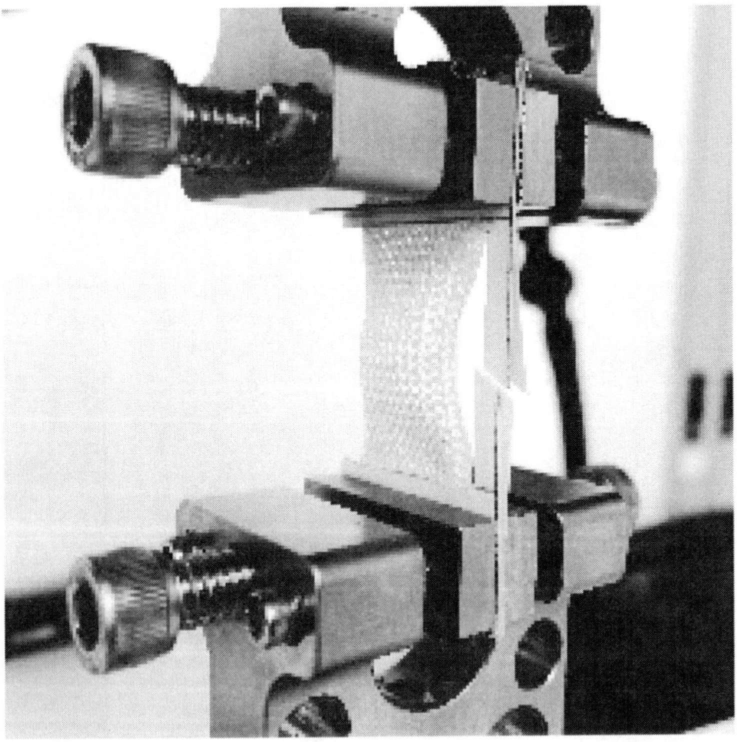

Figure 5. Sample placed in grips of the Bose Electro Force 3200 Series Axial Configuration tensile tester.

During testing, the temperature of the room was 22.9 °C, with a 21% humidity level. Once tightened between two clamps, a load was applied and samples were extended to a maximum displacement of 6.5 centimeters. Data for time, displacement and loading were recorded. Using MATLAB and Excel software, graphs were created showing stress vs. strain curves (Figure 6) and Young's modulus comparisons (Figure 7) for the range of samples. Samples are designated first by the angle of the transition line, and then by the percentage of purl stitches in the sample, which dictates the length of the transition line (45p23 indicates a 45 degree angle transition, in a sample that consists of 23% purl stitches and 77% knit stitches).

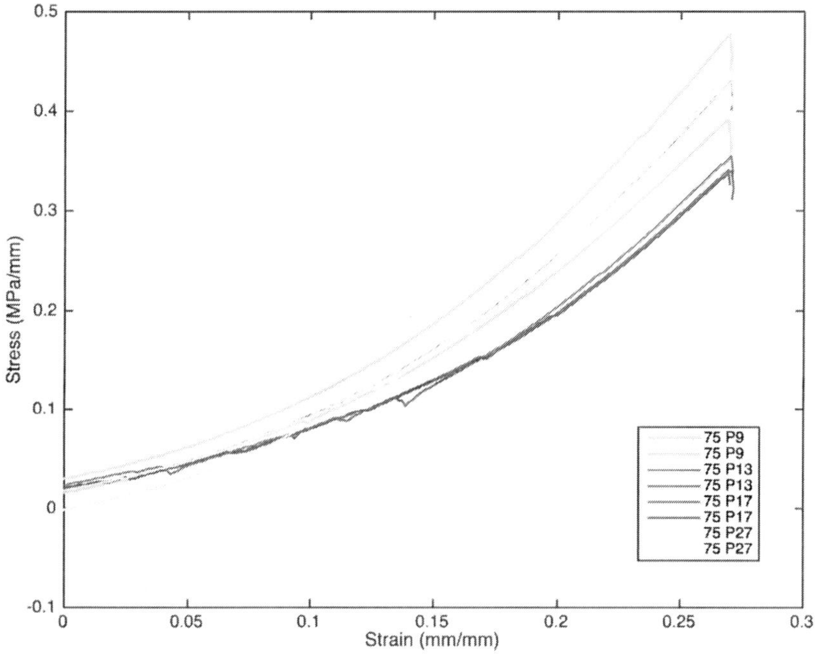

Figure 6. Stress vs. strain graph of samples with a 75-degree transition line, MATLAB.

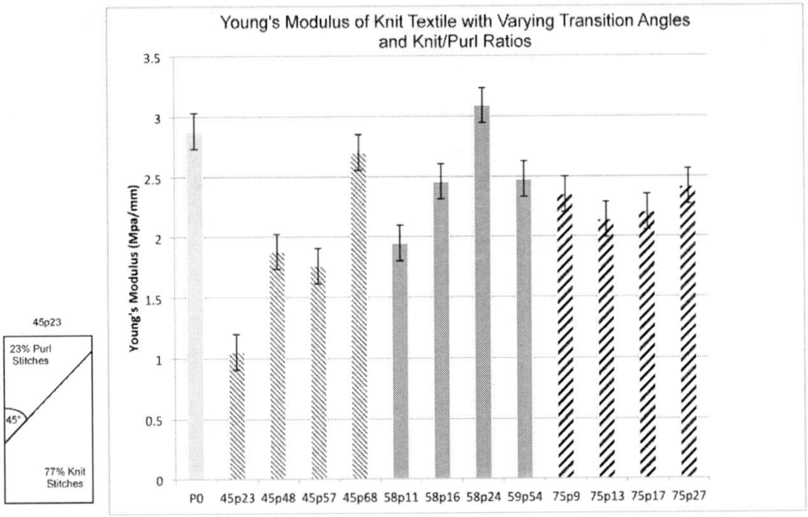

Figure 7. Young's modulus comparisons between transition angles and lengths, Excel.

3. RESULTS

3.1. Self-Folding Angle Behaviors

From these knit samples, it was observed that that:

- With transition lines between 0 and 45 degrees, purl stitches are dominant; the self-folding mechanisms create a strong fold that pushes the knit stitches down below.
- With transition lines between 75 and 90 degrees, knit stitches are dominant; the self-folding mechanisms create a strong fold that pushes the purl stitches down below.
- With transition lines at angles greater than 45 but less than 57 degrees, the strength of the purl-dominant folds diminishes as the angle approaches 57 degrees.
- With transition lines at angles less than 75 but greater than 59 degrees, the strength of the knit-dominant fold diminishes as the angle approaches 59 degrees.
- With transition lines at the angle of 57–59 degrees, neither the knit nor the purl stitches are dominant: the self-folding mechanisms are in equilibrium and a flat fabric is produced.

3.2. Mechanical Testing Data

By looking at the results of the tensile test, we can see several trends. At transition angles of 45 and 58 degrees, we can see a possible positive correlation between the Young's modulus and an increase in the length of the transition line (which occurs with an increase in the percentage of purl stitches) however, this will require further validation through statistical analysis. For transition angles of 75 degrees, however, we see less change in the Young's modulus occurring with an increase in the length of the transition line.

This result of the tensile tests helps to support the hypothesis that the demonstrated self-folding angle effects are a result of the folding forces along the line of transition, which will vary with an increase or decrease in the length of the line. By the time the knit-dominant force in the folding line is strong enough to overtake the purl (at angles greater than 75 degrees), the ratio of purl stitches is so low that the horizontal folding forces are finally overcome, allowing a more vertical folding action to occur (Figure 4). Once the horizontal effect of the purl stitches is this low, there is no longer an observable trend in the effect on the Young's modulus of the fabric.

4. DISCUSSION

Through understanding the self-folding mechanisms created by knit and purl stitches, and the angles at which these are most effective, it was observed that the angle of stitch geometries may be used to consistently control the self-folding behavior of knit textiles. Though material characteristics of the yarn may result in variance of properties in the fabricated textiles, mechanical properties demonstrate consistent effects created by these angles. Ideally, using this formula, a wide range of textiles may be designed for smart textile applications, including those with auxetic characteristics, buckling mechanisms, cell-like structuring, high levels of articulation, and transforming mechanisms. Future designs will be explored in terms of scalability and performance under stress and continued fatigue. Using the defined parameters for angular folds, we may use the results of this study to develop stitch architecture designs that will give auxetic characteristics to smart textile garments and devices.

5. FUTURE WORK

While a polyester yarn was primarily used in this research, we found that similar structures could be produced using the same structure patterns paired with a variety of yarn materials. Moving forward we will work on developing a statistical model that can be used to validate our observations and help us further predict folding behavior, as well as characterize the specific effects of yarn type on the self-folding behavior and on the material and mechanical properties of the textile.

ACKNOWLEDGMENTS

This material is based upon work supported by the National Science Foundation Graduate Research Fellowship under Grant No. DGE-10028090/DGE-1104459. Any opinion, findings, and conclusions or recommendations expressed in this material are those of the authors(s) and do not necessarily reflect the views of the National Science Foundation. This material is also based on work supported by Drexel COE Engineering Design Teaching Fellowship awarded to RMS.

Special thanks to Kristy Jost, Kevin Egan, Robert Lehrich, Christina Kara and all of my colleagues at the Shima Seiki Haute Technology Laboratory and the Natural Polymers and Photonics Group.

AUTHOR CONTRIBUTIONS

Chelsea Knittel, Genevieve Dion, Caroline Schauer, and Diana Nicholas conceived and designed the experiments; Chelsea Knittel and Reva Street performed the experiments; Chelsea Knittel, Caroline Schauer and Genevieve Dion analyzed the data; Chelsea Knittel wrote the paper.

CONFLICTS OF INTEREST

The authors declare no conflict of interest.

REFERENCES

1. Zhang, Y.; Nan, K.; Xiao, D.; Liu, Y.; Luan, H.; Fu, H.; Wang, X.; Yang, Q.; Wang, J.; Ren, W.; et al. A mechanically driven form of Kirigami as a route to 3D mesostrucure. PNAS **2015**, 112, 11757 11764.

2. Kuribayashi, K.; Tsuchiya, K.; You, Z.; Tomus, D.; Umemoto, M.; Ito, T.; Sasaki, M. Self-deployable origami stent grafts as a biomedical application of Ni-rich TiNi shape memory alloy foil. Mater. Sci. Eng. A **2006**, 419, 131–137.

3. Felton, S.; Tolley, M.; Demaine, E.; Rus, D.; Wood, R. A method for building self-folding machines. Science **2014**, 345, 644–646.

4. Hoff, E.V.; Jeong, D.; Lee, K. OrigamiBot-I: A Thread-Actuated Origami Robot for Manipulation and Locomotion. In Proceedings of the IEEE/RSJ International Conference on Intelligent Robots and Systems, Chicago, IL, USA, 14–18 September 2014.

5. Gao, W.; Huo, K.; Seehra, J.S.; Ramani, K.; Cipra, R.J. HexaMorph: A reconfigurable and Foldable Hexapod Robot Inspired by Origami. In Proceedings of the IEEE/RSJ International Conference on Intelligent Robots and Systems, Chicago, IL, USA, 14–18 September 2014.

6. Yao, S.; Georgakopoulos, S.V.; Cook, B.; Tentzeris, M. A Novel Reconfigurable Origami Accordion Antenna. In Proceedings of the EEE/RSJ International Conference on Intelligent Robots and Systems, Chicago, IL, USA, 14–18 September 2014.

7. Trease, B.P.; Thomson, M.W.; Sigel, D.A.; Walkemeyer, P.E.; Zirbel, S.; Howell, L.; Lang, R. Origami Inspired Folding of Thick, Rigid Panels; NASA Technical Reports Server, National Aeronautics and Space Administration: Washington, DC, USA, 2014.

8. Shim, J.; Perdigou, C.; Chen, E.R.; Bertoldi, K.; Reis, P.M. Buckling-induced encapsulation of structured elastic shells under pressure. Proc. Natl. Acad. Sci. USA **2012**, 109, 5978–5983.

9. Alderson, K.; Alderson, A.; Anand, S.; Simkins, V.; Nazare, S.; Ravirala, N. Auxetic warp knit textiles structure. Basic Solid State Phys. **2012**, 249, 1322–1329.

10. Ugbolue, S.C.; Kim, Y.K.; Warner, S.B.; Fan, Q.; Yang, C.-L.; Kyzymchuk, O.; Feng, Y.; Lord, J. The Formation and Performance of Auxetic Textiles. Part II: Geometry and Structural Properties. J. Text. Inst. **2011**, 102, 424–433.

11. Glazzard, M.; Breedon, P. Weft-knitted auxetic textile design. Basic Solid State Phys. **2013**, 251, 267–272.

12. Liu, Y.; Jimmy, H.H.; Lam, K.C.; Liu, S. Negative Poisson's Ratio Weft-knitted Fabrics. Text. Res. J. **2009**, 80, 856–863.

13. Hu, H.; Wang, Z.; Liu, S. Development of auxetic fabrics using flat knitting technology. Text. Res. J. **2011**, 81, 1493–1502.

14. Alderson, A.; Kim, A. Expanding materials and applications: Exploiting auxetic textiles. Tech. Text. Int. **2005**, 14, 29–34.

15. Wang, Z.; Hu, H. Auxetic materials and their potential applications in textiles. Text. Res. J. **2014**, 84, 1600–1611. [

16. Wang, Z.; Hu, H. 3D auxetic warp-knitted spacer fabrics. Basic Solid State Phys. **2014**, 251, 281–288.

17. Dion, G. Garment Device: Challenges to Fabrication of Wearable Technology. In Proceedings of the 8th International Conference on Body Area Networks, Brussels, Belgium, 30 September–2 October 2015.

18. Yuskel, C.; Kaldor, J.M.; James, D.L.; Marschner, S. Stitch meshes for modeling knitted clothing with yarn-level detail. ACM Trans. Graph. Proc. ACM SIGGRAPH **2012**, 31. Available online: http://www.cs.cornell.edu/projects/stitchmeshes/stitchmeshes-highres.pdf (accessed on 14 September 2014).

19. Underwood, J. The Design of 3D Shape Knitted Preforms. Ph.D. Thesis, RMIT University, Melbourne, Australia, 2009.

20. Jackson, P. Folding Techniques for Designers: From Sheet to Form; Laurence King Publishing Ltd.: London, UK, 2011. [

21. American Society for Testing and Materials International. Standard Specification for Tensile Testing Machines for Textiles; ASTM D76/D76M-11; ASTM International: West Conshohocken, PA, USA, 2006.

22. Luminita, C.; Florin, F. Experimental Study on the Mechanic Behaviour of Weft Knitted Fabrics. Fibers Text. East. Eur. **2012**, 20, 34–39.

23. American Society for Testing and Materials International. Standard Test Method for Breaking and Elongation of Textile Fabrics (Grab Test); ASTM D5034-09; ASTM International: West Conshohocken, PA, USA, 2013.

CHAPTER 3

Systems thinking in designing automotive textiles

Pammi Sinha*, Subramanian Senthilkannan Muthu, Iain Taylor, Rita Schulze, Katharine Beverley, Claire Day and Nicoleta Tipi

Sinha, Beverley, Day, Schultz: Department of Fashion and Textiles,University of Huddersfield, Queensgate, Huddersfield HD1 3DH, UK

ABSTRACT

We present the complexities in terms of designing automotive exterior seating materials (seat coverings and interior linings) at Sage Automotive Interiors (UK), which is a division of a global international automotive textile supplier with headquarters in the US. Sustainability and innovation are emphasized in documents communicating the company's vision. Using a case study approach, we consider the current design, development and manufacture process and examine it for the potential for feedback loops, identify potential leverage points to effect change and how the process could divert wastes from disposal. We will highlight where sustainable decisions can be incorporated and the difficulties in achieving true sustainability. We argue that a systems approach is needed from conception to final product to ensure economic recycling of textiles and fibres used in automotive seating. Without which, the reality is at best incineration for energy and/or landfill, thus losing important, finite resources forever from a diminishing resource pool of raw materials.

KEYWORDS

Automotive textiles; Sustainability Systems; Recycling; Incineration; Waste Landfill

1. BACKGROUND

1.1 Systems thinking

Systems may be defined as a set of independent but interlinked phenomena; identifying and using beneficial relationships and linkages between different parts of a system is key to optimizing the whole (Charnley et al. 2011; Mingers and White 2010). Meadows (1997) identified nine places of intervention to influence and change systems: (9) numbers - subsidies, taxes and standards, (8) material stocks and flows - substituting with more sustainable material, (7) regulating negative feedback loops - control undesirable factors by legislating to minimize harmful materials or processes, e.g. REACH, (6) driving positive feedback loops - encourage desirable factors through promotion and aid for sustainable practices, (5) information flows - working towards transparency in the supply chain, (4) rules of the system - develop existing or new practices and policies beneficial to all stakeholders, (3) the power of self-organization - internal intervention akin to organizational or self-learning, (2) the goals of the system - profit to survive and compete, but now incorporate social and environmental concerns, and (1) the mindset or paradigm out of which the system arises - the accepted status quo, influencing this can bring about the biggest changes. All systems have inputs, outputs and feedback mechanisms; for this paper, we examine the design process for automotive textiles and the system within which it operates in to outline the implications for environmental product design.

1.2 Sustainability in the automotive industry

The European Union's end-of-life vehicles (ELV) directive stipulates that, by 2015, 95% of new vehicles need to be recyclable (EURAPA 2010) and that 'vehicle manufacturers and material and equipment manufacturers must: (a) endeavour to reduce the use of hazardous substances when designing vehicles, (b) design and produce vehicles which facilitate the dismantling, reuse, recovery and recycling of end-of-life vehicles, (c) increase the use of recycled materials in vehicle manufacture and (d) ensure that components of vehicles placed on the market after 1 July 2003 do not contain mercury, hexavalent chromium, cadmium or lead, except in the applications listed in Annex II of the Directive (Normand 2008). Manufacturers are requesting textiles that disassemble and are recyclable; suppliers are being requested to recall their products after use to recycle them (Normand 2008), potentially forcing readjusting their operations to support this new additional element to their business. Many original equipment manufacturers (OEMs) currently use the design for disassembly processes using compatible chemicals, components and similar fibre types to increase their recycling and reuse efficiencies. The automotive industry is interesting because environmental sustainability of textiles is not (yet) the main focus of the car companies' sustainability efforts, which centre on fuel efficiency and lightweight cars. OEMs are therefore much more hesitant to make improvements regarding EPD for those 'peripheral' parts of the car which customers are less likely to ask about and where the influence on sales and reputation is perceived as uncertain. On the other hand, seating textiles are visible to the customer, and so, it is important to understand the internal and external pressures which the design team faces. Therefore, the design process is analysed to get a detailed understanding of the factors driving the design decisions.

1.3 The automotive assembly system

Often described as tiered, up to four tiers excluding the OEMs, tier 1 supplies interior compartments, e.g. seats, engines, floors; tier 2 carries out subassembly activities such as cut and sew and supply parts such as headliners; tier 3 supplies the manufacturing for the upholstery whilst tier 4 supplies the materials (Shishoo 2008). Tiers 1 and 2 companies may be integrated, and 'cut and sew' operations may be carried out by the same company. Fabric laminators supply their services the cut and sew operations (Tier 2). The fabric formation companies may also conduct fabric lamination in-house or out-source. There are many permutations; companies may supply yarn directly to the fabric formation company, foam and scrim directly to a fabric lamination company or only yarn to the fabric formation company and the fabric formation company supplies the scrim to the fabric lamination company. Moreover, the three companies may be integrated (Taylor 2012). An example is the consolidation by tier 1 companies. The OEMs, frustrated by the increasingly longer lead times and operational constraints, only worked with fabric suppliers with cut and sew operations. Tier 1 companies (e.g. Johnson Controls) have bought out fabric companies and become vertically integrated companies with control of upstream *and* downstream activities making the full range of products from seats, instrument panels, headliners, etc. With 25% of the market, Johnson Controls vertical integration captures increasing amount of value along the chain but loses flexibility and speed. The independent status of Sage AI is advantageous because it affords flexibility and speed within the system (Taylor 2012).

2. METHODS

A single case study, according to Yin (2009), was conducted using a number of semi-structured interviews from November 2011 until July 2014 (on site and off site), with the managing director, design and development director, designers and members of the logistics and technical team. The interviews were recorded, transcribed and triangulated with the company's documents and literature regarding product development process. The analysis was validated by the design and development director. The commercial sensitivity, the detailed nature of the analysis and associated time requirements necessitated a single case study. Sage AI (UK) was chosen as a case study company due to their strong interest in environmental product development and existing close working relationships with the University.

To analyse the design process, we used Integration Definition for Function (IDEF) modelling, a standardized and widely used method to model manufacturing processes (Rashid and Ismail 2007). Used as a visual tool, it helps describe, analyse and communicate ideas (Yu and Wright 1997). It has proven effective for detailed system analysis (Knowledge Based Systems 2010). The modelling often starts from a description of the process at the highest level, which is then broken down hierarchically to the level of detail required for a meaningful analysis. This allows the user to describe the details of complex processes whilst keeping a good overview of the activities (Sugiyama et al. 2008).

IDEF-0 has also been used in the past to support corporate decision-making processes and to support the redesign for the improvement of processes (Hernandez-Matias et al. 2008). In recent years, IDEF-0 has been used to help introduce environmental criteria into business processes. For example, IDEF-0 models were used to illustrate new process designs, considering environmental and health criteria (Hirao et al. 2008) or to show how risk assessment and life cycle assessment can be facilitated using IDEF-0 to help understand the process and information requirements (Kikuchi and Hirao 2009). Kikuchi et al. (2012) used IDEF-0 to illustrate a systematic assessment process for environmental, health and safety aspects in the process development phase. Sugiyama et al. (2008) used IDEF-0 as a modelling tool to support the integration of LCA in process design - the method helped them to highlight information requirements for the newly required activities.

According to the IDEF-0 standard, the purpose and viewpoint, system boundaries and meaning of the arrows should be defined. For this study, the purpose of the model was to understand the design development process of the case study company and to identify where environmentally relevant decisions are made in this process. The model takes the perspective of the design and development director, with the intention to focus on the steps in the design and product development process stages. As a consequence, the model includes all processes which the design team are directly involved in, starting with research and development, design concept development and selection by client, up to the point where a master sample of the fabric has been selected as a template for production. Figure 1 demonstrates a basic structure of the IDEF-0 (National Institute of Standards and Technology 1993; Kikuchi et al. 2012). The control arrows in the IDEF-0 model represent factors which need to be taken into account in the respective process step. They are the most relevant factors for the identification of the barriers.

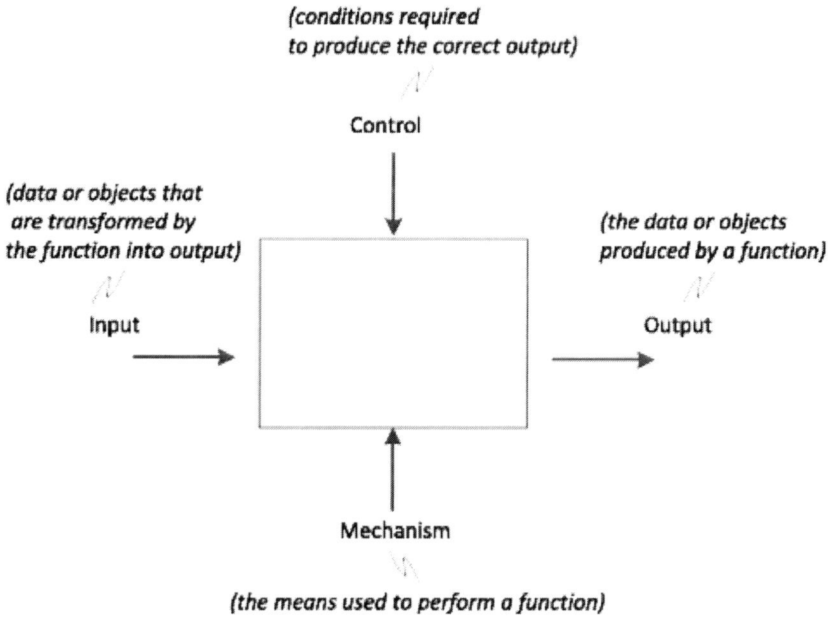

Figure 1 Basic structure of IDEF-0, based on (National Institute of Standards and Technology **1993**).

The purpose of the IDEF-0 modelling was to obtain a detailed examination of the product design/development process at the case study company and was used as a basis to identify the following:

1. where in the design process the most environmentally relevant decisions are being made

2. the crucial factors affecting these decisions, and
3. barriers to and enablers for environmental product development in the current design process, using the introduction of recycled yarn as a particular case study example.

3. RESULTS AND DISCUSSION

3.1 Sage AI (UK) company background

Sage AI, originally a part of the former Milliken Automotive and the Viktor Achter Groups. Sage Automotive Interiors (Sage AI) business is headquartered in the US, and solely in the automotive sector with bases in Poland, Japan, India, South America and China. The Sage AI business is entirely related to the automotive industry. Their fundamental business is design, incorporating engineered and technical qualities to add value to their products. Sustainability and innovation is paramount to their vision.

Sage AI produces textile interiors based on two manufacturing routes: woven or knitted. Most developments are related to enhancing these processes, e.g. through embossing, stitch and HF welding. The samples are developed through their plants overseas; thus, knowledge of their capabilities is essential. This knowledge accrues through experience, and the creative or technical designers visit the partner manufactures every month. Most innovations are for enhancing the fabrics, e.g. rotary embossing, 12 head embroidery machine and HF welding. Various OEMs now produce shorter runs and offer special editions to add more interest to the colour and design of the automotive Powell NB (2006). The use of embroidery is a new technical capability, enabling mass customization service. When developing initial ideas (for either internal or external clients), Sage AI liaise with their suppliers' technicians to establish feasibility with partner companies; where, if it is not possible, they will develop new supplier relationships and run feasibility trials.

Sage AI (UK) own manufacturing plants in Poland (weave) and Egypt (knit). The creative or technical designers visit the manufacturers monthly to maintain communication and address any design/production issues. Costs limit fibre choice usually to polyester which is then designed for variety of looks, e.g. highly lustrous, soft matt, textured and novelty yarns to create interest in the woven and knitted fabrics. Wool can be used but is expensive, therefore limited (Sage used wool for the BMWi3). Yarns are supplied by Autofil Worldwide Ltd., who also piece- or yarn-dye and texturize to a very wide variety of looks and feel; the blend, has to be visually appealing to the customer (the OEM), pass various rigorous tests (such as abrasion and light fastness), be durable (up to 10 years) and most importantly, cost effective.

3.2 Application of textiles to seating

The manufactured polyester fabrics are usually laminated with foam and a lightweight fabric 'scrim'. These are manufactured to a schedule and then shipped to a cut and sew company. These companies, often part of the final seat manufacturer, receive the fabric in roll form and mechanically cut parts which are then sewn together to create a seat cover. The covers are then usually shipped, often over long distances, to where the seat itself is being made, where they are incorporated into the final seat. The seat is then sequenced in trackside as the car they are destined for is being made just in time; the seat assembler receives notification of which type, colour and style manufacture of the seat is required a few hours before it is required to meet the rest of the car on the assembly line. Other parts, e.g., engines, headliners, instrument panels, doors etc. are equally sequenced in.

3.3 The design and product development process

The product development process may take between 3 and 5 years from brief setting to final vehicle production. This is extended by the additional 7 years approximate length of time that the textiles will be in vehicle. Figure 2 outlines the process from concept to master sample; for ease of describing the process, stages have been assigned various developmental stages of the process (stages A1 to A4).

Product development or innovation begins with a brief, either from the customer/client (OEM, which designers interviewed termed as *client based*) or in-house through their own product/consumer knowledge (e.g. designer, manager, etc. - the designers interviewed termed this as *open-line brief*). Sage AI developed FXC for Honda from a brief set by Honda, whereas YES Essentials was developed through an open-line brief based on internal knowledge and consumer research. It is not uncommon for the OEMs to tender briefs to more than one supplier at a time. OEMs distribute requirements via a brief to specify requirements in terms of objectives for the textiles; this process may commence up to 3 to 5 years in advance of the final vehicle production dependent on whether incremental or innovative design is involved. Dependent on the client's requirements, the brief may focus on pattern requirements and technological developments, whilst others may be more responsive to emotive desires of the consumer (Figure 3a,b).

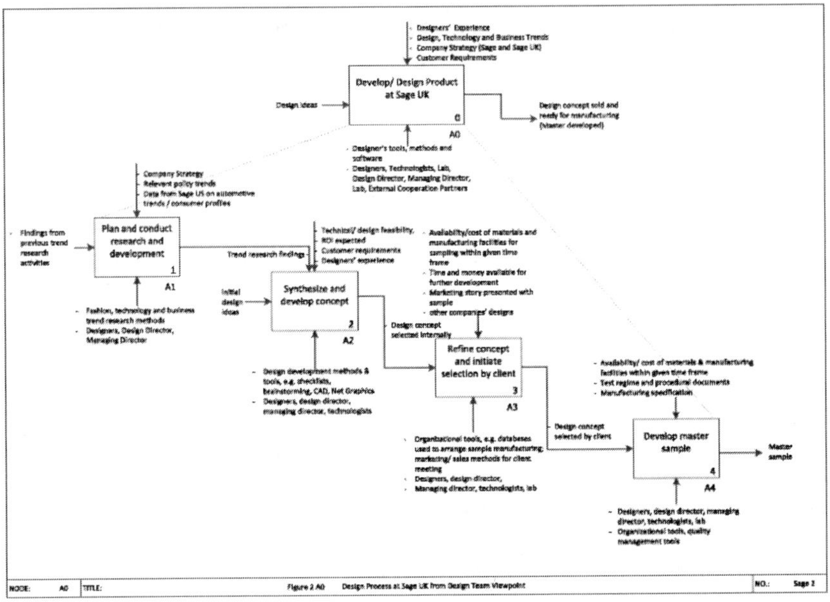

Figure 2 . Product development at Sage AI (concept to master sample).

It may be noted that an extra consideration needs to take place when developing a concept by open-line brief, in that the technical feasibilities and standards for the concept need to be researched; moreover, internal briefs are no longer common at Sage as the resource implications and risks to approval by a potential client are high and raise too many complex questions when considering how copyright is assigned and to whom (Sage or their client).

Stage A1: plan and conduct research and development

This initial stage A1 can take between 2 weeks and 2 months dependent on the client's needs. According to Powell, a brief will consist of performance [standards] and cost parameters, colour, pattern, scale of pattern, lustre and handle which is then taken forward by the designers (Powell 2008a). Timing and major stage-gates will also be confirmed. If the initial stage-gate is short, the design team will show concepts already *in situ* to support quick turnaround of sampling; when the stage-gate is long, the designers will develop new concepts for a particular brief, and this may include two stages prior to sampling.

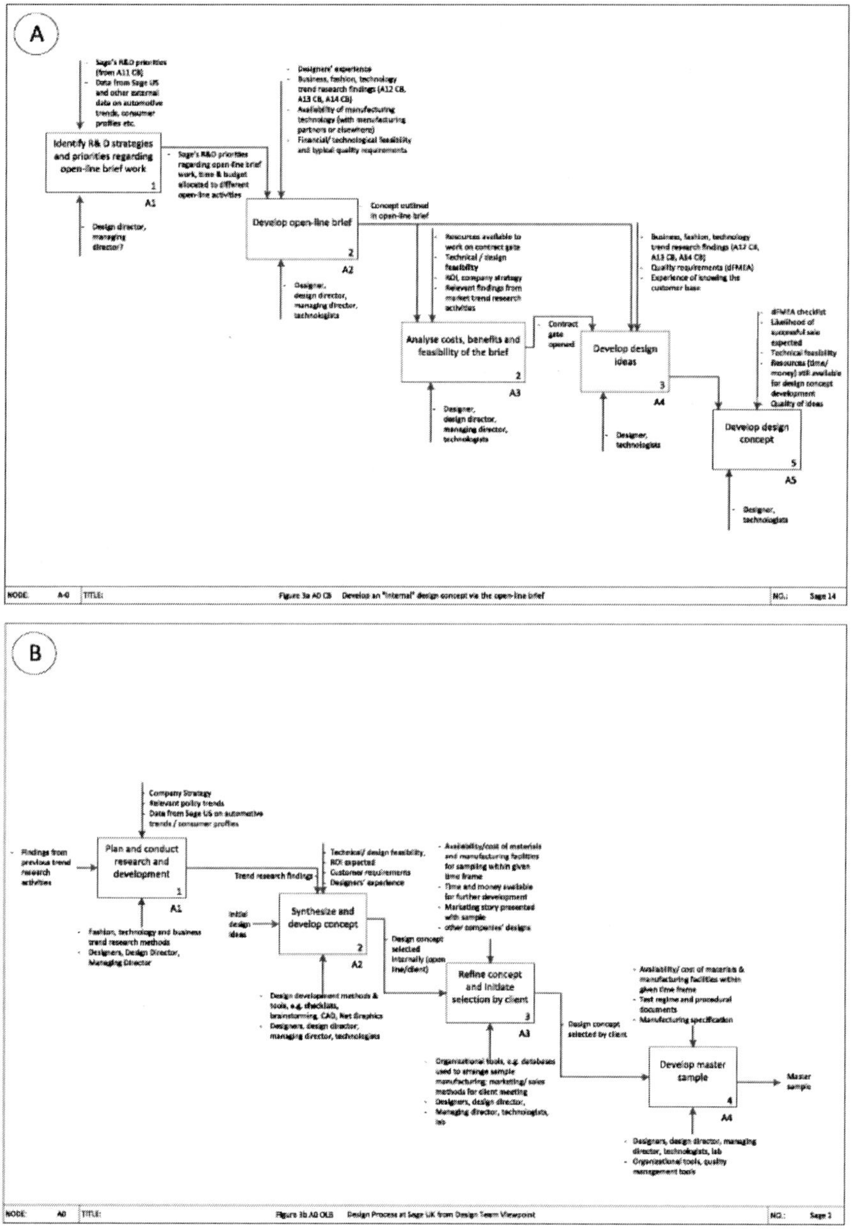

Figure 3. Open-line brief. (A) Open-line brief product development. **(B)** OEM set brief product development.

Vehicle interiors are mainly plain, striped, checked or small-scale ditsies in neutral colouring with the addition of a small percentage of a brighter accent colour to add interest (Powell 2008a). Various OEMs who produce shorter runs have started to offer special editions to add more interest into the colour and design of the automotive (Powell et al. 2008b citing Powell 2006). As already stated, vehicles are developed up to 5 years in advance; alongside this is the timeframe of 7 years that the textiles will be in vehicle production before an updated model is developed and ultimately replaces it. Thus, the fabric has to remain on trend for a much lengthier amount of time than fashion or home interiors in terms of appearance and colour.

Due to the long lead times, a global macro-trends study is projected every 4 years and micro-trends analysed such as ‘luxury’, ‘sports’ and ‘comfort’ to focus the design team. Sage AI used to profile geographically but now references according to age and lifestyle, e.g. ‘youth’ and ‘active’. Other elements to be considered are environmental trends, colour, texture and consumer profile. The designers produce a global colour show consisting of four themes to maintain constant fabric design and development and visit trade shows to keep abreast of latest yarn or fabric development techniques and suppliers. Figure 4a illustrates the open-line brief route whilst 4b illustrates the OEM set route.

The development of an internal brief provides an opportunity for Sage to develop environmentally friendly innovations. Whether those are taken up by the car manufacturers depends on trends and politics, company strategies and product cost. The risks of the product being accepted are high at this stage; therefore, it is important for Sage to research trends, politics, potential clients’ company strategies and estimate the product’s cost and feasibility; all of this is resource intensive.

Stage A2: synthesis and development of concept

The initial idea is transferred into a CAD software. Figure 4c illustrates that the open-line brief route requires an examination of the return on investment of resources spent on developing the concept and its fit with the company’s strategic priorities before they proceed to developing it into design ideas for the client.

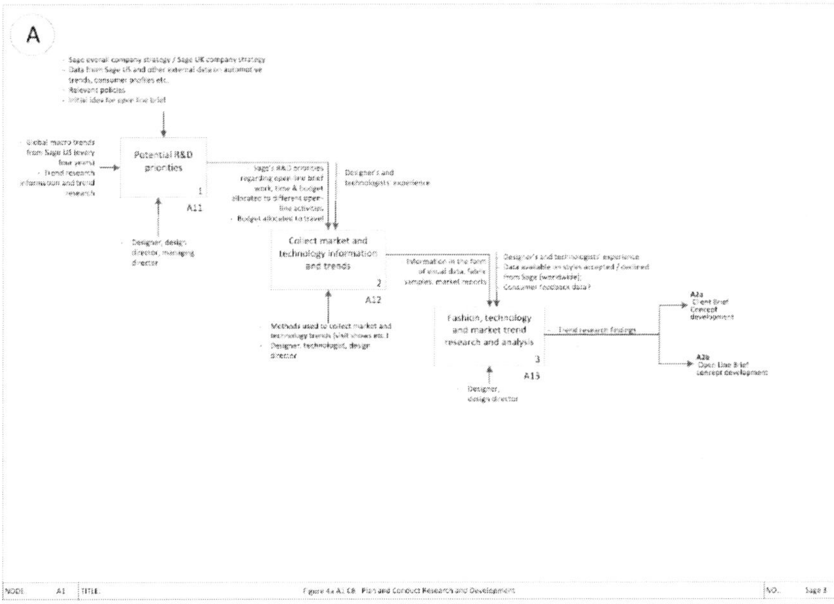
A
Relevant policies
Potential R&D priorities
1
A11
Designer's and technologists' experience
Collect market and technology information and trends
2
A12
Fashion, technology and market trend research and analysis
3
A13
Designer, design director
A2a
A2b
NODE A1 TITLE
Figure 4a A1 CB Plan and Conduct Research and Development
NO. Sage 3

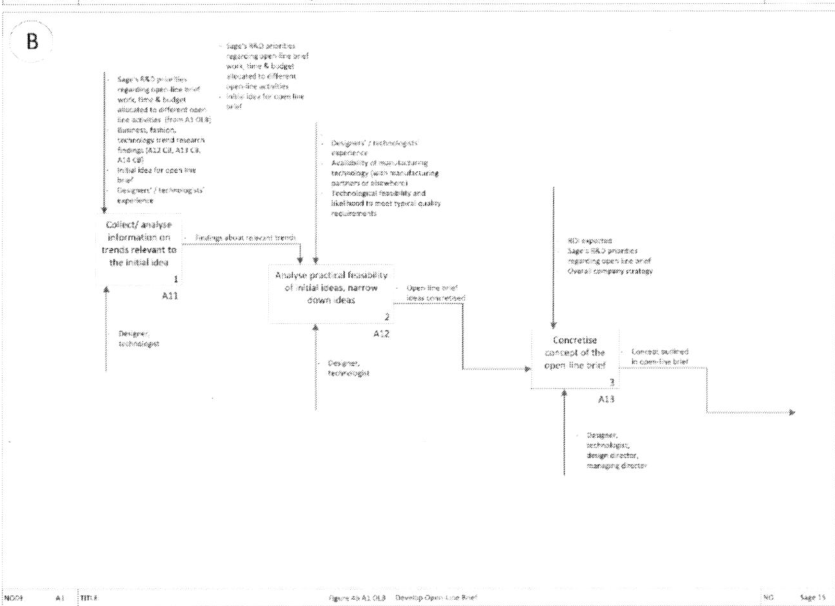
B
Collect/ analyse information on trends relevant to the initial idea
1
A11
Designer, technologist
Analyse practical feasibility of initial ideas, narrow down ideas
2
A12
Concretise concept of the open line brief
3
A13
Designer, technologist, design director, managing director
NODE A1 TITLE
Figure 4b A1 OLB Develop Open Line Brief
NO Sage 15

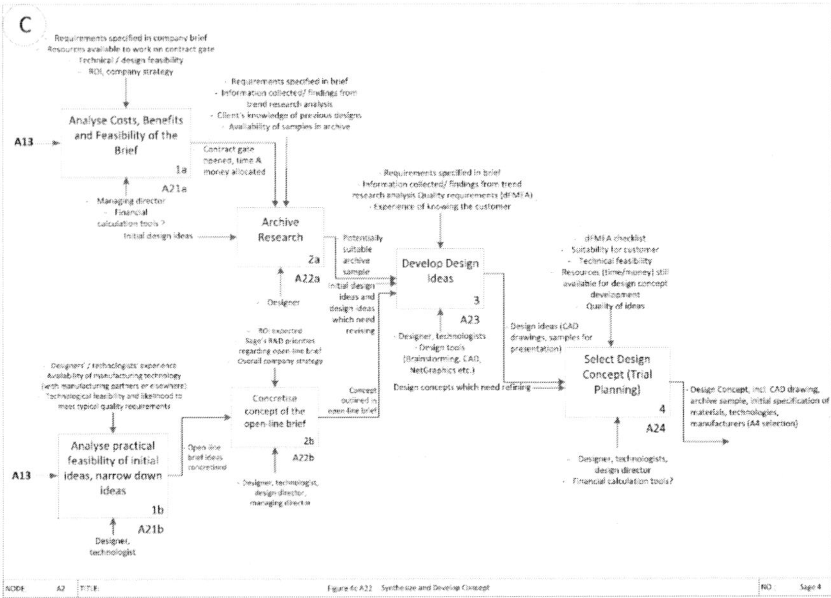
C
Requirements specified in company brief
Resources available to work on contract gate
Technical / design feasibility
ROI, company strategy
A13
Analyse Costs, Benefits and Feasibility of the Brief
1a
A21a
Managing director
Financial calculation tools ?
Contract gate opened, time & money allocated
Requirements specified in brief
Information collected/ findings from trend research analysis
Client's knowledge of previous designs
Availability of samples in archive
Initial design ideas
Archive Research
2a
A22a
Designer
Potentially suitable archive sample
Initial design ideas and design ideas which need revising
Requirements specified in brief
Information collected/ findings from trend research analysis Quality requirements (dFMEA)
Experience of knowing the customer
Develop Design Ideas
3
A23
Designer, technologists
Design tools (Brainstorming, CAD, NetGraphics etc.)
Design ideas (CAD drawings, samples for presentation)
dFMEA checklist
Suitability for customer
Technical feasibility
Resources (time/money) still available for design concept development
Quality of ideas
Design concepts which need refining
Select Design Concept (Trial Planning)
4
A24
Designer, technologists, design director
Financial calculation tools?
Design Concept, incl. CAD drawing, archive sample, initial specification of materials, technologies, manufacturers (A4 selection)
ROI expected
Sage's R&D priorities regarding open-line brief
Overall company strategy
Concretise concept of the open-line brief
2b
A22b
Designer, technologist, design director, managing director
Concept outlined in open-line brief
Designers' / technologists' experience
Availability of manufacturing technology (with manufacturing partners or elsewhere)
Technological feasibility and likelihood to meet typical quality requirements
A13
Analyse practical feasibility of initial ideas, narrow down ideas
1b
A21b
Designer, technologist
Open-line brief ideas concretised
NODE A2 TITLE: Figure 4c A22 Synthesize and Develop Concept
NO: Sage 4

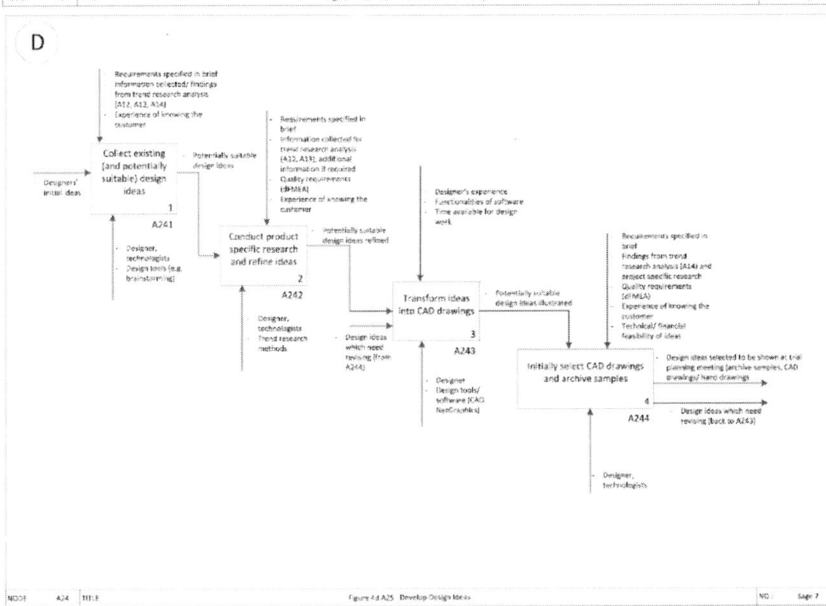
D
Requirements specified in brief
Information collected/ findings from trend research analysis (A12, A13, A14)
Experience of knowing the customer
Designers' initial ideas
Collect existing (and potentially suitable) design ideas
1
A241
Designer, technologists
Design tools (e.g. brainstorming)
Potentially suitable design ideas
Requirements specified in brief
Information collected for trend research analysis (A12, A13), additional information if required
Quality requirements (dFMEA)
Experience of knowing the customer
Conduct product specific research and refine ideas
2
A242
Designer, technologists
Trend research methods
Potentially suitable design ideas refined
Designer's experience
Functionalities of software
Time available for design work
Transform ideas into CAD drawings
3
A243
Design ideas which need revising (from A244)
Designer
Design tools/ software (CAD, NetGraphics)
Potentially suitable design ideas illustrated
Requirements specified in brief
Findings from trend research analysis (A14) and project specific research
Quality requirements (dFMEA)
Experience of knowing the customer
Technical/ financial feasibility of ideas
Initially select CAD drawings and archive samples
4
A244
Designer, technologists
Design ideas selected to be shown at trial planning meeting (archive samples, CAD drawings/ hand drawings
Design ideas which need revising (back to A243)
NODE A24 TITLE: Figure 4d A25 Develop Design Ideas
NO: Sage 7

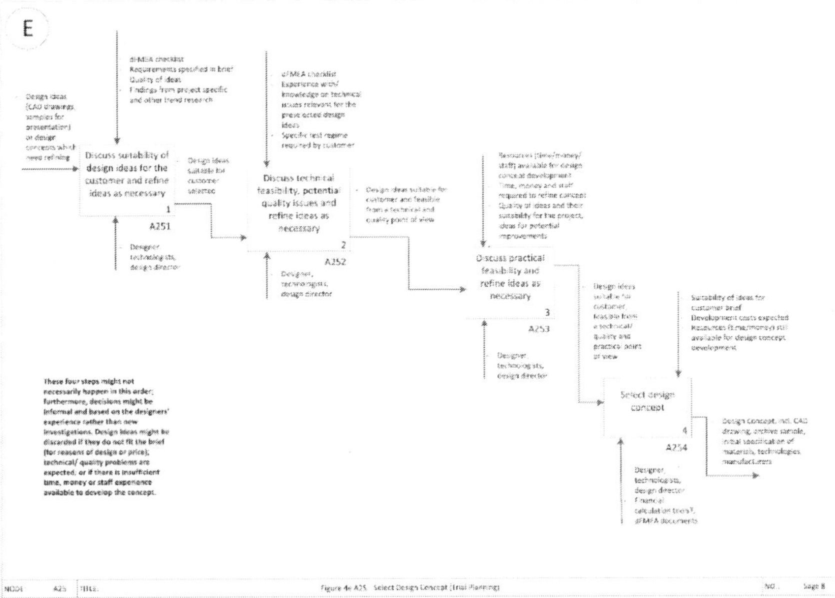

Figure 4 . Stages A1 to A25. (A) Client based brief route for stage A1: plan and conduct research and development. **(B)** Open-line brief route for stage A1: plan and conduct research and development. **(C)** Stage A2: synthesis and development of concept. **(D)** Stage A24: developing the design idea. **(E)** Stage A25: select design concept (trial planning).

During stage A2, the design idea is developed into a concept (that will be discussed at a meeting with potential clients). At stage A24, the design idea is examined for selection with regard to whether to continue to refine or not. The design ideas might not be good enough (in which case, they need to go back one step). A design functional modes and effects analysis (dFMEA) review is undertaken which outlines the technical requirements (included in dFMEA) to narrow down the selection of potential fabrics and material; much of this depends on the design team's knowledge of what is likely to work. Technological requirements are manufacturer dependent. Figure 4d illustrates the activities surrounding the decision to select the design concept to develop further.

At stage A25, the concept is refined towards developing a sample. The design concept (perhaps in the form of a CAD illustration, or a selection of archive samples), includes specification of materials, technologies required for the making, sources of laminators, etc.

Figure 4e illustrates the activities involved in refining the idea into a concept that will be developed for discussion in a client meeting. The concept is produced on a CAD programme (in this case, NED Graphics); this is emailed to the specified manufacturing location and processors and (dependent on colourways) will take 4 to 6 weeks for sampling to materialize. If the colour of the yarn is already available (e.g. black), this would be a 4-week process; however, if the yarn needs to be dyed specifically to match other elements of the interior such as vinyls and leathers, this can take approximately 6 weeks.

Stage A3: refine concept and initiate selection by client

Concepts considered feasible for client discussions are developed into a sample representative of the actual fabric that may be used by the client company (Figure 5a).

Samples to be considered may be new samples designed or they may be archival samples or samples that did not quite achieve the appropriate standards in a previous round of testing. The overriding considerations are return on investment (likelihood to get the job, likelihood to get good money out of job if accepted), availability of machinery and raw material within specified time frame, time availability and cost of sampling. The initial decision-making regarding whether a sample should be put into manufacture is illustrated in Figure 5b.

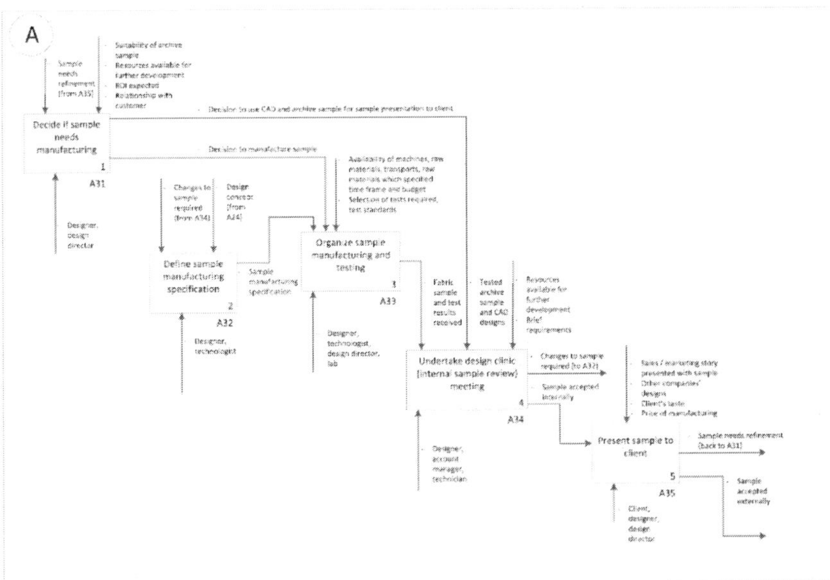

Figure 5a A3 Refine Concept and initiate Selection by Client

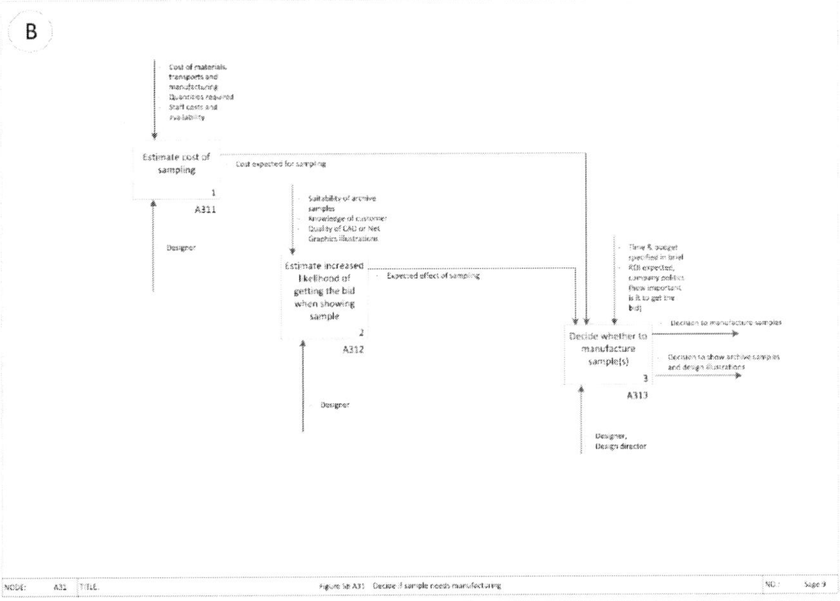
B
Cost of materials, transports and manufacturing
Quantities required
Staff costs and availability
Estimate cost of sampling
1
A311
Cost expected for sampling
Designer
Suitability of archive samples
Knowledge of customer
Quality of CAD or Net Graphics illustrations
Estimate increased likelihood of getting the bid when showing sample
2
A312
Expected effect of sampling
Designer
Time & budget specified in brief
ROI expected, company politics (how important is it to get the bid)
Decide whether to manufacture sample(s)
3
A313
Decision to manufacture samples
Decision to show archive samples and design illustrations
Designer, Design director
NODE: A31 TITLE: Figure 5b A31 Decide if sample needs manufacturing NO.: Page 9

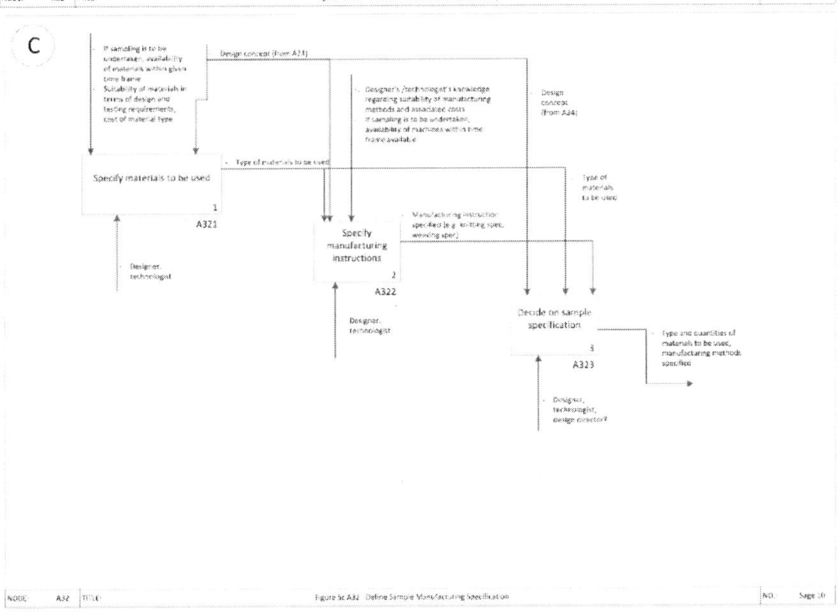
C
If sampling is to be undertaken, availability of materials within given time frame
Suitability of materials in terms of design and testing requirements, cost of material type
Design concept (from A24)
Designer's /technologist's knowledge regarding suitability of manufacturing methods and associated costs
If sampling is to be undertaken, availability of machines within time frame available
Design concept (from A24)
Specify materials to be used
1
A321
Type of materials to be used
Type of materials to be used
Designer, technologist
Specify manufacturing instructions
2
A322
Manufacturing instruction specified (e.g. knitting spec, weaving spec)
Designer, technologist
Decide on sample specification
3
A323
Type and quantities of materials to be used, manufacturing methods specified
Designer, technologist, design director?
NODE: A32 TITLE: Figure 5c A32 Define Sample Manufacturing Specification NO.: Page 10

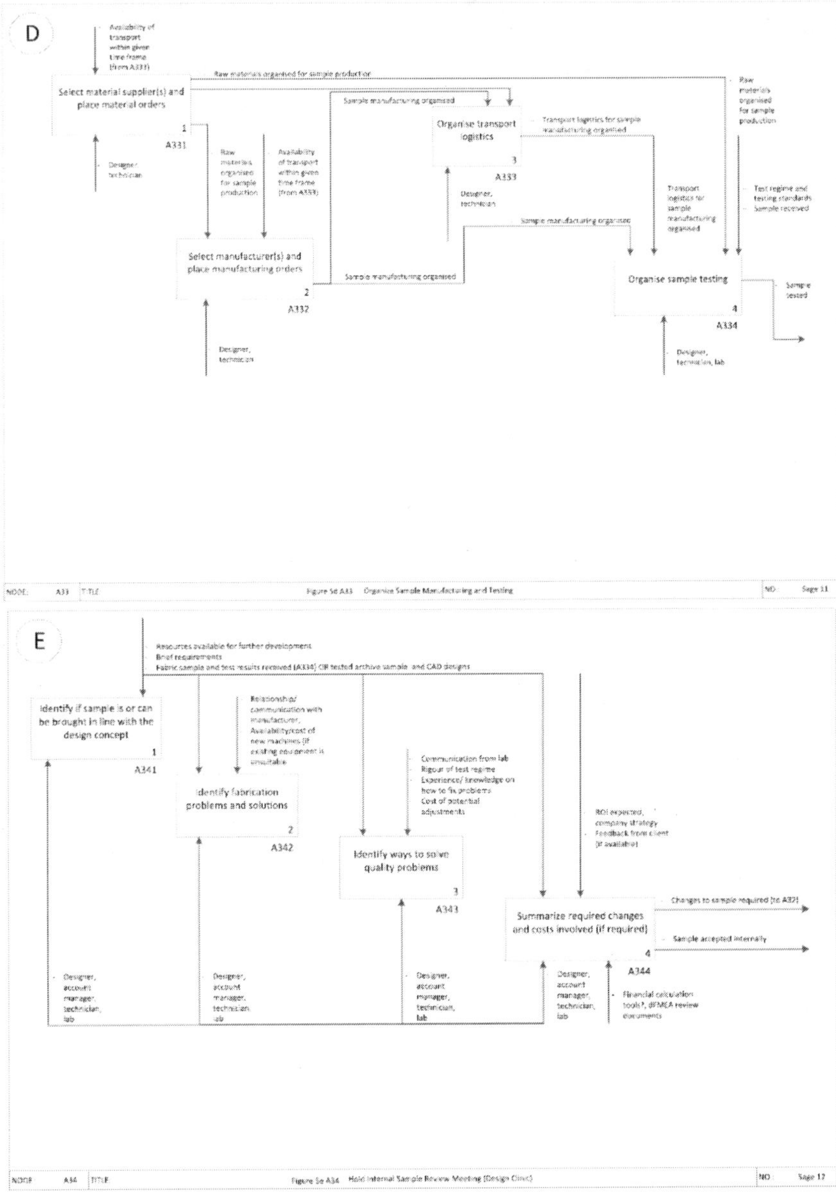

Figure 5 Stages A3 to A34. (A) Stage A3: refine concept and initiate selection by client. **(B)** Stage A31: initial decision if a sample should be manufactured. **(C)** Stage A32: defining sample manufacturing specification. **(D)** Stage A33: organizing sample manufacturing and testing. **(E)** A34: hold an internal review meeting (design clinic).

Once the decision is made to take a concept to the manufacturing level, specifications for manufacturing need to be developed and defined, as illustrated in Figure 5c.

As the sample specifications are being defined, there is also the need to organize the logistics for the transportation of the samples and their testing. Samples in automotive textiles are many metres of fabrics, representing bulk and hence the requirement for cost and time efficiencies to be met, as illustrated in Figure 5d.

Once tested, the sample is then examined by the design team for quality of manufacture, standard of testing and financial implications. Any problems are identified with appropriate rectification routes. Samples accepted through this internal design review are presented to the client company (stage A35) who will make the decision to either review the sample again after specified modifications or to agree that the sample is further developed for manufacture.

Stage A4: develop master sample

Once the clients have reviewed the samples, decisions are made regarding the merits of developing the sample for production; the first stage of this is the 'master sample'. Approved samples from the client review at stage A35 undergo a second set of testing and colour-tuning (Figure 6); if not approved, design alterations and further sampling will occur. When samples and testing are approved, the design process is finalized, a term described by Sage AI as 'mastering', and this is whereby the whole product is fixed (yarns, design, etc.), contracts are raised, the client firm is financially committed and the design is then assigned to production.

3.4 Decision making through the design process regarding recycled textiles

Sage AI has maintained sustainability as core to their corporate values from Milliken and normally use polyester yarns for their textile products that are visible within the automotive (because it is cost-effective, recyclable and meets the rigorous test requirements). The polyester may be coloured and texturized to a very wide variety of looks and feel that are cost-effective, visually appealing to the customer (the OEM) and pass a number of rigorous tests (such as abrasion and light fastness) so that it will last for up to 10 years.

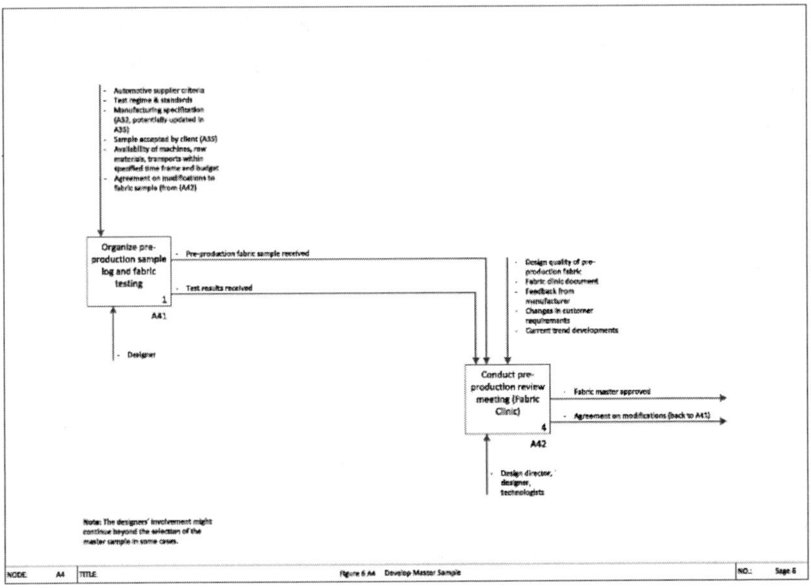

Figure 6 . Stage A4: developing the master sample.

In line with their sustainability vision, Sage AI also use a percentage of post-industrial recycled yarns and post-consumer (PTC) recycled yarns blended with virgin yarns. Post-industrial recycled yarn is easier for Sage AI to use as properties and provenance are known. Most yarn manufacturing takes place in Asia, but the suppliers of polyester yarns are based in wider locations, e.g. Sage AI (UK) have a supplier in the UK in Nottingham. As the tests that automotive yarns and fabrics have to pass are so rigorous, there are few automotive yarn suppliers. Automotive textile interior fabrics and yarns are therefore sourced from a small pool of yarn suppliers, easing the complexity of the provenance of the yarns.

PTC recycled yarns are made from 100% plastic bottles, and the benefits of using post-consumer recycled yarns is the reduction in greenhouse gas emissions, the reduced energy consumption compared to making virgin yarns from raw materials and associated landfill charges and land use for storing waste plastic bottles (Collora 2012). The benefits are somewhat counterbalanced by the considerations of another supply chain loop; post-consumer yarns are currently 100% from plastic bottles which require processing into yarns, and there are the collection of plastic bottles, cleaning, transportation and storage of bottles before manufacturing processes, raising issues about transparency within this recycling route (European Environmental Agency 2012).

The use of recycled textiles used in production is a dilemma: the recycled polyester is blended with virgin polyester and so the resultant fabric does not look different from the non-recycled fabrics, thus raising questions from OEM customers and final vehicle consumers alike about the validity of the sustainable claims and extra costs. The cost of using recycled yarns appears to be in the processing of the yarn: there is a surcharge/premium in terms of the raw materials which is based on the polyester and not the plastics making it difficult for the OEM customer to understand the costing. The cost implications are, for example, 7% mark up on fabric that is 40% recycled and 11% on fabric that is 100% recycled; when this is taken back to the yarn, the cost is double or three times that on the normal price of the yarns. According to Ford's research shared with Sage AI, the premiums charged on recycled yarns varied across the world: 10% to 15% (USA), 15% to 20% (China, India, Brazil), 20% to 25% (Europe), 20% to 30% (Korea), and 50% to 100% (Thailand).

As well as the yarn composition, the application of textiles within the visible areas of automotives presents a problem regarding sustainability. Between seven to nine million vehicles are disposed of annually in the EU; only 75% of this is recyclable. Textiles form part of the remainder non-recyclable materials (Collora 2012). Textiles from ELV seating make up about 15% of automotive shredder residue, (ASR - possible to process into vehicular sound proofing insulation and then finally incinerated) (Erth et al. 2008a, 2008b).

Seat covering textile is laminated to impart waterproofness, increased abrasion, stain, flame and UV resistance, retro-reflection properties which the OEM requires testing standards to be achieved. Lamination involves backing onto foam, usually with adhesive. Currently, the most cost-effective lamination process renders textile products to be non-recyclable as lamination involves placing adhesives onto the fabric. It is possible to replace adhesive and foam-based lamination process with three-dimensional knitted or non-woven textiles products stitch-bonded or needle-punched onto the exterior fabric enabling separation of layers and recyclability. It is expensive though and currently only in high-value cars.

As costs appear to be an overriding consideration, we examined the design process to identify where and how decisions were made that were directly related to considerations about using recycled yarns (and by implication the lamination process used). This is presented in Table 1.

From Table 1, it can be seen that there are few, if any, decisions that are under Sage AI's control, save the decision to develop an open-line brief. This is a route that is not now very common for Sage AI as it is risk-laden and demands resources in terms of time and money with no guarantees of developing a sample that wins a contract. However, having developed fabric samples in this way to show at various industry shows did place them as companies who are not only interested in sustainability but also actively seeking solutions.

Table 1 . Decision-making for using recycled yarn

Decision	Decision explained	Relevance for environmental sustainability	Relevance in recycled yarn scenario?
Determination of R&D priorities (A11)	R&D priorities (including the level of detail at which this is defined), time and budget allocated to different open-line activities	Level of priority given to EPD in R&D has direct implications for both innovative and incremental EPD activities (incl. level of detail about specific EPD aspects)	N/A
Decision whether to work on a company brief or not (contract gate opening) (A21a)	Decision whether to open a contract gate is made; a certain amount of time and money is allocated to the project	a) The decision whether to work on a brief which requires research into EPD is made, and the time and money allocated to the brief is determined (impacts on the likelihood of the bid being successful). b) The amount of time/ money allocated to a 'conventional bid' affects the time left for work on specific EPD projects. c) For any bid, the availability of resources (amount of time/ number of staff members allocated to working on the bid) could influence whether there is scope for exploring EPD as part of the design development	Yes
Decision on which open-line briefs to take forward (internal product development work) (A21b and A21b)	The decision which open-line brief ideas to work on is taken/ open-line brief ideas are concretized	Taking the R&D priorities into account, the process of narrowing down design ideas requires a decision on whether to work on EPD as part of the R&D work - and what priority they are given over other R&D ideas - this may be an implicit or explicit process	N/A
Design idea development - decision which ideas to present (A23)	Design ideas are selected (CAD drawings, samples for presentation)	A decision whether to include design ideas which require sustainable materials and production methods is made.	Yes (combined analysis with design concept selection)
Design concept selection (A24)	Design concept selected, incl. CAD drawing, archive sample, initial specification of materials, technologies, manufacturers (A4 selection)	The design concept selection will largely determine what the actual product will look like and how it will be made, which has direct implications for the environmental impact of its production.	Yes (combined analysis with design idea development)
Organization of sample manufacturing and testing (A33)	a) Suppliers and manufacturing partners are selected for the product sample (if successful, they are likely to be commissioned with the production of the product once it goes into production as well). b) Testing is organized (test regime specified by the respective OEM)	The choice of suppliers and manufacturing partners will affect the environmental impact of the product. The importance of the choice of the manufacturing partners depends on the availability of alternatives and the variability between the manufacturing processes with regards to their environmental impact (e.g. the age of the machines used, materials used, energy sources etc.) - the differences are likely to be greater if the market is global.	Yes

Decisions around developing sustainable textiles solutions for automotives are through negotiation with the lead firm, the OEM that is seeking to manufacture the model. A study of incorporating sustainability into the supply chain management at Volkswagen (Koplin et al. 2007) noted essentially two forms of environmental supply management: (1) the integration of environmental criteria/standards into product and production-related decisions along the whole supply process ('greening the supply chain') and (2) the optimization of the environmental compatibility of purchased goods ('product-based green supply') (Vermeulen et al. 2011).

In the first route, the supplier is evaluated for their environmental performance; achieving and providing certificates for environmental management systems such as ISO 14001 are regarded as minimum qualification criteria for discussions to commence about supplying a product. Sage AI (UK) has ISO 14001 and will also work with suppliers of their raw materials who have either ISO 14001 or QS-9000 (an automotive standard developed by General Motors, Chrysler and Ford based on ISO 9000). In the second route, all stages of the life cycle are considered from packaging, recycling and disposal. As products are developed in accordance to specifications set by the OEM in the brief, measurable sustainable criteria should help to evaluate the environmental credentials of the product. However, as the case study with Sage AI (UK) attests, the interest in green procurement of sustainable textile products for the automotive sector in Europe has not gained enough ground to warrant all or majority of textile design to be from recycled polyester or the use of non-adhesive lamination. A recent study of sustainable supply chain management within the Indian textile industry not only noted that it was important to identify the most influential drivers within the region as these enablers were observed to be context based (Diabat et al. 2014) but they also observed that within the automotive sector, the customer was a very influential driver to sustainable textile design.

4. CONCLUSION

This paper presents the dichotomies in developing sustainable textile materials for the automotive industry. Recycled and renewable raw materials are two main drivers in the growth of the Mobiletech markets which accounts for about 22% of the technical textiles base, making it the largest single market and a growth area. There is, however, a disparity in the use of recycled yarns in automotives: in the US market, the Ford Mondeo Fusion Hybrid has been reported as having 100% recycled yarn whilst the model in the Asian and EU markets is reported to have up to 43% recycled yarn (Green Supply Chain News 2013). According to Messe Frankfurt (2013), the use of textiles within automotive has risen from 20 kg per car in 2,000 to 26 kg in 2012 and is set to rise to 35 kg in 2020, driven partly by the use of textiles composites and substitution of heavier materials for textiles to cover fuel consumption and CO2 emissions. Sustainability is important for consumers in the mature automotive markets of the USA and EU; of the top ten decisions driving purchase of automobiles, fuel efficiency was number one and that environmental considerations was number four (KPMG 2013).

As the use of recycled fabrics (and the use of any recyclable textile component) push up the final costs of the car, the consumer has to be interested to buy recycled/sustainable, and this is not easy to predict: there is a large interest in the North Americas, but not so much of a concern within Europe. Research conducted by (European Environmental Agency 2012) identified that in their sample of 1,000 men and women in USA, 70% of them felt that sustainability was now a mainstream issue for them and that 9 out of 10 of these 'green consumers' wanted to know where recycled materials are used and what recycled fibre brands are used in their cars. The decision to use recycled content within the product therefore rests on cost and questions that need to be addressed by Sage AI (UK) which include:

- Is it a 'deal-breaker', i.e. would the use of recycled textiles parts secure a sale?
- What are the cost implications on the total supply chain?
- Does legislation present any advantages in the use of recycled textiles in production?

Our study demonstrates how the design for automotive textiles is a system, whereby design decisions are a series of negotiations with the client OEM. Our interviews found that although Sage AI headquarters did provide worldwide trends information to all locations, there was some concern regarding consumer knowledge about the EU markets; the implications for Sage AI (UK) are most notable in their ability to persuade the client to work with them towards the most sustainable textiles solutions, rather than achieving lowest cost. With reference to Meadows (1997) and her nine interventions for influencing and changing systems, the automotive sector already have governmental policy drivers and material stocks and flows in that the technology is currently available for recyclable and sustainable textiles to be developed for the automotive sector. Given the strength of consumer interest in the US for sustainable textiles within the vehicle, it would be interesting to research how the US consumer became engaged with this. Some of the questions raised from this case study are as follows:

- How are the costs of recycled yarns determined?
- How can equality in costs of recycled/virgin yarns be achieved?
- How can/should the sustainability message for recycled textiles be delivered?

- What new technologies may be developed to ensure that end-of-life considerations can be taken into account when developing laminated textiles for automotive interiors?

COMPETING INTERESTS

The authors declare that they have no competing interests.

REFERENCES

1. Charnley, F., Lemon, M., Evans, S. (2011), Exploring the process of whole system design, Design Studies 32 (2011) 156–179, doi:10.1016/j.destud.2010.08.002 156, Elsevier Ltd.

2. Collora, J. (2012) Recycling empty PET bottles, Bottled Water Reporter, December 2011-January 2012, pp8-14, International Bottled Water Association http://www.nxtbook.com/ygsreprints/IBWA/g23199_IBWA_DecJan12/#/10

3. Diabat, A, Kannan, D, & Mathiyazhagan, K. (2014). Analysis of enablers for implementation of sustainable supply chain management - a textile case. Journal of Cleaner Production, 83(2014), 391–403.

4. EEA (European Environmental Agency) (2012) Movements of waste across the EU's internal and external borders, EEA, ISBN 978-92-9213-337-5

5. Erth, H, Gulich, B, & STFI. (2008a). Three-dimensional textiles and nonwovens for polyurethane foam substitution in car seats. In R Shishoo (Ed.), Textile advances in the automotive industry (pp. 140–149). Cambridge: Woodhead Publishers Ltd.

6. Erth, H, Gulich, B, & STFI. (2008b). Three-dimensional textiles and nonwovens for polyurethane foam substitution in car seats. In R Shishoo (Ed.), Textile advances in the automotive industry (p. 147). Cambridge: Woodhead Publishers Ltd.

7. EURAPA. (2010). Summaries of EU legislation: end-of-life vehicles. Retrieved 07-11-2011, from http://europa.eu/legislation_summaries/environment/waste_management/l21225_en.htm

8. Green Supply Chain News (2013) Ford Motor Company keeps pushing the green envelop as it continues to expand use of recycled materials

inside its cars, http://www.thegreensupplychain.com/news/13-05-20-1.php?cid=7093

9. Hernandez-Matias, JC, Vizan, A, Perez-Garcia, J, & Rios, J. (2008). An integrated modelling framework to support manufacturing system diagnosis for continuous improvement. Robotics and Computer-Integrated Manufacturing, 24(2), 187–199. http://dx.doi.org/10.1016/j.rcim.2006.10.003.View Article

10. Hirao, M., Sugiyama, H., Fischer, U., & Hungerbühler, K. (2008). IDEFO activity modeling for integrated process design considering environmental, health and safety (EHS) aspects. In B. Bertrand & J. Xavier (Eds.), Computer Aided Chemical Engineering (Vol. Volume 25, pp. 1065–1070): Elsevier.

11. Kikuchi, Y., & Hirao, M. (2009). Activity and information modeling of comprehensive assessment for sustainable process design. In J. Jacek & T. Jan (Eds.), Computer Aided Chemical Engineering (Vol. Volume 26, pp. 1123–1128): Amsterdam, London:Elsevier.

12. Kikuchi, Y., Papadokonstantakis, S., Banimostafa, A., Sugiyama, H., Hungerbuhler, K., & Hirao, M. (2012). Analysis and modeling of information required for process assessment on environment, health, and safety by IDEF0 and UML. In A. K. Iftekhar & S. Rajagopalan (Eds.), Computer Aided Chemical Engineering (Vol. Volume 31, pp. 1392–1396): Amsterdam, London:Elsevier.

13. Knowledge Based Systems, I., (2010). IFED0 - function modelling method, Retrieved 17/09/2012, 2012, from http://www.idef.com/IDEF0.htm

14. Koplin, J, Seuring, S, & Mesterharm, M. (2007). Incorporating sustainability into supply management in the automotive industry - the case of the Volkswagen AG. Journal of Cleaner Production, 15, 1053–1062.

15. KPMG. (2013). KPMG's global automotive executive survey 2013. https:// www.kpmg.com/KZ/ru/IssuesAndInsights/ArticlesAndPublications/ Documents/KPMGs-Global-Automotive-Executive-Survey-2013.pdf

16. Meadows, D., H, (1997) Places to intervene in a system, Whole Earth, Winter, http://www.wholeearth.com/issue/2091/

17. Messe Frankfurt (2013) Sustainability Recycling concepts and natural fibres in motor vehicles, http:// www.messefrankfurt.com/frankfurt/en/media/textiletechnologies/ techtextil/frankfurt/texte/ autositze-und-dachhimmel-aus-holz--flachs--mais-und-recycel-tem-.html

18. Mingers, J, & White, L. (2010). A review of the recent contribution of systems thinking to operational research and management science. European Journal of Operational Research, 207, 1161. Elsevier.

19. National Institute of Standards and Technology. (1993). Draft federal information processing standards publication 183 - standard for integration definition for function modeling (IDEF0).

20. Normand, X. (2008). In R Shishoo (Ed.), Textile advances in the automotive industry (pp. 86–109). Cambridge: Woodhead Publishing Ltd. pp. 87–88.

21. Powell, NB. (2006). New textile product development: processes, practices, and products. The Journal of The Textile Institute, 97(2), 155–166.

22. Powell, NB. (2008a). Design of automotive interior textiles. In R Shishoo (Ed.), Textile advances in the automotive industry (pp. 113–139). Cambridge: Woodhead Publsihing Ltd.

23. Powell, NB, Handfield, R, & Barnhardt, R. (2008b). Mapping the automotive textile supply chain. In R Shishoo (Ed.), Textile advances in the automotive industry (pp. 17–42). Cambridge: Woodhead Publishing Ltd.

24. Rashid, MM, & Ismail, H. (2007). Generic tool for measuring the reliability of product development processes. Journal of Modelling in Management, 2(1), 71–93.

25. Shishoo, R. (2008). Textile advances in the automotive industry: Woodhead Publishing.

26. Sugiyama, H, Hirao, M, Fischer, U, Hungerb, U, & Hler, K. (2008). Activity modeling for integrating environmental, health and safety (EHS) consideration as a new element in industrial chemical process design. Journal of Chemical Engineering of Japan, 41(9), 884–897.

27. Taylor, I. (2012). Interview with design and product development director at Sage AI, December 2011.

28. Vermeulen, I, Van Caneghem, J, Block, C, Baeyens, J, & Vandecasteele, C. (2011). Automotive shredder residue (ASR): reviewing its production from end-of-life vehicles (ELVs) and its recycling, energy or chemicals' valorisation. Journal of Hazardous Materials, 190(1–3), 8–27.

29. Yin (2009), Case study research: design and methods, p. 13, California:Sage

30. Yu, B, & Wright, DT. (1997). Software tools supporting business process analysis and modelling. Business Process Management Journal, 3(2), 133–150.

CHAPTER 4

Textile-Based Weft Knitted Strain Sensors: Effect of Fabric Parameters on Sensor Properties

Ozgur Atalay [1,*], William Richard Kennon [1] and Muhammad Dawood Husain [2]

[1] School of Materials, University of Manchester, Manchester, M60 1QD, UK

[2] Department of Textile Engineering, NED University of Engineering and Technology, University Road 75270, Karachi, Pakistan

ABSTRACT

The design and development of textile-based strain sensors has been a focus of research and many investigators have studied this subject. This paper presents a new textile-based strain sensor design and shows the effect of base fabric parameters on its sensing properties. Sensing fabric could be used to measure articulations of the human body in the real environment. The strain sensing fabric was produced by using electronic flat-bed knitting technology; the base fabric was produced with elastomeric yarns in an interlock arrangement and a conductive yarn was embedded in this substrate to create a series of single loop structures. Experimental results show that there is a strong relationship between base fabric parameters and sensor properties.

KEYWORDS

strain sensor; yarn input tension; elastomeric yarn; conductive yarn; contact pressure; contact area; conductive textiles

1. INTRODUCTION

Electro-textiles can be defined as textiles with unobtrusively built-in electronic and photonic functions [1]. They are mostly used for electromagnetic shielding, anti-static and heating purposes, and also for soft circuits: electric circuits or sensors made out of a combination of special fabrics, threads, yarns and electronic components [2]. Electrical functions can be embedded in textiles by using weaving, knitting and embroidery or nonwoven production techniques. The integration of electronic properties directly into the clothing environment carries some advantages such as increased comfort, mobility, usability and aesthetic properties. However, there are some challenges to be addressed. Yarns that are used for making cloth should be fine and elastic in order to ensure the wearer's comfort. The fibres have to be able to withstand handling and fabrics should have low mechanical resistance to bending and shearing which means they can be easily deformed and draped [3].

The creation of textile-based strain sensors has attracted researchers' attention so many investigators have studied this area and numerous different kinds of technique have been used in order to create strain sensing structures. These sensors have been used to measure human body movements or respiratory activity [4–10]. De Rossi et al. [11,12] created strain sensing fabrics by coating Lycra/cotton fabrics with polypyrrole and carbon loaded rubbers. Polypyrrole-coated fabrics showed an average gauge factor of about −13. These strain sensing fabrics exhibited a strong variation of strain-resistance with time and they showed a high response time to applied mechanical stimulus. Fabrics coated with carbon loaded rubber had a gauge factor of approximately 2.5 and fabric sensors made with this type of material showed good strain sensing properties between 1% and 13% strain. Xue et al. [13] also created strain sensing structures by coating nylon 6 and polyurethane fibres with polypyrrole. According to this research, polypyrrole-coated nylon 6 fibres showed good sensing performance, whereas polypyrrole-coated polyurethane fibres did not produce promising results as a strain sensing structure. Also, Mattmann et al. [14] created a strain sensor by using a thermoplastic elastomer and carbon particles and they were able to recognize upper body postures with an accuracy of 97%.

Another method for the creation of textile based strain sensing structures is the embedding of conductive yarns into knitted or woven structures. As a distinction from the coated sensors, they offer an integration of the sensing part during the manufacturing stage of the fabric. Thus, this approach reduces the production stage to one step. Since woven fabrics are generally characterised by their dimensional stability, poor skin contact and limited elastic recovery, knitted structures are more suitable for strain sensor applications as it is easier to create flexible structures which fit closely against the human body. Zhang et al. [15,16] created knitted strain sensors by using stainless steel yarns and carbon yarns and identified that the contacting electrical resistance between overlapped fibres is the primary factor in the sensing mechanism. They also considered electrically conductive fabrics as pure resistive networks and found a solution for the plain fabric circuit network. Yang et al. [17] modelled 1 × 1 conductive rib fabrics as resistive networks and found a relationship between extension and the equivalent resistance both experimentally and theoretically. Li et al. [18] also investigated the relationship between the electrical resistance and textile force which includes the length related resistance of conductive yarns and the contact resistance of two overlapped yarns. However, the effects of base fabric knitting parameters on sensor characteristics were not studied in the earlier research.

The primary objective of this new study was to develop fabric-based strain sensors using the knitting route. Elastomeric yarns with different linear yarn density have been used to create interlock based structures, because the interlock structure has the highest dimensional stability among the basic weft knitted structures. Thus, this characteristic enables the creation of more reliable sensors in terms of repeatability. Silver coated polymeric yarn was used as a sensing element and this was knitted over the interlock base structure as a series of single loops of the fabric which were arranged to help reduce the conductive yarn structural deformation during long term force loading. Different elastomeric yarn input tensions were applied to produce interlock structures with different fabric compactness in order to investigate the effect of contact pressure on the electrical resistance as well as sensor characteristics. The following section describes the electro-mechanic theory and production of knitted strain sensors followed by the testing method for the electromechanical properties of the sensors. The third part reports the results obtained from the experimental procedure and discussion of the electro-mechanical properties of the sensor.

2. MATERIALS AND METHODS

2.1. Production of Knitted Strain Sensing Fabrics

A single design of knitted strain-sensing fabric was devised and three different variations of the basic knitted sensor were created using 800 decitex (the mass in gram of 10,000 m of yarn), 570 decitex double covered elastomeric yarns and silver coated nylon yarn. The three groups of sensing structures, each with different characteristics, were manufactured with a Shima Seiki SES 122-S ten gauge computerised flat-bed knitting machine by varying the elastomeric yarn input tension and linear yarn density. Table 1 presents the manufacturing parameters of the knitted strain-sensing fabrics.

Table 1. Manufacturing parameters of knitted strain-sensors.

Sample Group	Elastomeric Yarn Input Tension (cN/Tex)	Elastomeric Yarn Linear Density	Effective Conductive Area	Number of Conductive Wales	Number of Conductive Courses
Group 1	0.125	800	23 mm × 42 mm	36	12
Group 2	0.062	800	31 mm × 44 mm	36	12
Group 3	0.125	570	20 mm × 41 mm	36	12

At the technical face of the fabric, silver plated nylon yarn was used to create conductive loops; it has 235 decitex fineness and 200 Ω/m linear resistance. The three groups of strain sensing fabrics were manufactured using an interlock arrangement and the conductive yarn was embedded into this interlock structure in a series of single loops. It should also be noted that the same amount of silver yarn was used for each group during the manufacturing process, so the length of conductive yarn was kept equal for each sample. The needle notation, carrier locations and front side photograph of the knitted samples are shown in Figure 1 and a magnified image of a sample is shown in Figure 2. Five specimens were prepared for each group. Samples belonging to the first two groups were produced using 800 decitex elastomeric yarns with an accurately controlled run-in tension of 0.125 cN/Tex. The second batch of samples was knitted with an accurately-controlled run-in tension of 0.062 cN/Tex on the elastomeric yarn. The Third group of samples was produced using 570 decitex elastomeric yarns with run-in tension of 0.125 cN/Tex during knitting. Thus, samples with three different tightness factors were created and this varied the compactness of the knitted structures. The tightness factor in the metric system may be expressed as *Tex*——√/1, where the yarn linear density is in Tex and the loop length (l) is expressed in mm [19]. In this study, Knapton's [20] structural knitted cell (SKC) concept was adopted for calculation of interlock fabric loop length. According to this concept, the effective loop length is the length of yarn in one SKC which consists of four single loops. The schematic diagram in Figure 3 shows one structural knitted cell in the interlock fabric. Compactness is an important fabric property which affects fabric properties including dimensional stability, strength, drape, handle and shrinkage. Normally, structures with high a tightness factor have higher wale and course stitch density values. Since course and wale spacing decreases, higher contact pressure occurs between adjacent courses and wales and this study will reveal the effect of the tightness factor on fabric electrical resistance and the sensing characteristics.

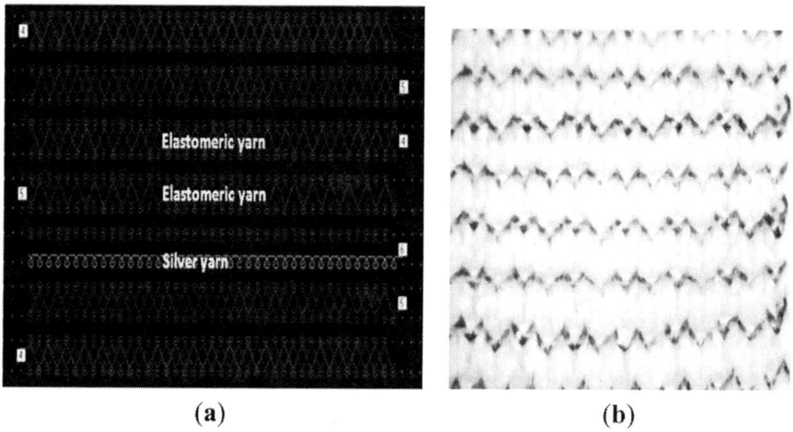

Figure 1. (**a**) Needle notation for knittedstrain-sensing fabric shows arrangement of elastomeric and silver yarn in the structure; (**b**) Technical face of knitted sampleshows conductive loops in the structure.

Figure 2. Magnified image of knitted sensor showing a conductive course.

2.2. Electro-Mechanical Theory

2.2.1. Gauge Factor

In strain sensors, the gauge factor (GF) is an important parameter and it gives information about the sensitivity of the sensor. The GF is calculated as follows:

$$GF = \frac{\frac{\Delta R}{R}}{\varepsilon} \tag{1}$$

where:

ΔR = the change in the resistance;

R = the initial resistance (the resistance before extension);

ε = the strain value.

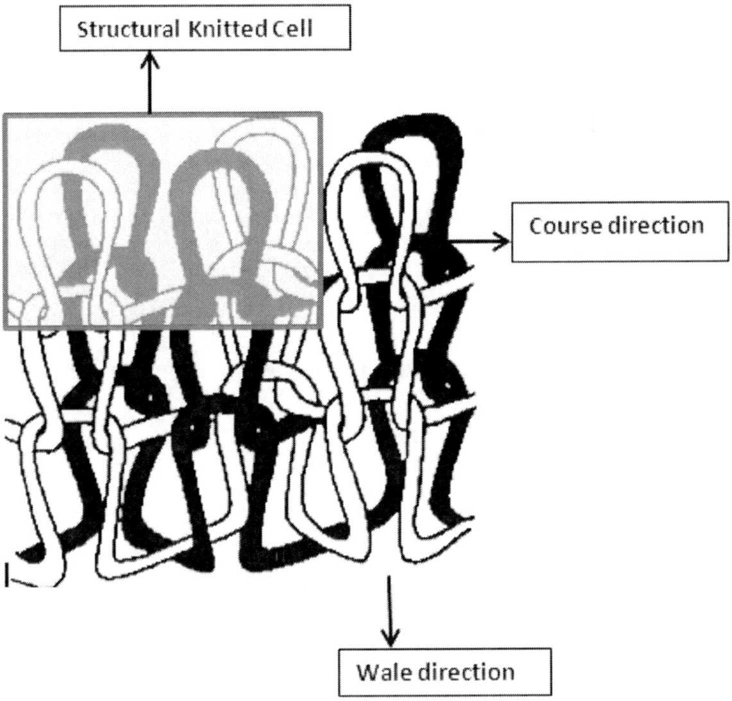

Figure 3. Schematic diagram of Interlock fabric showing a structural knitted cell.

2.2.2. Structural Design

As may be seen from Figure 4, due to the usage of elastomeric yarn in the structure, conductive yarn loops make contact with adjacent loops at their heads and limbs also at their sinker loops which are pressed together. In addition to this, conductive loops are located within the interlock structure in the form of a zigzag arrangement in which they are alternately located in a higher or lower position relative to each other. This is due to the modification of the conductive plain loops which are used in the interlock structure. This feature improves contact area significantly and enables sensor measure up high strain levels. It is not clear from the needle notation of the knitted sensor shown Figure 1 that adjacent loops of conductive yarns can be designed to touch each other. Hence, this type of sensor derives predominantly from practical knitting experience. These are functions of the design of the knitted sensor and are enhanced by the incorporation of elastomeric yarn and interlock structure.

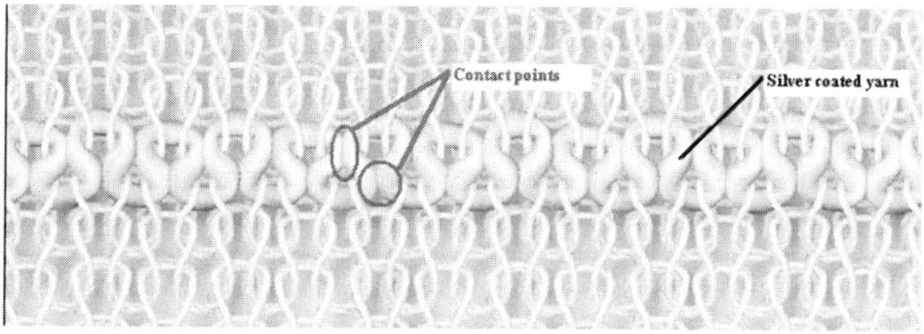

Figure 4. Schematic diagram of sensor design showing the geometry of the conductive yarn.

2.2.3. Contact Theory

According to Holm's [21] contact theory:

$$R_c = \frac{\rho}{2}\sqrt{\frac{\pi.H}{nP}} \qquad (2)$$

where:

R_c = contact resistance;

ρ = electrical resistivity;

H = material hardness;

n = number of contact points;

P = contact pressure.

From Equation (2), it can be see that the electrical resistivity and material hardness are constant for a given material, but the number of contact points and the contact pressure are variable depending on the sensor design. Thus, higher contact pressure and increased number of contact points between conductive parts lower the contact resistance. In this sensor design, contact pressure between the conductive loops has a maximum value before extending the fabric, but during the force loading stage, uniaxial tensile force reduces the level of contact between the conductive loops. Hence, contact pressure between conductive loops lessens depending on the level of applied strain, so the overall electrical resistance of the proposed sensor increases with strain. The three groups of sensors with different compactness have been produced In order to study the effect of contact pressure between the conductive loops and to characterise the sensor behaviour.

2.3. Test Procedure for Knitted Strain Sensors

In order to calculate the GF and to investigate the sensor characteristics of the knitted strain sensing fabrics, electro-mechanical measurements of the various samples were performed under multi-cyclic tensile stress using a Zwick/Roell BTC-FR2.5TS.D09 tensile testing machine to apply repeated mechanical extension and deformation. The change of resistance was measured simultaneously with applied strain by the tensile tester in combination with a Wheatstone bridge arrangement. Experimental data was recorded using the Testexpert software. The samples were subjected to levels of up to 40% extension in the course direction and the change of electrical resistance was recorded over the time. The extension level of 40% was chosen to mirror typical human body extensions, as the proposed sensor can be used for monitoring human body movements. The tensile testing machine has a fixed and a moveable crosshead which may be driven at a range of speeds. In this research, samples were tested with a constant rate of extension of 120 mm/min with a full test comprising 20 repeats. In addition to the multi-cyclic tensile test, two further different tests were performed. Firstly, fabric samples were subjected to conditioning extension with a two minute dwell time at 40% strain and a two minute dwell time at 0% in order to study the relaxation behaviour of the sensor. Thereafter, each knitted strain sensing fabric was extended up to the 40% strain level at 120 mm/min speed with 1,000 repeats so as to investigate the effect of long term cycling.

3. RESULTS AND DISCUSSION

3.1. Comparison of Average Electrical Resistance and Average Base Fabric Parameters

The initial electrical resistance of the knitted strain sensors was measured before any external applying tensile forces in order to see the effect of base fabric knitting parameters on the electrical resistance of five samples. Table 2 shows the average electrical resistance values and the base fabric knitting parameter values.

Table 2. Comparison of average electrical resistance and average base fabric parameters.

Sample Group	Elastomeric Yarn Input Tension (cN/Tex)	Elastomeric Yarn Linear Density (dtex)	Wale Density (wales per·cm)	Course Density (courses per·cm)	Structural-Cell Stitch Length (mm)	Tightness Factor	Electrical Resistance (Ω)
Group 1	0.125	800	8.67	15.55	15.36	1.84	166.47
Group 2	0.062	800	8.27	12.20	19.75	1.43	242.02
Group 3	0.125	570	8.88	17.15	10.97	2.17	206.58

When the first two groups of samples are compared, samples belonging to group 1 demonstrate a lower average electrical resistance value. A number of aspects of the work have been considered which may help to explain this situation. Samples belonging to group 1 have higher average wale density, so more pressure is exerted on the touching points of the conductive yarn loops. According to Holm's contact resistance theory, when the contact pressure and number of contact points increase, then the contact resistance decreases. In addition to this, more contact points are generated by higher wale density and this also causes the resistance values to be reduced. As seen in Table 2, group 3 samples were produced by applying same elastomeric yarn input tension per Tex as those in group 1. Since lower linear yarn density was used, group 3 samples have highest wale density as well as the highest course density. In the light of this information, group 3 samples were expected to have the lowest resistance values. However, in this case the elastomeric linear yarn density affected the measurements significantly. As some of the conductive yarn loop contact points are located between the wales of interlock fabric, the finer elastomeric yarn which was used for group 3 created fewer contact areas in comparison to using thicker yarn. Thus, the average electrical resistance of the group 3 samples was higher than group 1 samples. However, since the amount of silver yarn was kept constant, the conductive loops created an enhanced ridge effect at the lowest stitch length values as conductive yarn loops located on the interlock stitch.

3.2. Sensor Characterisation

The graphs in Figure 5 show the relative change in resistance versus strain for the three groups of samples whilst they are being subjected to cyclic tensile testing. The samples were cycled between 0% and 40% strain at a speed of 120 mm/min, with 20 repeats and there were no dwell times at either the lowest or the highest strain levels. 20 sets of measurement were averaged for plotting each graph. It can be seen in the graphics of Figure 5 that there are in fact two hysteresis loops described by the curves and these extend from 0% to around 8% strain and from 8% to 40% strain respectively for group 1 sensors; from 0% to around 5% strain and from 5% to 40% strain for group 3 sensors. The reason for this behaviour is that the applied strain creates textile deformation over the fabric to such an extent that group 1 and 3 sensors starts to experience time-depended recovery. Thus, during the cyclic tests, this level of deformation stretches the fabric to its elastic recovery limit and buckling is apparent in the samples when the strain is released. It does not appear that has been inflicted on the sensors, but recovery takes an extended period of time. When loading is applied to a buckled sample, the fabric is initially pulled flat and this causes the touching points of the conductive knitted loops to make enhanced contact with each other so there is a slight decrease in resistance in the very early stage of straining a distorted sample. This phenomenon occurs only up to the 2.8% level of strain for group 1 sensors and 2.4% level of strain for group 3 sensors. Hence, the working range of sensors can be considered as being between the finishing strain values of first hysteresis loops and the 40% strain limit. Also, the maximum hysteresis values of sensors are 3%, 5.8% and 3.4% for group 1, group 2 and group 3 sensors, respectively. As seen from the graphs, the three groups of sensors demonstrate different behaviour and different gauge factors with applied strain. When the first two groups which are produced with the same linear yarn density were compared, group 1 showed more linear response over the 40% strain range. Actually both of the sensors can be characterised by two linear regions at their working range. Group 1 sensors have an initial linear region up to 19% strain, then a second linear region between 19% and 40% strain. The gauge factors values of sensors have been calculated to be close 3.75 for strains below 19% then the gauge factor falls to 2.16 for strains between19% and 40%. Group 2 sensors have their first linear region up to 9% strain and their second linear region between 9% and 40% strain and the gauge factor values are approximately 4.3 and 0.9 respectively for the given regions. As group 1 samples are more compact than those in group 2 they create more contact regions and more contact pressure on the touching points of adjacent conductive loops. Thus, higher strain rates are needed to separate conductive contact points from each other in tightly knitted structures. Different levels of tightness during the knitting process can cause the different mechanisms to start and finish at different strain levels. When the group 1 sensor in the Figure 5 is considered, the two slopes reflect two different effects. Increasing strains up to 19% cause the upper parts of the limbs of adjacent loops to separate. Above 19%, the conductive loops start to separate from their sinker loops. The mechanism is the same in every case, but different levels of compactness of knitting cause the effects to occur at different levels of strain. When the group 3 samples are considered, it would be appear that there is just one linear range through the whole 40% strain range and the effect has a 0.75 gauge factor value. Group 3 comprises those samples with the most tightly knitted structure and they prove to have the highest linearity with the lowest gauge factor. Table 3 shows the R^2 (coefficient of determination) values of the best fitted linear curves of resistance-strain data of knitted sensors. While the compactness of structure endowed more linearity, the fineness of the elastomeric yarn resulted in the lowest gauge factor value due to the reduced contact area of each of the conductive yarns compared with other samples which have been made using thicker yarns.

As seen from Figure 6, there is also a contribution to the change of the resistance measurement which derives from strain of the conductive yarn itself. This is felt to be more significant at higher strain rates, as it is at higher strain rates that the knitted loops start to be distorted when the fabric is stretched. Hence, the length of conductive yarn increases and as a result of this, the overall resistance of the sensor continues to increase even though the knitted loops have virtually ceased to make contact at their touching points and Figure 7 shows magnified images of a part of one conductive course during various level of textile extension.

When the sensor is intended to be used for a longer period of time and for more operational cycles proposed sensor needs to show stable properties. Figure 8 shows 1,000 cycle of knitted sensors extended up to 40% strain value. The very small deviation from the horizontal of the top and bottom lines shown in Figure 8 indicate that the sensors are extremely stable.

Over the duration of the test, the unloaded (starting) resistance of the samples increased by just 7 Ω, 6.28 Ω and 2.50 Ω for group 1, group 2 and group 3 sensors, respectively, and the peak (40%) resistance increased by 4 Ω, 0.70 Ω and 2.7 Ω for group 1, group 2 and group 3 sensors, respectively. Thus, the sensor can be considered to be stable over long-term usage. This design of a sensor offers a distinct solution to the drift problem which has been widely reported in textile-based sensors. The base fabric is a modified interlock structure and has been shown to provide particularly stable knitted base. The conductive yarn loops are located onto the technical face of the interlock structure and during tensile testing the applied force gradually separates the fabric wales. Hence, the conductive yarn itself affected minimally by the applied force and it retains its structural properties even longer usage.

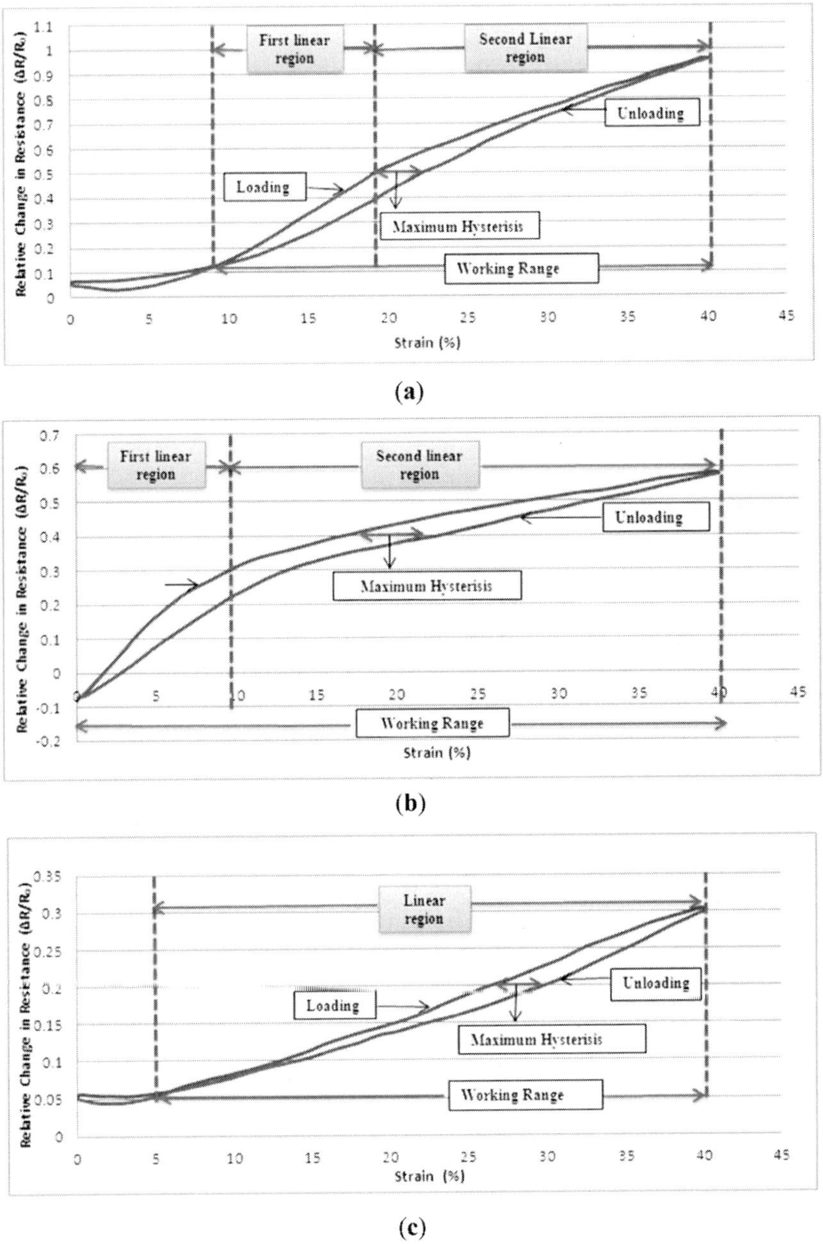

Figure 5. Relative change in resistance-strain graphs of three groups. (**a**) Group 1; (**b**) Group 2; (**c**) Group 3.

Table 3. R^2 values of knitted sensors.

Sample Type	First Linear Region	Second Linear Region	Whole working range
Group 1	0.996	0.997	0.980
Group 2	0.985	0.985	0.868
Group 3	-	-	0.997

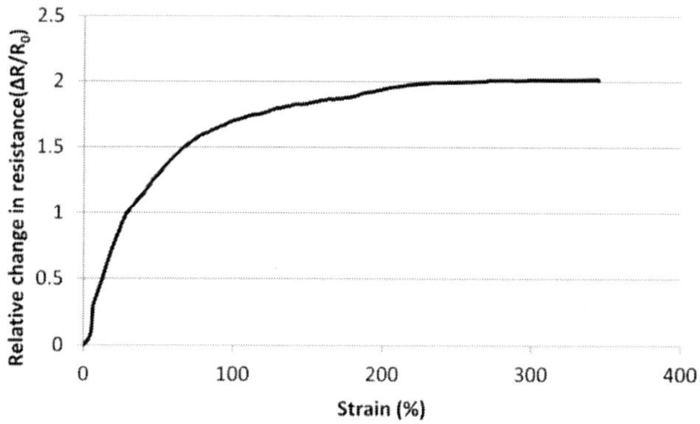

Figure 6. Resistance versus Strain up to breaking point of group 1 sample.

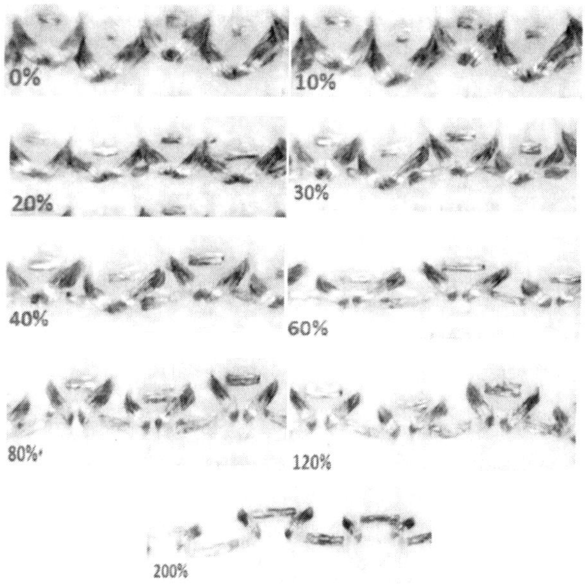

Figure 7. Magnified images of one conductive course during various levels of textile extension from 0% to 200%.

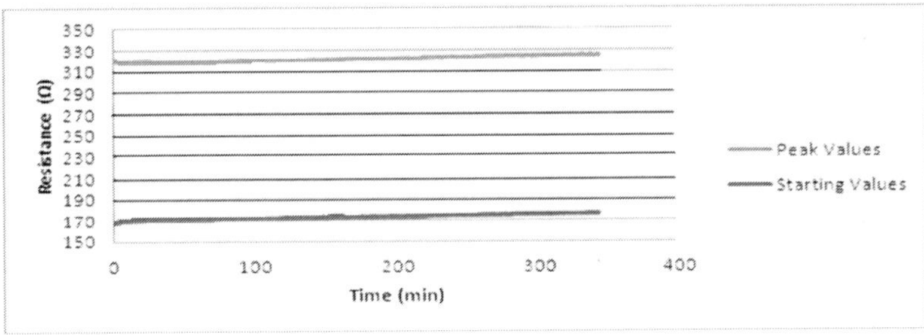

(a)

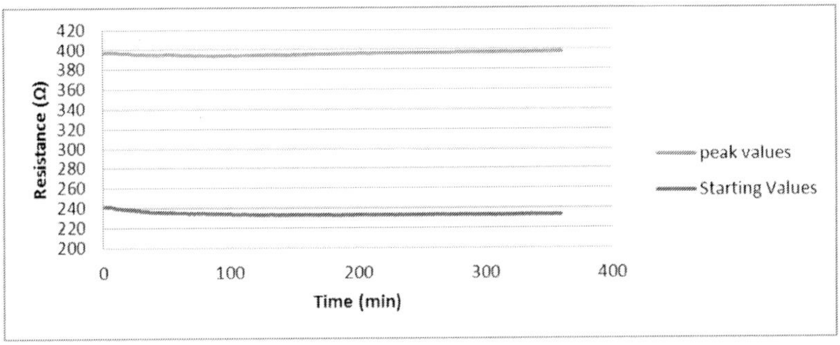

(b)

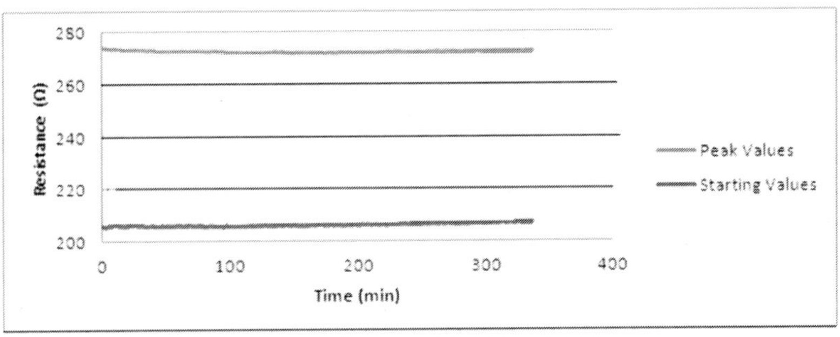

(c)

Figure 8. Long term cycling behaviour graphs of three group sensors: **(a)** Group 1; **(b)** Group 2; **(c)** Group 3.

In Figure 9 relaxation behaviour of the sensors has been shown. The dwell times at maximum strains are 2 min. When the strain is kept constant at the 40% strain level, the group 1 sensor relax an average of 14.9 Ω, the group 2 sensors relax by 6.1 Ω and the group 3 sensors relax by 11.4 Ω. These relaxation levels caused inaccuracy of 9.3%, 4.3% and 16.6%, respectively, for each sensor and it seems that when the tightness factor of the structure increases, inaccuracy caused by relaxation increases for a given strain value which is kept equal for each group.

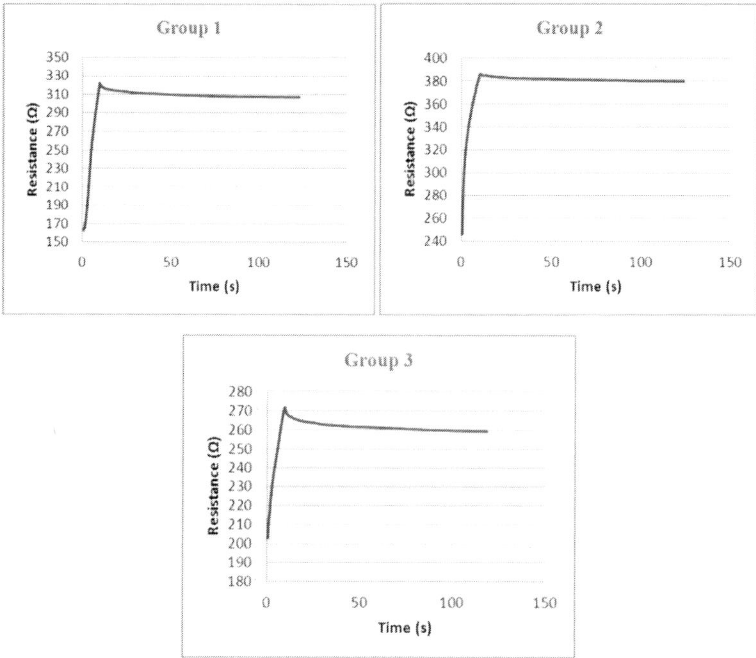

Figure 9. Relaxation behaviour of sensors at 40% strain.

4. CONCLUSIONS

In this paper, a novel textile-based strain sensor and the effect of base fabric parameters on its sensor properties have been presented. A strong relationship has been established between the sensor characteristics and the base fabric parameters. Variations in elastomeric yarn input tension have greatly affected the sensors' linear range and gauge factor values. More compact structures showed higher linearity in respect of one specific sensor design. Also, the elastomeric yarn linear density affected the contact resistance of the conductive loops by decreasing the number of contact points. The tightness factor values of the base fabric also affected the relaxation behaviour sensors. Less compact structures showed improved accuracy during the relaxation period. The sensor design offers a good solution to the problem of drift which is usually seen in textile- based sensors. Future work includes alternating of conductive yarn input tension and using of different types of conductive yarns in order to see the effect of these variables on the sensor properties.

ACKNOWLEDGMENTS

We are thankful to Edward Lay for the cooperation in the preparation of samples.

CONFLICTS OF INTEREST

The authors declare no conflict of interest.

REFERENCES

1. Tao, X. IntroductionIntroduction. In Wearable Electronics and Photonics; Tao, X., Ed.; CRC Press: Boca Raton, FL, USA, 2005; pp. 1–12.

2. Electrotextiles. Conductive materials, textiles. Available online: http://openmaterials.org/2011/03/27/materials-101-electrotextiles/ (accessed on 15 April 2013).

3. Kirstein, T.; Cottet, D; Grzyb, J; Tröster, G. Wearable Computing Sytems—Electronic Textiles. In Wearable Electronics and Photonics; Tao, X., Ed.; CRC Press: Boca Raton, FL, USA, 2005; pp. 177–197. [

4. Huang, C.T.; Tang, C.F.; Shen, C.L. A Wearable Textile for Monitoring Respiration, Using a Yarn-Based Sensor. Proceedings of the 2006 10th IEEE International Symposium on Wearable Computers, Montreux, France, 11–14 October 2006; pp. 141–142.

5. Kim, K.; Lee, I.K.; Choi, S.S.; Kim, S.S.; Lee, T.S.; Cha, E.J. Wearable Transducer to Monitor Respiration in a Wireless Way. Proceedings of 6thInternational Special Topic Conference on Information Technology Applications in Biomedicine (ITAB 2007), Tokyo, Japan, 8–11 November 2007; pp. 174–176.

6. Pacelli, M.; Loriga, G.; Paradiso, R. Flat knitted sensors for respiration monitoring. Proceedings of the IEEE International Symposium on Industrial Electronics (SIE 2007), Vigo, Spain, 4–7 June 2007; pp. 2838–2841.

7. Mitchell, E; Coyle, S.; O'Connor, N.E.; Diamond, D.; Ward, T. Breathing feedback System with Wearable Textile Sensors. Proceedings of the 2010 International Conference on Body Sensor Networks (BSN), Sigapore, 7–9 June 2010; pp. 56–61.

8. Guo, L.; Berglin, L.; Wiklund, U.; Mattila, H. Design of a garment-based sensing system for breathing monitoring. Text. Res. J. **2013**, 83, 499–509.

9. Li, X.; Zhang, R; Yu, W; Wang, K; Wei, J; Wu, D; Cao, A; Li, Z; Cheng, Y; Zheng, Q. Stretchable and Highly Sensitive Graphene-on-Polymer Strain Sensors; Scientific Reports; Nature Publishing Group: London, UK, 2012.

10. Kannaian, T.; Neelaveni, R.; Thilagavathi, G. Development and characterisation of elastomeric tape sensor fabrics for elbow angle measurement. Indian J. Fibre Text. Res. **2011**, 36, 436.

11. Scilingo, E.P.; Lorussi, F.; Mazzoldi, A.; De Rossi, D. Strain-sensing fabrics for wearable kinaesthetic-like systems. Sens. J. **2003**, 3, 460–467.

12. Lorussi, F.; Scilingo, E.P.; Tesconi, A.; Tognetti, A.; De Rossi, D. Wearable Sensing Garment for Posture Detection, Rehabilitation and Tele-Medicine. Proceedings of 4th International IEEE EMBS Special Topic Conference on Information Technology Applications in Biomedicine, 24–26 April 2003.

13. Xue, P.; Tao, X.M.; Kwok, K.W.; Leung, M.Y.; Yu, T.X. Electromechanical behavior of fibers coated with an electrically conductive polymer. Text. Res. J. **2004**, 74, 929–936.

14. Mattmann, C.; Clemens, F.; Tröster, G. Sensor for measuring strain in textile. Sensors **2008**, 8, 3719–3732.

15. Zhang, H.; Tao, X.; Wang, S.; Yu, T. Electro-mechanical properties of knitted fabric made from conductive multi-filament yarn under unidirectional extension. Text. Res. J. **2005**, 75, 598–606.

16. Zhang, H.; Tao, X.; Yu, T.; Wang, S. Conductive knitted fabric as large-strain gauge under high temperature. Sens. Actuat. A: Phys. **2006**, 126, 129–140.

17. Kun, Y.; Guang-li, S.; Liang, Z.; Li-wen, L. Modelling the Electrical Property of 1 × 1 Rib Knitted Fabrics Made from Conductive Yarns. Proceedings of the Second International Conference on Information and Computing Science, 2009 (ICIC'09), Machester, UK, 21–22 May 2009; pp. 382–385.

18. Li, L.; Au, W.M.; Li, Y.; Wan, K.M.; Wan, S.H.; Wong, K.S. Electromechanical Analysis of Conductive Yarn Knitted in Plain Knitting Stitch under Unidirectional Extension. Procedings of Textile Bioengineering and Informatics Symposium (TBIS 2008), Hong Kong, 14–16 August 2008; pp. 793–797.

19. Horrocks, A.R.; Anand, S.; Anand, S.C. Handbook of Technical Textiles; CRC Press: Cambridge, UK, 2000; Volume 12.

20. Gravas, E.; Kiekens, P.; Langenhove, L. Predicting fabric weight per unit area of single and double-knitted structures using appropriate software. Autex Res. J. **2006**, 6, 223–237.

21. Holm, R.; Holm, E. Electric Contact Theory and Application, 4th Ed. ed; Springer-Verlag: New York, NY, USA, 1967.

CHAPTER 5

Memory Bandage for Functional Compression Management for Venous Ulcers

Bipin Kumar [1,*], Jinlian Hu [2,*] and Ning Pan [1]

[1] Division of Textiles, Biological & Agricultural Engineering, UC Davis, CA 95616, USA

[2] Institute of Textiles and Clothing, The Hong Kong Polytechnic University, Kowloon 999077, Hong Kong

ABSTRACT

In current compression practice for the treatment of chronic venous disorders, there has always been a challenge of controlled compression by a bandage to achieve a particular pressure range in the affected region of the limb. The challenges in compression in the products could be solved if there were the possibility of stress control in fabric. Herein, we are exploiting the newly discovered phenomena, i.e., stress memory, in a memory polymer (MP) for the design and investigation of a smart bandage for functional compression benefits. A memory bandage is developed using a blend yarn consisting of MP filaments (segmented polyurethane) and nylon filaments. Results showed the possibility to control or manage the internal stress developed in the bandage in wrapped position by simple heating, and thus allowing pressure readjustment externally. Extra pressure generated by the bandage increases with increasing the level of temperature and strain ($p < 0.05$). The pressure variations also depend on the number of layers and limb circumference ($p < 0.05$). The memory bandage could have a great potential over existing conventional compression products, as they could give more freedom to govern pressure level whenever needed during the course of compression therapy as a novel wound care management system.

KEYWORDS

bandage; compression therapy; memory polymer; venous ulcers

1. INTRODUCTION

Compression therapy provides a means to treat venous leg ulceration, edema, venous hypertension, and other chronic venous disorders [1]. The objectives of compression therapy are to reduce the venous hypertension in the affected area, reduce the swollen limb to minimize the size and maintain a uniform pressure gradient in the leg from toe to knee to improve the venous return to the heart [2]. External compression is given to the affected leg portion during compression therapy using compression materials like bandage, stockings, pressure garments, etc. [3]. The pressure is generated in the interface between bandage and skin because of compression by the bandage during its application over the limb by the application of external force. This pressure is called interface pressure [4,5]. The success of this treatment depends to a great degree on the level of pressure applied by the bandage at the affected portion on the limb, and sustenance of this pressure during the course of the treatment. This interface pressure has to be quite accurate within certain limits and should not be below or above the prescribed level, otherwise it can lead to certain complications during treatment.

Several researchers have used Laplace's law to find the relationship among several factors affecting the interface pressure [6]. In the context of compression therapy by a bandage, the relevant equation derived from the above law is expressed as:

$$P = \frac{2\pi \times N \times F}{C} \tag{1}$$

where P is the pressure, F is fabric tension per unit width, N is number of bandage layers and C is the limb circumference. The interface pressure produced by a bandage depends on the complex interaction of many factors—the tension in the fabric, the physical structure and elastomeric properties of the fibers from which it is manufactured, the size and shape of the limb to which it is applied, the application technique (number of layers wrapped) and the nature of any physical activity undertaken by the patient [7]. Achieving desired compression range is difficult as the pressure applied by the bandage is governed by leg attributes (shape or size) and the elasticity or stiffness of the fabric [8]. Moreover, pressure drop over time is a cause of concern. Experimental studies showed that the pressure decreases over time due to reduction in swelling of leg, and also decrease in the internal stress in the fibrous structure of the bandage [9,10]. Replacement of the bandage is needed once the pressure fall below a target level. Clearly, pressure management and its sustenance is a huge challenge to provide effective treatment.

Currently, no external means exists to govern pressure exerted by the bandage at wrapped position on the leg. Controlling or readjusting pressure on the leg would provide many advantages to health practitioners, and would make treatment more effective. The challenges in compression products could be solved if there were the possibility of stress control in fabric materials. In this regard, smart polymer fibers could offer choices with internal stress responsive to an external stimulus. Among stimulus-responsive polymers, shape memory polymers (SMPs) have high relevance to the present needs and have been solving many real life problems in the areas of aerospace, biomedicine, transport, construction, electronic, textile and consumer products [11,12]. SMPs can sense the environment and/or their own state, make a judgment and then change their functions according to predetermined purpose. They can memorize the original shape and allow the materials to recover the original shape from a temporarily deformed shape under appropriate stimuli, such as heat, light, magnetic field, water, etc. The shape memory effect (SME) of a SMP involves two aspects: fixability and recoverability. Fixability refers to the capability of the SMP to change from the original un-deformed shape to a temporary deformed shape through a suitable programming process, i.e., shape fixing, while the recoverability indicates its ability to recover the original shape. In the programming process, SMPs are mechanically deformed above a particular temperature (also known as transition temperature, T_g) and the deformed shape is temporarily fixed by cooling the SMP below T_g. The most important characteristic of SME is the stability of this temporarily deformed shape in which the internal stress is frozen and does not change in the absence of suitable stimuli. The deformed shape is actively triggered to recover the original permanent shape by exposure to an appropriate stimulus. Owing to this unique property, SMPs have found increasing applications in engineering and medical fields where the potential of shape fixity or recovery is primarily exploited for desired functions [13,14,15,16]. Recently, however, several other memory behaviors, including stress memory [17], temperature memory [18], chrome memory [19] and electric memory [20], have also been reported in such polymers, and it seems more appropriate therefore to call them memory polymers (MPs) in general.

Similar to shape memory, a new phenomenon, i.e. stress memory, has been discovered recently in MPs [17]. In stress memory, the stress in a polymer can be programmed, stored and retrieved reversibly with an external stimulus (e.g., heat), similar to the shape memory effect. Herein, the stress is stored upon inducing certain strain level beyond the T_g and cooling below this point. The stress is further released under constraint upon triggering by an external stimulus such as heat. This stress memory might be mistaken as recovery stress, which has been studied quite extensively [21], but investigation has confirmed that they are quite different phenomena. Although stress freezing can be obtained via a normal shape fixing process of the MP, where the internal stress is also frozen or stored at lower temperature ($<T_g$) in a deformed MP, but upon triggering the MP under constraint, the recovered stress does not remain stable but decreases over time. This is primarily because of the viscous stress that causes stress relaxation. This adds limitations to the use MP for some applications where the stress level has to be maintained, or a stable cyclic stress variation is required, such as pressure bandage, massage devices, etc. It has been confirmed that the recovery stress contains multi-components including elastic, thermal and viscoelastic forces in addition to possible memory stress [17,22]. In order to obtain stress memory phenomenon, the recovery stress should be pure so that it can be repeated, analyzed, and a meaningful signal for different applications. If only the memory stress is properly identified and exploited in a MP using suitable programming methods, then this could serve the special functions in stress control [17]. This opens several new research domains for MPs where stimuli-responsive forces are required/involved, including in sensors, sportswear, compression garments, massage devices, nerve conduits, bone tissue engineering, artificial muscles and dynamic mattress. Herein, this novel stress memory phenomenon has been used for smart and functional compression benefits by a memory bandage fabric.

2. EXPERIMENTAL SECTION

2.1. Preparation of SMP filaments

The MP filament was prepared using segmented polyurethane (PU) via melt spinning process. For filament spinning, the MP polymer chips were initially prepared using bulk polymerization method. The MP was synthesized using polytetramethylene ether glycol (PTMEG; Aldrich Chemical Company, Atlanta, GA, USA; M_n = 650) as soft segment, and 4,4′-methylene diphenyl diisocyanate (MDI; Aldrich Chemical Company, Atlanta, GA, USA) and 1,4-butanediol (BDO; Acros Organics, Geel, Belgium) as hard segment. First, the PTMEG and MDI was mixed and stirred for 2 h to form a pre-polymer at 80 °C. Thereafter, the BDO was added for cross-linking to complete the reaction. The ratio (weight) of hard to soft segments was 13:12. The MP chips was obtained after curing the mixture for 24 h at 100 °C. These chips were used for melt spinning to spin the filaments. The filaments were spun in pure nitrogen environment using single screw extruder with a temperature range from 175 to 202 °C [23]. The winding speed was set at 500 m/min, and the obtained linear density of the filaments was 18.6 tex. The glass transition temperature of the MP filaments chosen in the study was around 30 °C, as can be observed in Figure 1. The shape fixity and shape recovery of the filaments were 23.7% and 95%, respectively. More information on thermo-mechanical characterization of the MP can be found in the [23].

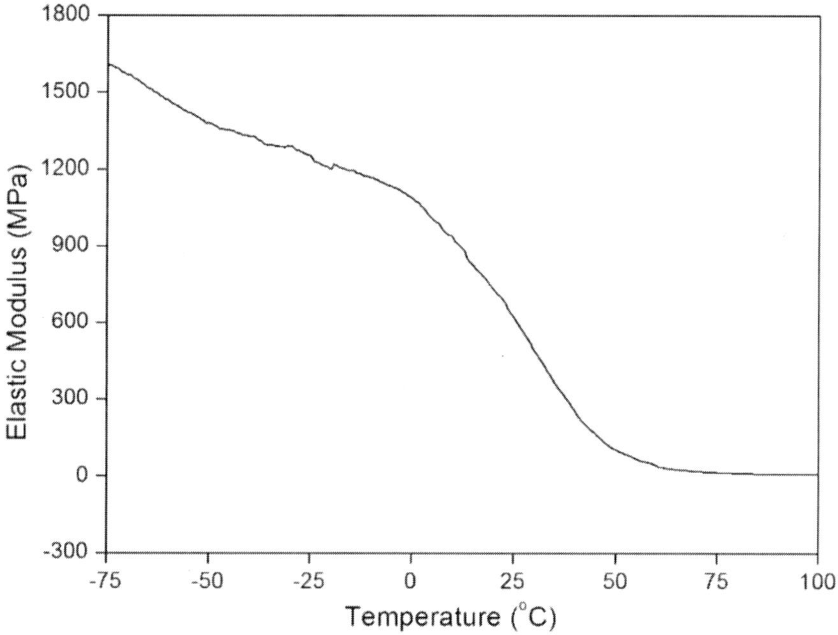

Figure 1. Variation of elastic modulus vs. temperature of the MP using DMA.

2.2. Development of Memory Bandage

The memory bandage was prepared from a yarn blend consisting of the MP and Nylon filaments using V-bed double jersey flat weft knitting machine. The fabric characteristics are given in Table 1. As an example to demonstrate the pressure control using memory bandage, the measurement was conducted using a Kikuhime™ pressure sensor (Figure 2). After the application of the bandage on the cylindrical tube, the entire set-up was placed in a heated chamber for the MP activation, and the variation in the interface pressure was obtained. Different cylindrical tubes were used to stimulate different circumferences of a human leg. Extra pressure generated by the bandage was obtained at different levels each for the factors—temperature (T), number of layers (N), cylinder circumference (C) and applied strain (ε).

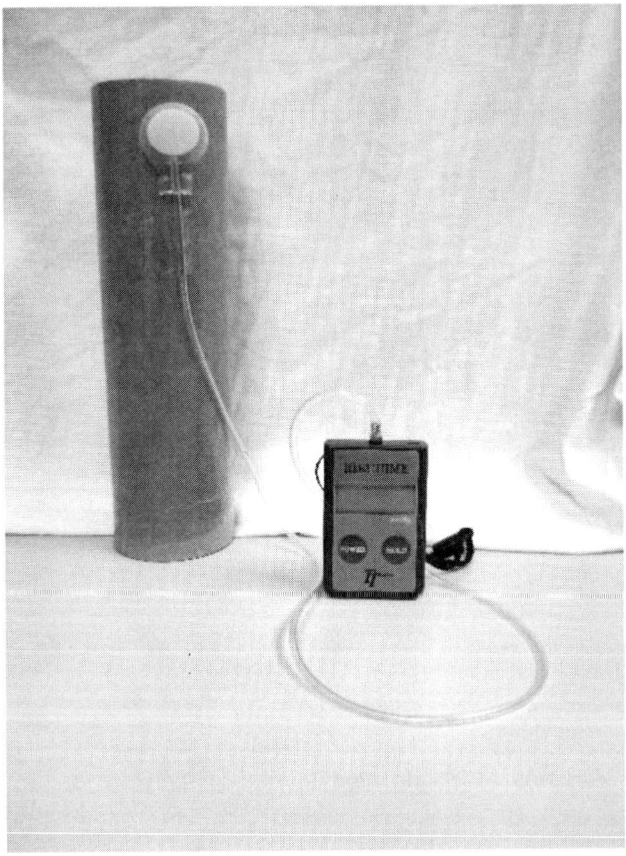

Figure 2. Set-up for interface pressure measurement.

Table 1. Characteristics of the memory bandage.

Weave Structure	Weft knitted
Filament density, tex	18.9 (Nylon); 13 (SMPU filament)
Composition, %	59.2 (Nylon); 40.7 (SMPU filament)
Mass Density, g/m^2	341.2
Fabric Thickness, mm	0.85
Filaments per unit length	8 (wales/cm); 12 (courses/cm)

3. RESULTS AND DISCUSSION

As the recovery in a MP is largely governed by entropy elasticity, the change in material stress σ could be expressed by temperature T as [24,25]:

$$\left(\frac{\partial \sigma}{\partial T}\right)_{l,V} = \frac{\sigma}{T}$$
$$\Rightarrow \sigma = f(T) = a + bT$$
$$\Rightarrow \Delta\sigma = b \times \Delta T \quad (2)$$

where the sample length l and volume V are kept constant. The equation indicates the possibility of controlling the stress in the MP by external temperature control. The stress in the bandage fabric could be changed if MP filaments in its structure exist. To confirm this, thermo-mechanical tensile testing was employed to examine the stress variation in a deformed memory bandage specimen upon activating under constraint. The measurement was done using a tensile tester (Instron 5566) for loading and unloading, anchored with temperature chamber for heating and cooling. Figure 3 shows the result of stress increase (Δσ) in the specimen under a fixed tensile strain. It can be inferred that the stress increase (Δσ) in the fabric is due to presence of MP filaments. The Δσ increased rapidly for the initial time and finally saturated to a fixed value for the given temperature. Figure 3b shows the results of stress variation with temperature. The increase in stress varies linearly with the temperature, as shown in Equation (4).

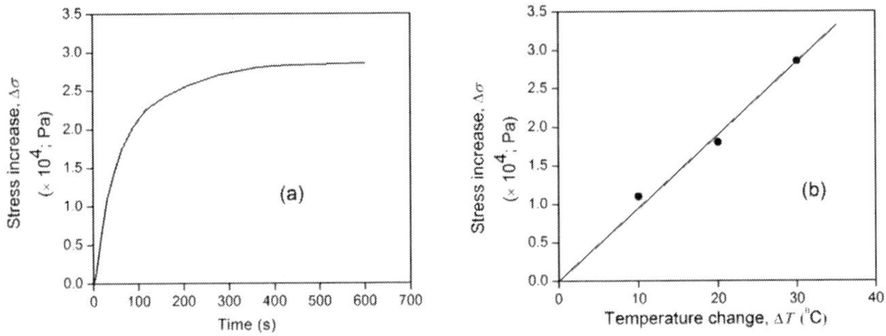

Figure 3. (**a**) Stress increase Δσ in the bandage upon activation under constraint deformation (ε = 30%; T = 50 °C). (**b**) Results of stress increase Δσ with increase in temperature ΔT (Equation (4)).

If such MP bandage is applied over a limb surface (here it is applied to a cylindrical tube), there is an external constraint imposed to the bandage by the interface surface. As the interface pressure depends on the stress in the bandage as expressed by Laplace's law (Equation (1)), the bandage could allow pressure change directly over the limb by simple temperature modulation. Figure 4 shows the pressure variation by the bandage once the temperature is increased. Increasing the temperature ($>T_g$) allows the activation of memory stress in MP filaments (similar to Figure 3), and the overall stress is increased, which allows the pressure to change. At lower temperatures ($T < T_g = 30$ °C), the memory stress in the MP filaments is not activated, and a constant pressure is maintained, which is present only due to elastic stress present in the structure of bandage while wrapping on the cylindrical surface.

For a given T ($>T_g$), the P increases rapidly for the initially period (600 s) and then finally reaches saturation. The rate and amount of extra pressure generated during activation is greater at higher temperatures ($p < 0.01$). The extra pressure increases (ΔP) are 8 mmHg and 4 mmHg, respectively, at 50 °C and 35 °C. The MP filaments used in the fabric has a transition range (~20–50 °C; Figure 1), because of which, it is highly unlikely that all switching segments would activate together with increasing temperature to a fixed level. It is expected that more switching segments are activated at higher T (~50 °C) as compared to lower T (~35 °C). Therefore, more memory stress is generated in the MP filaments at higher temperatures, which therefore results in greater pressure increases. No significant pressure changes were observed beyond 50 °C. This is due to activation of all switching segments in the MP network around 50 °C.

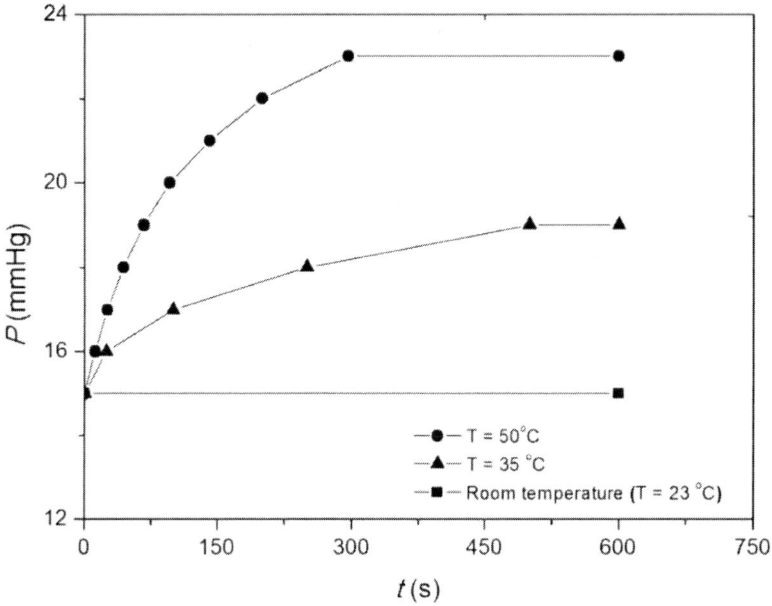

Figure 4. Pressure variation at different levels of temperatures (ε = 30%; C = 23.6 cm; N = 1).

For practical application, the temperature should not be increased too much above body temperature, as this could cause patient discomfort. This is why the transition range of MP actuation needs to carefully optimized to obtain the desired pressure variation in the achievable and allowable temperature conditions. The T_g could be optimized to a particular range by the proper choice of different components (type and molecular weight of switching segment and hard segment content) involved in the formation of the MP [23,26,27]. Herein, we developed MP filament using segmented polyurethane with low T_g (~30 °C) and the activation domain was 20–50 °C.

Before recommending the SMP and suggesting it as a credible compression product, there are many aspects that need to be thoroughly examined, such as how much extra pressure could be achieved, the temperature range, other external factors that could decide pressure variation, the repeatability, etc. Apart from temperature change, several other variables that could affect the pressure results exist. Figure 5 summarizes the effect of important variables on the amount of ΔP generated by the bandage upon activation. Similar to temperature, strain has the same effect on the MP bandage (Figure 5b). More ΔP is observed at high strain level. This is due to differences in the extent of recovery of memory polymer at different levels of strain. Inducing more strain in the MP results in significant entropy reduction because of orientation of polymer chains, and upon heating the stretched MP, the shrinkage is more favorable to increase the entropy by coiling of polymer chains, as explained by rubber elasticity. Therefore, high strain in the MP is expected to have more stress upon heating. The effect of number of layers and circumference can be derived from the equation of Laplace's law. The pressure is inversely related to the limb circumference due to which ΔP observed at higher circumference (31.2 cm) is lower than pressure observed at 23.6 cm circumference (Figure 5d). Increasing the number of wrapping layers will also have an additive effect (Figure 5c), although the ΔP is not doubled if the number of layers (N) is increased from one to two. Primarily, the compression provided by the top layer is absorbed by the inner layers; thus, the total pressure exerted by multiple layers is less than the sum of pressure exerted by individual layers. The relation of P and N, as described by Laplace's law (Equation (1)), overestimates the pressure values. This relation is valid for thin-wall cylinder vessel and for continuum materials. Assuming fabric as a continuum and neglecting its thickness may also cause this discrepancy. Table 2 shows all the ΔP results of different levels of influencing factors.

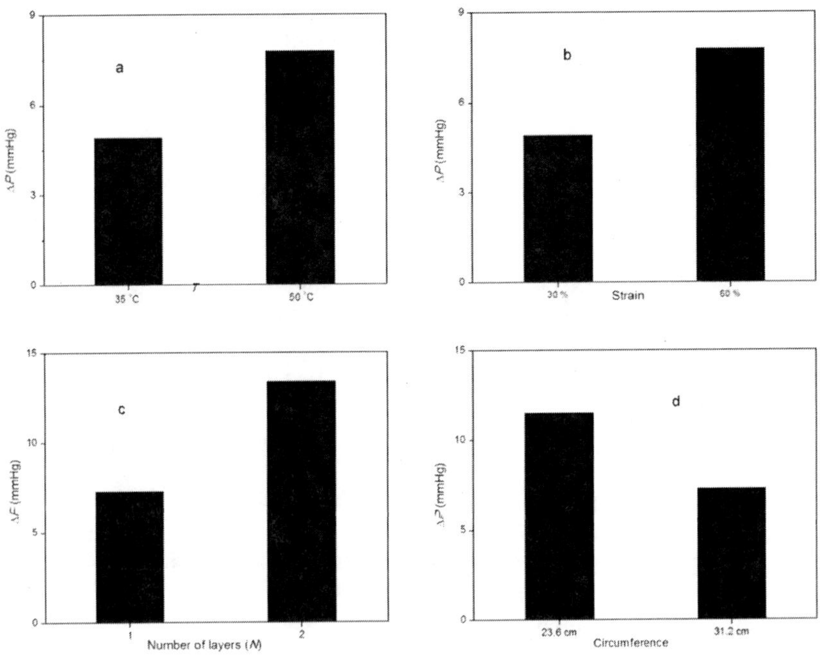

Figure 5. Result of extra pressure ΔP: (**a**) by changing temperature; (**b**) by changing strain; (**c**) by changing number of wrapping layers; and (**d**) by changing cylinder circumference.

Table 2. Results of extra pressure (ΔP, in mmHg) by the MP bandage at different levels of factors.

Temperature (°C)	Circumference							
	23.6 cm				31.2 cm			
	Applied strain				Applied strain			
	30%		60%		30%		60%	
	No. of layers		No. of layers		No. of layers		No. of layers	
	1	2	1	2	1	2	1	2
35	4.9	7.5	7.8	11.9	3.1	4.9	4.2	7.4
50	7.8	15.1	11.5	21.6	5.8	10.5	7.3	13.4

Notes: *T*: Temperature; *C*: Cylinder circumference; *N*: Number of layers; ε: Strain.

Finally, we also tested the repeatability of pressure change by the bandage under continuous heating and cooling cycles. The result is summarized in Figure 6. After an initial adjustment for the first four cycles, the pressure variation show similar results. This indicates that the bandage has the potential to reproduce the same pressure results, even after multiple uses, and thus allow readjusting pressure levels whenever needed.

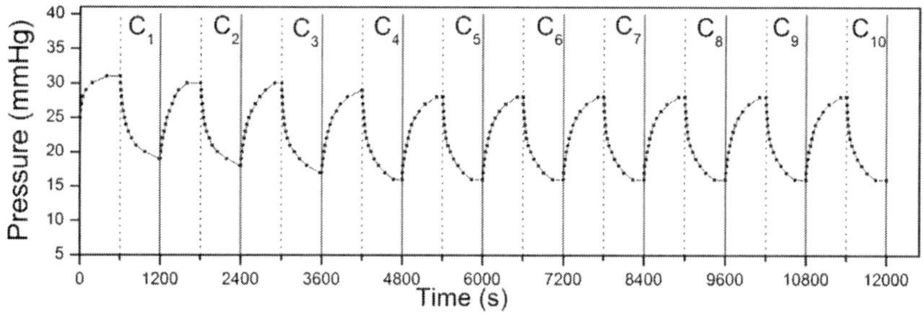

Figure 6. Cyclic pressure variation by the bandage upon continuous heating and cooling cycles. Each cycle (C_1 to C_{10}) includes heating of the bandage for 10 min in heated chamber (T = 50 °C), and then cooling down for another 10 min at room temperature (T = 23 °C).

The memory bandage could offer several advantages in compression management. First, this could give more freedom to nurses to control or readjust pressure at wrapped position, and to achieve an appropriate level of pressure. Second, it would be possible to counter the effect of pressure drop during the course of compression treatment. Earlier studies suggested that the pressure drop is inevitable for most of the compression bandages due to fiber's stress relaxation or due to limb reduction [9,28]. However, the advantage of memory bandage over the existing conventional product would be that, when the pressure drops below a targeted level, it would be possible to readjust pressure level even without replacing the bandage, and therefore could provide sustained compression. Third, it would be possible to extract multi-functionalities of static and dynamic compression benefits from the memory bandage. Most of the current means for compression are passive, meaning that they only provide a fixed level of initial pressure, incapable of having massage effect. For dynamic treatment (massaging of leg), the equipment like intermittent pneumatic compression (IPC) is recommended, which is too costly, noisy, bulky, and, once attached, requires immobile patients [29]. This novel bandage could be used for multiple compression benefits (static and dynamic) via a programmed heating stimulus (Figure 7).

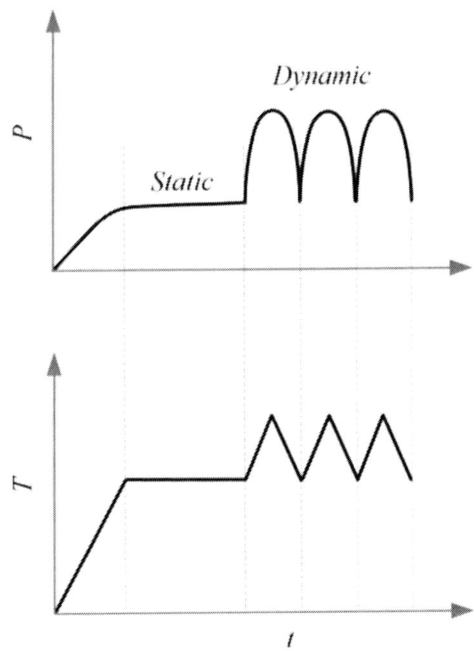

Figure 7. Exploiting controlled static and dynamic compression benefits from a memory bandage using a programmed heating stimulus.

4. CONCLUSIONS

The present work introduces a novel application of MP filaments in medical bandage for the functional compression management for the treatment of chronic venous disorders. A memory bandage was developed using MP filaments in the structure. The bandage was able to change the internal stress in its structure by external means, i.e., using heating stimulus. It was possible to control or manage the pressure exerted by the bandage in wrapped position, and extra pressure (ΔP) was generated by simple heating. Increasing the temperature and strain stimulated more ΔP. Other factors, i.e., circumference and number of layers, also affect the magnitude of ΔP. The bandage could offer the advantage of controlled compression to meet the desired needs for a given patient.

It is envisioned that the memory bandage could revolutionize compression treatment, and this goal could be achieved with more scientific effort, sharp vision, tremendous creativity and huge investment. There are many studies that are still needed to ensure the memory bandage is a credible compression product. In real practice, heating of MP on the leg to a particular temperature is a challenge, and simple and efficient heating arrangements need to be assured to have precise control on the pressure. The use of flexible heating systems (conductive fabrics) could be a good alternative. Another method is to develop body-temperature activated memory filament, where the activation could be performed by body heat. All this could be a part of future study where an integrated system (heating device and memory bandage) will be designed and examined. The other work is to conduct the pressure study on a soft surface. Naturally, the pressure on a soft surface (such as a real body) may vary compared to the hard surface presented in this study. The soft tissue compressibility could lead to changes in circumference C, and the parameters in Equation (1) will have to be adjusted accordingly for accurate pressure predictions. Future work should also focus on the effect of several other factors, e.g., different MPs, hard segment content, proportion of MP, fabric structure, time dependent study, etc.

ACKNOWLEDGMENTS

The authors acknowledge the support of team members who helped directly or indirectly during the preparation of the work, including Han Jianping and Lv Jing who helped in polymer synthesis and filament preparation. We also thank Harishkumar Narayana and Wu You for initial discussions.

AUTHOR CONTRIBUTIONS

The main author, Bipin Kumar, initiated the project, contributed in design of experiments, observed the key results of the paper and wrote up the manuscript. Jinlian Hu participated in experimental design, paper discussion and revision. Ning Pan helped in proof-reading and final editing of the paper.

CONFLICTS OF INTEREST

The authors declare no conflict of interest.

REFERENCES

1. Partsch, H. Compression therapy in leg ulcers. Rev. Vasc. Med. **2013**, 1, 9–14.

2. Ramelet, A.A. Compression therapy. Dermatol. Surg. **2002**, 28, 6–10.

3. Partsch, H. Compression for the management of venous leg ulcers: Which material do we have? Phlebology **2014**, 29, 140–145.

4. Kumar, B.; Das, A.; Alagirusamy, R. Study on interface pressure generated by a bandage using in vitro pressure measurement system. J. Text. Inst. **2013**, 104, 1374–1383.

5. Kumar, B.; Singh, J.; Das, A.; Alagirusamy, R. Comfort and compressio nal characteristics of padding bandages. Mater. Sci. Eng. C **2015**, 57, 215–221.

6. Basford, J.R. The Law of Laplace and its relevance to contemporary medicine and rehabilitation. Arch. Phys. Med. Rehabil. **2002**, 83, 1165–1170.

7. Farah, R.S.; Davis, M.D.P. Venous Leg Ulcerations:A Treatment Update. Curr. Treat. Options Cardiovasc. Med. **2010**, 12, 101–116.

8. Kumar, B.; Das, A.; Alagirusamy, R. Analysis of sub-bandage pressure of compression bandages during exercise. J. Tissue Viability **2012**, 21, 115–124.

9. Kumar, B.; Das, A.; Alagirusamy, R. Effect of material and structure of compression bandage on interface pressure variation over time. Phlebology **2013**, 29, 376–385.

10. Kumar, B.; Das, A.; Alagirusamy, R. Study of the effect of composition and construction of material on sub-bandage pressure during dynamic loading of a limb in vitro. Biorheology **2013**, 50, 83–94.

11. Zhao, Q.; Qi, H.J.; Xie, T. Recent progress in shape memory polymer: New behavior, enabling materials, and mechanistic understanding. Prog. Polym. Sci. **2015**, 49–50, 79–120.

12. Hu, J.L.; Zhu, Y.; Huang, H.H.; Lu, J. Recent advances in shape-memory polymers: Structure, mechanism, functionality, modeling and applications. Prog. Polym. Sci. **2012**, 37, 1720–1763.

13. Lendlein, A.; Behl, M.; Hiebl, B.; Wischke, C. Shape-memory polymers as a technology platform for biomedical applications. Expert Rev. Med. Devices **2010**, 7, 357–379.

14. Mather, P.T.; Luo, X.F.; Rousseau, I.A. Shape Memory Polymer Research. Annu. Rev. Mater. Res. **2009**, 39, 445–471.

15. Leng, J.S.; Lan, X.; Liu, Y.; Du, S. Shape-memory polymers and their composites: Stimulus methods and applications. Prog. Mater. Sci. **2011**, 56, 1077–1135.

16. Zhang, D.W.; George, O.J.; Petersen, K.M.; Jimenez-Vergara, A.C.; Hahn, M.S.; Grunlan, M.A. A bioactive "self-fitting"shape memory polymer scaffold with potential to treat cranio-maxillo facial bone defects. Acta Biomater. **2014**, 10, 4597–4605.

17. Hu, J.L.; Kumar, B.; Narayan, H.K. Stress Memory Polymers. J. Polym. Sci. Part B Polym. Phys. **2015**, 53, 893–898.

18. Kratz, K.; Madbouly, S.; Wagermaier, W. Temperature-Memory Polymer Networks with Crystallizable Controlling Units. Adv. Mater. **2011**, 23, 4058–4062.

19. Wu, Y.; Hu, J.; Huang, H.; Li, J.; Zhu, Y.; Tang, B.; Han, J.; Li, L. Memory Chromic Polyurethane with Tetraphenylethylene. J. Polym. Sci. Part B Polym. Phys. **2014**, 52, 104–110.

20. Yuan, J.K.; Zakri, C.; Grillard, F.; Neri, W.; Poulin, P. Temperature and Electrical Memory of Polymer Fibers. In Proceedings of the 7th International Conference on Times of Polymers (TOP) and Composites, Melville, NY, USA, 22–26 June 2014.

21. Liu, Y.P.; Gall, K.; Dunn, M.L.; Greenberg, A.R.; Diani, J. Thermomechanics of shape memory polymers: Uniaxial experiments and constitutive modeling. Int. J. Plast. **2006**, 22, 279–313.

22. Narayana, H.; Hu, J.; Kumar, B.; Shang, S. Constituent Analysis of Stress Memory in Semicrystalline Polyurethane. J. Polym. Sci. Part B Polym. Phys. **2016**.

23. Hu, J.; Meng, Q.H.; Zhu, Y.; Lu, J.; Zhuo, H.T. Shape Memory Fibers Prepared via Wet, Reaction, Dry, Melt, and Electro Spinning. US20090093606 A1, 9 October 2007.

24. Rosen, S. Rubber Elasticity. In Fundamental Principles of Polymeric Materials; Wiley-Interscience Publications: Hoboken, NJ, USA, 1982; pp. 189–198.

25. Kumar, B.; Hu, J.L.; Pan, N. Smart medical stocking using memory polymer for chronic venous disorders. Biomaterials **2016**, 75, 174–181.

26. Meng, Q.H.; Hu, J.; Yeung, L.Y.; Hu, Y. The Influence of Heat Treatment on the Properties of Shape Memory Fibers. II. Tensile Properties, Dimensional Stability, Recovery Force Relaxation, and Thermomechanical Cyclic Properties. J. Appl. Polym. Sci. **2009**, 111, 1156–1164.

27. Zhu, Y.; Hu, J.; Yeung, L.-Y.; Liu, Y.; Ji, F.; Yeung, K. Development of shape memory polyurethane fiber with complete shape recoverability. Smart Mater. Struct. **2006**, 15, 1385–1394.

28. Kumar, B.; Das, A.; Alagirusamy, R. An Approach to Determine Pressure Profile Generated by Compression Bandage Using Quasi-Linear Viscoelastic Model. J. Biomech. Eng. Trans. Asme **2012**, 134.

29. Feldman, J.L.; Stout, N.L.; Wanchai, A.; Stewart, B.R.; Cormier, J.N.; Armer, J.M. Intermittent Pneumatic Compression Therapy: A Systematic Review. Lymphology **2012**, 45, 13–25.

CHAPTER 6

Nanomaterials for Functional Textiles and Fibers

Pedro J. Rivero[1*], Aitor Urrutia[2], Javier Goicoechea[2] and Francisco J. Arregui[2]

[1]Institute for Advanced Materials (InaMat), Materials Engineering Laboratory, Department of Mechanical, Energy and Materials Engineering, Public University of Navarre, Campus Arrosadía S/N, 31006 Pamplona, Spain.

[2]Institute of Smart Cities (ISC), Nanostructured Optical Devices Laboratory, Department of Electrical and Electronic Engineering, Public University of Navarre, Campus Arrosadía S/N, 31006 Pamplona, Spain.

ABSTRACT

Nanoparticles are very interesting because of their surface properties, different from bulk materials. Such properties make possible to endow ordinary products with new functionalities. Their relatively low cost with respect to other nano-additives make them a promising choice for industrial mass-production systems. Nanoparticles of different kind of materials such as silver, titania, and zinc oxide have been used in the functionalization of fibers and fabrics achieving significantly improved products with new macroscopic properties. This article reviews the most relevant approaches for incorporating such nanoparticles into synthetic fibers used traditionally in the textile industry allowing to give a solution to traditional problems for textiles such as the microorganism growth onto fibers, flammability, robustness against ultraviolet radiation, and many others. In addition, the incorporation of such nanoparticles into special ultrathin fibers is also analyzed. In this field, electrospinning is a very promising technique that allows the fabrication of ultrathin fiber mats with an extraordinary control of their structure and properties, being an ideal alternative for applications such as wound healing or even functional membranes.

KEYWORDS

Nanoparticles; Coatings; Functional textiles; Functional nanowebs; Electrospinning; Antibacterial; Flame retardant; UV-protection; Superhydrophobic; Membranes; Wound dressing

1. INTRODUCTION

Natural fibers and textiles have been used for humans since ancient times. Our antecessors firstly used fur and animal skin for dressing and protection from the environment, but very soon they started to use vegetal fibers to make rudimentary fabrics. There are evidences of the use of dyed flax fibers into clothes more than 30,000 years ago [1]. For centuries, humans have used vegetal fibers (such as flax or cotton) and animal fibers (such as wool or silk) to produce yarns and then weave them into textiles using handmade processes. In the eighteenth century, there was a revolutionary industrial development [2] with the invention of machines that speeded up the manufacture of fabrics and made them more available and affordable. This technological revolution changed the concept of textile manufacturing transforming it into a real industry. It could be said that in the twentieth century, there was a second technological revolution with the synthesis of artificial fibers such as rayon [3], nylon, or polyester with good quality and low cost production techniques that rapidly gave those fibers a significant market share because of their good properties such as low cost, chemical stability, and outstanding versatility (dyes, colors, fiber diameters, engineered weaving for special clothes).

Nowadays, there is a new revolution on the textile industry with the apparition of new technologies that could add special functions and properties to the fabrics. For example, there has been significant improvement in technologies for textile coloring [1], digital printing on textiles [4], smart fabrics [5], and high performance functional textiles [6]. In this sense, nanoparticles play a key and significant role in this technological evolution since they show outstanding surface properties that allow to multiply their effect in comparison with bulky traditional additives and materials.

This work reviews the most relevant contributions of the use of nanoparticles for functionalize textile materials. Firstly, the analysis is focused on traditional synthetic fibers (mainly nylon, polyamide, or blends) and the use of nanoparticles for providing new properties such as antibacterial activity, flame retardant properties, UV-protection, supehydrophobicity, and others. And secondly, the use of nanoparticles is analyzed for special ultrathin fibers functionalization. These fibers can be obtained by electrospinnig technique for applications such as wound dressing, functional membranes for water treatment and others.

2. TRADITIONAL SYNTHETIC FIBERS

The continuous advances in organic chemistry, especially with the apparition of polymer chemistry discipline, enable the fabrication of synthetic fibers which are made from polymers or small molecules, being the petroleum as the most important raw material. Among the high amount of different types of synthetic fibers, the most known are polyamide, polyester, acrylic, and polyolefin. Within these four types, the polyamide (especially PA6, known as polyamide 6 or nylon) and polyesters (especially PET, known as polyethylene terephthalate) are the most used in the textile industry. This section is focused on the development of functionalized synthetic fibers, concretely polyamide and polyester fabrics. In general, these fabrics show poor wettability properties and hydrophobicity in nature which can affect the polymer processability and also fiber dyebility [7]. With the development of the nanotechnology, different types of nanoparticles (mostly inorganic or metal oxide) have been successfully incorporated into synthetic fibers in order to obtain functionalized fibers with special properties [8–11]. The major effects such as flame retardancy, antibacterial activity, or superhydrophobicity are presented in detail using different deposition techniques. In addition, special emphasis is oriented to the surface modification fabrics in order to improve the adhesion of the nanoparticles and obtain functionalized synthetic fibers with the desired properties. Finally, it also discussed the importance of controlling particles size as well as their corresponding shape because both phenomena are of vital importance to the final and desired function imparted to the fibers.

3. Flame Retardant Applications

Synthetic fibers such as polyamide or polyester are highly inflammable, and continuous investigations are carried out in the development of flame retardant products because the legislation requires the use of low flammability products in the industry [12]. In this context, many of the available systems for flame retardancy are composed of halogen-containing additives. However, new regulations are devoted to the use halogen-free compounds, particularly phosphorous-based systems, which are being used as a promising alternative as effective flame retardants in textile industry [13–15]. Recently, the incorporation of nanoparticles in the finishing fabrics as flame retardant seems to be a valid and interesting approach. The main advantage is that a low amount of them can be employed, although not all the nanoparticles available in the market can be used in the flame retardancy field [16, 17]. It has been corroborated that several key points have to be into consideration in order to obtain better results. Among them, the resultant morphology (shape and size), chemical nature, concentration, as well as their distribution as a function of the procedure employed onto textile fabrics constitute the main ingredients to be an efficient flame retardant system.

El-Hady et al. proposed a novel flame retardant approach based on the use of zinc oxide (ZnO) nanoparticles for their application to cellulosic fabrics (cotton polyester blend) [18]. A pad-dry-cure method was selected to incorporate the ZnO nanoparticles onto the fabrics. The use of two different polycarboxylic acids such as succinic acid (SA) or 1,2,3,4-butane tetarcarboxylic acid (BTCA) together with sodium hypophosphite (SHP) as catalyst have been employed for cross-linking fabrics. In this work, firstly, the effect of curing temperature was investigated, and it has been demonstrated that an increase of the curing temperature from 160 to 180 °C made possible the formation of new ester cross-linking between cellulose chains and polycarboxylic acids [19]. And secondly, the measuring of char yield was performed to study the influence of flame retardant. The results indicated that both SA and BTCA are effective agents in reducing flammability of treated fabrics in the presence of SHP. However, better experimental results were observed for BTCA in comparison with SA in reducing flammability. In addition, it has been demonstrated that by increasing the nano ZnO concentration (from 0.25 to 0.5 %) as well as BTCA and SHP concentrations led to decrease the fabric flammability.

Novel studies about the influence of nanoparticles adsorption as fire protection for fabrics was presented by Alongi et al. [13, 17, 20]. This research group has studied the enhancement of the flame retardant properties for synthetic fibers such as polyester or cotton/polyester blends. The effect of impregnating different types of nanoparticles onto textiles as lamellar nanoparticles (montmorillonites CNa, hydrotalcites HT, or bohemites OS1), globular nanoparticles (titanium dioxide, silica, or octalpropylammonium POSS), or a mixing of both globular/lamellar nanoparticles has been evaluated [16]. In this sense, according to the use of lamellar nanoparticles, the use of dispersing agents such as sodium polyacrylate (B500) or binders such as dicarboxylate ester (PES) and an initial pre-treatment of the fabric by cold oxygen plasma makes possible an increase in time ignition and a reduction of heat release rate. As far as globular nanoparticles are concerned, the use of a silicon-based structure soluble in water as octalpropylammonium (POSS) shows promising results as flame retardant with an important decrease of CO_2 and CO production. More specifically, polyester fabrics were not burnt when a methane flame was applied, although these results were not observed for cotton or PET-to-cotton blends. According to this, an interesting approach is presented in Fig. 1, where the results of untreated fabrics and fabrics coated with a combination of a commercial phosphorous-based flame retardant with different POSS amounts after cone calorimetry tests are shown. The untreated fabrics are totally destroyed after combustion test, whereas the sample only treated with the phosphorous flame retardant consisted of a thin film. However, the fabrics treated with both phosphorous flame retardant and variable POSS amounts (5, 15, and 30 wt.%, respectively) showed a residue more coherent and compact as a function of the nanoparticle amount. In addition, by increasing the POSS amount, the resultant time to ignition (TTI) and the corresponding peak value of heat release rate (pkHRR) were decreased in a remarkable way as it can be observed in Fig. 2. In addition, this same research group proposed that among all the possible nanoparticles combinations, a mixing of HT lamellar with SiO_2 globular nanoparticles seemed to be the most efficient flame retardant system [21]. This aspect was corroborated using the fire performance index (FPI) which is calculated as TTI-to-pkHRR ratio. When a higher value of FPI is obtained, better results as flame retardant were observed. The highest improvements in fire performance were observed for the HT-SiO_2 combination as evidenced all FPI values. These results were also corroborated for the PET-to-cotton blend where the HT-silica combination improves the fire performance to this blend. As a conclusion of the experimental results, a combination of both lamellar/globular nanoparticles presented better results as efficient flame retardant system in comparison with a single type of nanoparticles [16].

Until now, pad-dry-cure method or impregnation processes have been the most used techniques for the finishing fibers. However, an alternative approach for the incorporation of the nanoparticles into thin films is the layer-by-layer (LbL) assembly. This method makes possible the fabrication of nanocoatings onto the textiles fibers with a good control over the resultant thickness and roughness [22]. This method is performed by sequentially exposing the fabrics to a cationic polyelectrolyte and an anionic polyelectrolyte during a specific period of immersion time. In addition, a rinsing step in deionized water is necessary between both polyelectrolytes baths, and a drying step is performed between each rinsing time [20, 23, 24]. It is important to remark that the times of immersion as well as the times necessary for the rinsing and drying steps vary from author to author, although the methodology is always the same. According to this, Apaydin et al. designed new intumescent flame retardant coatings made from both cationic and anionic polyelectrolytes mixed with titanium dioxide (TiO_2) nanoparticles using the LbL assembly [25]. These nanocoatings have been performed onto both polyamide and polyesters fabrics, showing significant differences in surface properties (i.e., wettability and surface energy) as it was corroborated by SEM images, EDX, and contact angle values. The resultant thin films covered homogeneously all external polyamide fabrics, whereas polyester fabrics were covered partially with the presence of aggregates. This important difference in morphology is the key point to understand the different behaviors to flammability tests and the reason that the coating only acts efficiently in reducing the peak heat release rate (pHRR) value for the polyamide fabrics. The incorporation of these TiO_2 nanoparticles into LbL films slightly improved the fire retardant in comparison with a previous work of this research group based on polymeric coatings without any kind of nanoparticles [23].

Carosio et al. coated polyester (PET) fabrics by LbL assembly using negative colloidal SiO_2 nanoparticles of two different sizes (10 and 30 nm, respectively) with positively charged alumina-coated SiO_2 [20]. The best results have been observed for the smallest nanoparticles which increased the time to ignition and reduced the pHRR using cone calorimetry tests. One of the most relevant and important aspects of this work was the design of an environmentally friendly nanocoating based on nanoparticles, showing very promising results of flammability.

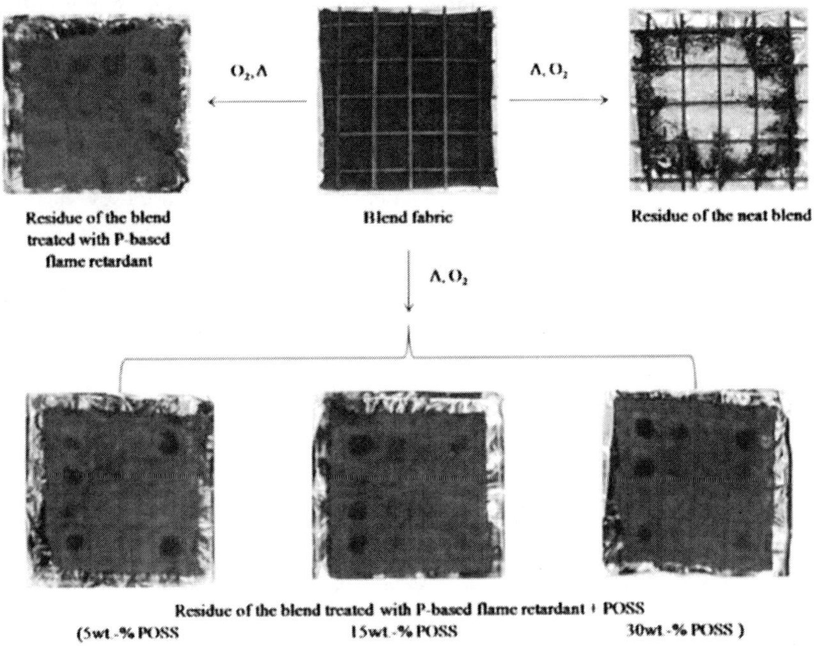

Figure 1 . Residues of untreated PET-to-cotton blend and fabrics treated with a phosphorous-based flame retardant and variable globular nanoparticle (octalpropylammomium, POSS) amounts after cone calorimetry tests. Reprinted with permission from [16]

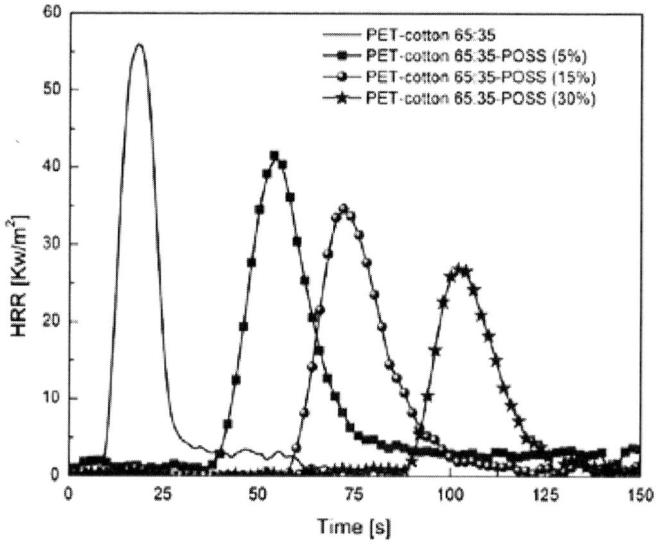

Figure 2 Heat release rate (HRR) curves of neat PET-to-cotton blend and of fabrics treated with a phosphorous-based flame retardant with different POSS amounts. Reprinted with permission from [16]

Finally, Conzalves et al. reported that multi-walled carbon nanotubes can be also used for improving the flame retardant properties when are applied onto polyester fabrics [26]. For this specific case, polyester fabrics functionalized with carbon nanotubes halved the burning rate and duplicated the burning time in comparison with raw polyester fabrics.

Finally, in order to have a better understanding of the different methodologies for obtaining flame retardant surfaces as well as the corresponding nanoparticles used for this purpose, a summary is shown in Table 1.

Table 1. Summary of the different types of nanoparticles with their corresponding size or the resultant thickness films for flame retardant applications as well as the deposition techniques used for a specific type of fabrics

Type of fabrics	Deposition process	Type of nanoparticles
Polyester and cotton/polyester blend	Impregnation process	Globular octalpropylammonium (POSS) nanoparticles, 2015 [16]
Cotton/polyester blend	Pad-dry-cure method	Zinc oxide (ZnO) nanoparticles (an average particle size of 30 nm), 2013 [18]
Polyester	Layer-by-layer assembly	Silica (SiO_2) colloidal nanoparticles (<10 nm average thickness), 2011 [20]
Polyester and cotton/polyester blend	Impregnation process	Mixture of silica (SiO_2) globular (spherical particles with an average size of 150 nm) and hydrotalcite (HT) lamellar nanoparticles, 2012 [21]
Polyamide and polyester	Layer-by-layer assembly	Titanium dioxide (TiO_2) nanoparticles, thickness film (approx. 500 nm), 2015 [25]
Polyester	Layer-by-layer assembly	Multi-walled carbon nanotubes (MWCNTs) (an average diameter of 9.5 nm and an average length of 1.5 μm), 2012 [26]

4. ANTIBACTERIAL PROPERTIES

Considering the special advantages of the application of nanostructured materials in textile industry, this section is focused on the modification of textile surface using different kinds of nanostructured materials which show excellent antimicrobial properties. Among all metallic materials, silver is of particular interest because it is a powerful antibacterial agent, showing antimicrobial efficacy against bacteria, viruses, and microorganisms [27, 28]. This excellent antibacterial activity makes possible their use in different and varied fields such as food preservation [29], safe cosmetics [30], medical devices [31, 32], water treatment [33], or textiles fabrics [34, 35]. An important aspect is that silver in both ionic and colloidal forms shows low toxicity to human cells and high biocompatibility [36]. However, the real mechanism of antimicrobial action related to the silver is not well understood, and different hypotheses are considered by the scientific community. Several studies indicate that the antibacterial effect is due to the release of silver ions from silver nanoparticles (Ag NPs), whereas other approaches are devoted to the adverse effect of the Ag NPs in the cell membrane, being these Ag NPs the responsible of the cell death. More details about these approaches can be found in [37].

As an example of multiple properties and their implementation in textile fabrics, Gerber et al. evaluated the effect of silver-tricalcium phosphate nanoparticles (Ag/TCP) onto polyamide 6 (PA6) fibers to build a reactive system against bacteria such as *Escherichia coli* and *Streptococcus sanguinis*, showing a killing efficiency of 99.99 or 100 % within 24 h contact time, respectively [38]. These great reactive properties can be implemented as self-disinfecting fibers in a broad range of applications. Other interesting approach is presented by Shastri et al. which evaluate the antimicrobial activity of nanosilver-coated sock fabrics against foot pathogens [39]. With the aim of the development of nanostructured silver production into fibers, Jiang et al. presented an effective method for preparing polyester fabrics via magnetron sputtering, showing excellent protection from ultraviolet protection, hydrophobicity, and good antibacterial performance [40]. SEM observations confirmed the presence of a thin layer of nanoscaled silver coating onto fiber surface after sputtering treatment. In addition, with an increase in sputtering time from 10 to 30 min, homogeneous deposition of silver nanoparticles (Ag NPs) with a particle size about 50 nm on the polyester fabrics is observed. In addition, the properties of the surfaces drastically change the resultant contact angle (CA) from hydrophilic (CA = 0°; untreated fabrics) to hydrophobic (CA = 132.2°) after modification with nanostructured silver. The silver-coated polyester has been tested against two kinds of bacteria such as *E. coli* and *Staphylococcus aureus*, showing an excellent antibacterial reduction ratio of 99.7 and 99.8 %, respectively. Interesting approaches were presented by Radetic et al. [41–43], which are devoted to the use of a corona treatment (electrical discharge at atmospheric pressure) or low-temperature radio frequency (RF) plasma for fiber surface activation which can facilitate the binding efficiency of colloidal Ag NPs onto polyester as well as polyamide fabrics, showing an enhancement of the antibacterial properties and better laundering durability.

As an important aspect to be in consideration is that there is a wide variety of synthetic methodologies to obtain Ag NPs. Most of them are focused on a strict control of several parameters such as shape, size, surface functionalization, or interparticle distance which affect their final properties [44]. Rivero et al. designed an experimental matrix where multicolored Ag NPs (Fig. 3a) were obtained as a function of the fine control of two parameters such as protective agent (polyacrylate (PAA)) and reducing agent (dimethylaminoborane (DMAB)) concentrations [45]. A great control of these parameters makes possible the synthesis of Ag NPs or silver clusters with different shapes and sizes.

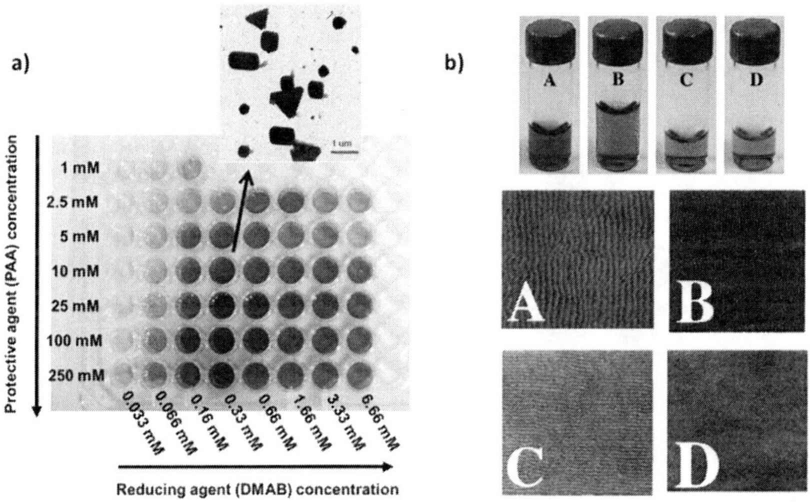

Figure 3 **a** Photograph of multicolor silver map obtained as function of variable protective agent and reducing agents. TEM micrograph shows the formation of Ag NPs with different shapes (*triangle*, *rod*, *cube*, *bar*) for 10 mM PAA and 0.33 mM of DMAB. Reprinted with permission from [45]. Copyright (2013) Springer. **b** Silver nanosols prepared at pH 8 and silver-nanosol-loaded polyester fabrics (*A*) NaBH4 reduction, M_w 1200, (*B*) NaBH4 reduction, M_w 15,000, (*C*) UV reduction M_w 1200, (*D*) UV reduction M_w 15,000. Reprinted with permission from [46]. Copyright (2008) American Chemical Society

In this sense, based on the preparation of silver-poly(acrylates) clusters, Falleta et al. proposed the synthesis of silver nanosols using silver nitrate ($AgNO_3$) as a source of silver ions, PAA as a protective agent at two different molecular weights (M_w 1200 and M_w 15,000) and two different reducing mechanisms (chemical reduction with sodium borohydride, $NaBH_4$ and UV radiation exposure), showing all dispersions different colorations (Fig. 3b, top) [46]. UV-vis spectra corroborated the presence of well-separated absorption bands. One was located at 400 nm which is related to the localized surface plasmon resonance (LSPR) absorption band of the silver, whilst another band with a broad signal was located at visible region which is associated with the formation of silver-polyacrylate complexes. These nanosols were applied onto polyester textiles by simply immersing, and the treated fabrics maintained the coloration after the rinsing and drying steps which corroborate the complete adhesion of the nanoparticles onto the fabrics surface (Fig. 3b, down). The resultant-treated polyester fabrics showed a strong inhibition effect on all microorganisms (*S. aureus, Staphylococcus epidermidis, Pseudomonas aeruginosa*, and *Candida albicans*) whereas the untreated fabrics did not show any inhibition zone, as it can be observed in Fig. 4.

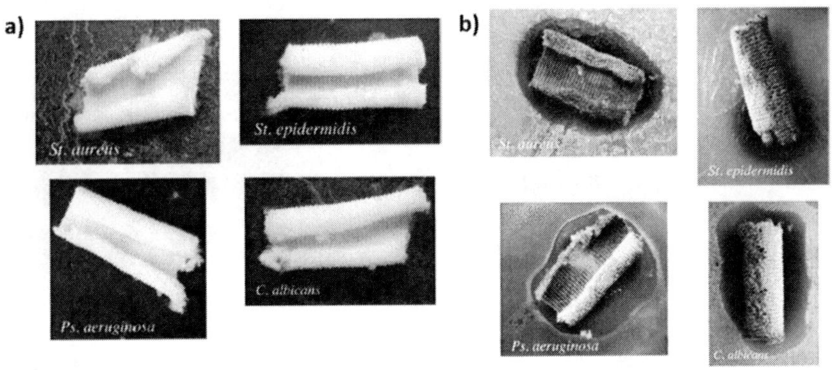

Figure 4 .**a** Effect of untreated polyester samples toward the growth of the investigated stains; **b** effect of Ag-treated polyester samples toward the growth of the investigated stains. Reprinted with permission from [46]. Copyright (2008) American Chemical Society

Other alternative approach is the synthesis of nanosilver on fabrics using silver ammonia complex [47–49]. Textor et al. proposed the modification of polyamide fabrics surface using a heterogeneous Tollens' reaction [47]. In a first step, the surface was modified by treatment with glutardialdehyde (GDA), making possible the establishment of aldehydes at the polymer surface. In order to corroborate the presence of this aldehyde layer, measurements of the contact angle of water for GDA-treated polyamide films were performed. An increase of contact angle from 50° (untreated) to 66° (the highest GDA concentration) verified the surface modification. Further, in a second step, polyamide fabrics were treated with Tollens' reagent, making possible the formation of a thin silver layer due to the chemical reduction of silver ions (Ag^+) to Ag NPs for the presence of aldehydes. A color change to brownish indicated a complete reduction of Ag^+ in solution, and it is assigned to the surface plasmon resonance of silver layers of a thickness in the range of several nanometers [45]. The experimental results showed an excellent antimicrobial activity of the silver finished polyamide fabrics against *E. coli* and even more an excellent durability of the coated fabrics even after 30 laundry cycles.

A similar approach based on the use of silver-ammonia complex is presented by Montazer et al. [48]. In this case, Ag NPs have been prepared using polyvinyl pyrrolidone (PVP) as a reducing/stabilizing agent through UV irradiation which reduce silver ions (Ag^+) to nanoparticles (Ag^0). These nanoparticles have been deposited onto nylon-knitted fabric surface using a simple dip-pad technique. The antibacterial activity of the treated fabrics has been tested against *E. coli* and *S. aureus*, showing a reduction percentage of the bacteria growth up to 99.2 % after 20 washes. This same research group proposed later the synthesis of Ag NPs within the polymeric chains of PA6 fabrics by using a silver-ammonia complex $[Ag(NH_3)_2]^+$ [49]. It was demonstrated that the polymeric chains of polyamide acted as both stabilizing and reducing agents for the in situ synthesis of the nanoparticles without using any other type of chemical compounds as stabilizing or reducing agents. The silver/ammonia complex could be oxidized through the polyamide chains, and consequently, the silver ions could be reduced to Ag NPs. Furthermore, the nitrogen atoms precedent of polyamide chains can stabilize Ag NPs through coordination. Due to this fixation of Ag NPs within polymeric chains of polyamide, the resultant-coated fabrics showed excellent antibacterial properties against *E. coli* and *S. aureus* even after repeated washing cycles.

Another different alternative method for the preparation of Ag NPs is the in situ synthesis process. The in situ synthesis of Ag NPs is based on chemical reactions in solution (often termed wet chemistry) that yields loading metal ions and a further synthesis of metal nanoparticles. The use of an adequate encapsulating and reducing agent is one of the keys to synthesize the metal nanoparticles with a specific morphology. In this contest, Babaahmadi et al. reported a novel and one-step method for the in situ synthesis of Ag NPs through reduction of silver nitrate ($AgNO_3$) with stannous chloride ($SnCl_2$) on polyamide (PA) fabrics, using cetyltrimethyl ammonium chloride (CTAB) as stabilizer [50]. The treated fabrics showed good antibacterial properties against both Gram-positive (*S. aureus*) and Gram-negative (*E. coli*) bacteria. The quantitative tests were performed based on the inhibition zone surrounding of the samples, as a function of inhibition or not bacterial growth. According to this same methodology, Allahyarzadeh et al. proposed the in situ synthesis of nanosilver on polyester fabrics using NaOH/Nano TiO_2 with self-cleaning and antibacterial properties [51]. For this purpose, the alkali hydrolysis makes possible the enhancement of the surface activity, improving the nanoparticle absorption and producing ethylene glycol. The Fourier transform infrared (FTIR) spectra corroborated the presence of ethylene glycol because the intensity of characteristic peaks attributed to –OH at 3054 and 3419 cm^{-1} and the peak attributed to C=O (carboxylic acid) at 1716 cm^{-1} were increased, proving the production of hydroxyl (–OH) and carboxyl (–COOH) functional groups on the fiber surface. After that, the oxidation of hydroxyl groups to aldehyde groups in ethylene glycol resulted in carboxylate ions and electrons, making possible the final reduction of silver ions to Ag NPs. In addition, X-ray diffraction (XRD) confirmed the presence of Ag NPs on the fabric surface. The synthesis of these Ag NPs onto polyester fabrics resulted in excellent antibacterial efficiency against *S. aureus* and *E. coli*. This methodology of in situ synthesis can be also used for synthesizing copper nanoparticles within the polymeric chains of nylon fabrics by using ascorbic acid as a reducing agent and CTAB as a capping agent [52] as well as for synthesizing nanometal oxides in nonwoven polyester fabrics [53].

As it has been previously commented, not only Ag NPs show antibacterial activity. It has been demonstrated that nanometal oxides can be used as antibacterial agents such as titania (TiO_2), silica (SiO_2), or zinc oxide (ZnO) nanoparticles can be also used as an antibacterial finishing for textiles [54–56]. Among all these nanometal oxides, it is well-known that TiO_2 nanoparticles show extraordinary photocatalytic activity, non-toxicity, high availability and biocompatibility, and being of great interest for manufacturing processes in order to impart the specific antibacterial properties. However, as it happened with Ag NPs, the exact mechanism of the bactericidal activity of the TiO_2 nanoparticles is not well understood and several approaches are used to demonstrate the TiO_2 action in the corresponding bacterial cell death [57].

Montazer et al. used an enzymatic pretreatment to improve the functionality of textile materials based on TiO_2 nanoparticles, showing great results as self-cleaning, antibacterial, and UV-protection [54]. The methodology is based on two well separately steps. In a first step, a polyester/wool fabric as a blend of fibers fabric was treated with two different types of enzymes such as lipase or protease to hydrolyze the wool and polyester surface, respectively, increasing the surface activity. In a second step, the fabrics were dipped into an ultrasonic bath containing nano-TiO_2 and a cross-linking agent, BTCA. This agent was used to enhance the nanoparticles adsorption onto the fabric surface. The treated samples exhibited excellent antibacterial activity against *E. coli* as the reduction of bacteria could reach to 100 % with 0.75 % nano-TiO_2 and 99 % with 0.25 % nano-TiO_2. In this contest, related to the use of TiO_2 nanoparticles, Mihailovic et al.

demonstrated the novel properties of polyester fabrics modified by corona discharge or RF plasma and colloidal TiO_2 nanoparticles [58–60]. This specific treatment provided a better binding efficiency of the TiO_2 nanoparticles onto the polyester fabrics. The results of the treated fabrics showed multifunctional properties such as antibacterial activity, self-cleaning effect with long-term stability, and good UV-protection.

El-Gabry et al. have treated polyester fabrics with SiO_2 nanoparticles to improve the properties of textile fabrics [55], showing good antibacterial properties and excellent durability of the treatment to repeat home laundering, thanks to the use binder such as an acrylate copolymer.

Farouk et al. proposed the design of hybrid polymers based on nanosized ZnO particles, being a great alternative as an antibacterial finishing for textiles [56]. These hybrid materials are based on glycidiltrimethoxisilane (GPTMS) and were applied onto cotton/polyester (65/35 %) fabrics, using a padding process. The antimicrobial activity of the modified fabrics with a constant mass of ZnO (10 wt.%) but with a varying particle size was tested against Gram-negative bacteria (*E. coli*) and Gram-positive bacteria (*Micrococcus luteus*). An important consideration of this work is that the fabrics treated with the hybrid polymer loaded with ZnO particles of the smaller particle size (30–60 nm) leads to a complete annihilation of both bacteria, while fabrics coated with hybrid polymers loaded with bigger ZnO particles (600–650 nm) achieve a 93 and 97 % reduction of the *E. coli* and *M. luteus* populations, respectively. This enhanced bioactivity related to the smaller particles is associated with the higher surface-area-to-volume ratio.

Other approaches for obtaining antibacterial activity can be performed with the design of nanocomposites (Ag/ZnO, Ag/SiO_2) [61, 62] or well with the combination of metal nanoparticles which show separately antibacterial properties [63, 64]. As it has been previously demonstrated, the use of ZnO nanoparticles with antimicrobial activity for textile treatment [56], this same research group proposed in a later paper the preparation of Ag/ZnO composite nanoparticles by the reduction of silver on the surface of ZnO nanoparticles [61]. The coating process was carried out on cotton/polyester (50/50 %) blend fabrics, and the antimicrobial tests were performed against the same kind of bacteria *E. coli* and *M. luteus*. As a conclusion of the results, the antibacterial activity of Ag/ZnO nanoparticles increases with a higher concentration of silver. In addition, this increasing of antimicrobial activity for the higher silver concentration in comparison with the other samples can be associated with decreased crystallite size with the corresponding increase of the active surface area, making possible an enhancement in the biocidal effect.

An interesting approach is presented by Shin et al. where silver-doped silica-complex nanoparticles have been synthesized for the design of antibacterial materials [62]. In this work, firstly, Ag NPs were synthesized using both liquid-phase and alcohol reduction methods, and secondly, silver-doped silica-complex nanoparticles were synthesized by using a sol-gel process, showing an excellent antibacterial activity against *S. aureus* and *E. coli*. As a conclusion obtained by silver-doped silica-complex nanoparticles, the average size of the nanoparticles decreased when the molecular weight (M_w) of the stabilizing agent PVP. In addition, the number of particles increased by a mercapto treatment (–SH group) from mercaptopropyl trimethoxysilane (MPTMS) which showed a narrower size distribution in comparison with the silica treated by an amino treatment ($-NH_2$ group) from aminopropyl trimethoxysilane (APTMS).

Ali et al. proposed the synthesis of both chitosan nanoparticles (CSN) and silver-loaded chitosan nanoparticles (Ag-CSN) and their further incorporation onto fabrics with the aim of designing bioactive polyesters [63]. An ionic gelation method using chitosan as a polycation and sodium tripolylphosphate (TPP) as a polyanion was used to prepare the CSN nanoparticles, whilst a further loading with silver ions was used to prepare the Ag-CSN nanoparticles. In addition, the minimum inhibitory concentration (MIC) against S. *aureus* bacteria was found to be 50 and 500 times less for CSN and Ag-CSN, respectively, as compared to bulk chitosan material. The Ag-CSN nanoparticles showed a further increase in activity due to the synergistic effect of Ag and chitosan nanoparticles. These results indicated that an antibacterial activity of 90 or 100 % were obtained for CSN and Ag-CSN nanoparticles, respectively, at very low concentration (0.2 %), whereas the bulk chitosan showed an antimicrobial activity of 58 % for the same concentration.

Another interesting approach based on the simultaneous combination of two different types of nanoparticles is presented by Mihailovic et al. [64]. In this work, it is shown the potential of combining both colloidal Ag and TiO_2 NPs onto polyester fabrics, making possible to obtain multifunctional properties such as antimicrobial activity, UV protection, and photocatalytic behavior onto the fabrics surface. According to the antimicrobial tests, it is important to remark that the fabrics treated with both types of nanoparticles (TiO_2 and Ag NPs) provided a higher antimicrobial efficiency in comparison with the fabrics treated only with Ag NPs. The modified fabrics were tested against Gram-negative bacteria (*E. coli*), Gram-positive bacteria (*S. aureus*), and fungus (*C. albicans*).

Mahltig et al. proposed the application of three different modified silica sols containing silver components onto polyamide fabrics, showing a lasting antimicrobial effect against *E. coli* even after 40 washing cycles [65]. The different sols (sol A, sol B, and sol C, respectively) were applied onto polyamide fabrics using a pad-dry-cure process. In order to determine the antimicrobial effect of the sol gel coatings, the number of colony forming units was evaluated as the samples are prepared as well the number of washing cycles are increased up to 40 washes. It was observed that after coating preparation, the antimicrobial effect related to both sols A and B was lower in comparison with sol C. The main reason of this different behavior is related to the chemical composition of the sols. Although both sols A and B presented the same silver source (silver nitrate, $AgNO_3$), the sol C presented a different one (commercial colloidal silver). In addition, both sols B and C consisted of a quaternary ammonium compound such as hexadecyltrimethylammonium-p-toluenesulfonate (HTAT) used as a dispersing/stabilizing agent which is also well-known because of its antimicrobial activity. Due to this, sols B and C showed better results due to the presence of HTAT in their chemical composition in comparison with sol A. However, a drastic change in their antimicrobial effect was observed after 40 washing cycles because the polyamide fabrics coated with sol C lost all their antimicrobial activity. This important difference could be associated with the silver source in sol C (commercial colloidal silver) which was not so tightly fixed to the fabric surface. SEM images of the coatings showed particles of bigger size (micrometer range) with a flake shape for polyamide fabrics coated with sol C whereas particles of smaller size (below 100 nm) with a spherical shape were observed for fabrics coated with sols A and B. Due to this, the particles of bigger size are easily abraded under conditions of repeated washing cycles as it happens for sol C. When sol A with sol B was compared, smaller and more regular distribution of the in situ synthesized nanoparticles onto polyamide fabrics were observed for sol B, due to the use of a surfactant such as HTAT, and it is the reason of their higher durability under the conditions of repeated washing cycles with a better antimicrobial efficiency after 40 cycles.

It is important to consider that there are other types of synthetic fibers such polypropylene (PP) or polypropylene/polyethylene (PP/PE) which can be also modified to show antibacterial activity [66, 67]. As an example, it has been corroborated the antibacterial efficacy onto nonwoven polypropylene using gold nanoparticles [66] or well onto padded PP/PE nonwoven by using different nanosilver colloids [67].

Finally, a summary is shown in Table 2 of the different nanoparticles used for antibacterial activity and the possibility of obtaining high durability antimicrobial effect after successive washing cycles. As it has been previously commented, the vital importance of controlling the corresponding size of the nanoparticles plays a key role in the resultant antibacterial activity of the nanoparticles [56, 61, 63, 65]. A decrease in the corresponding size show better results in the antibacterial tests due to the higher surface-area-to-volume ratio in comparison with bulky materials [63] or particles of bigger size [65]. In addition, Table 2 also summarizes the fabrics which show laundering durability as a function of an adequate surfactant or a specific pretreatment to enhance the adhesion of the nanoparticles onto the fabrics surface.

Table 2 . Summary of the type of nanoparticles used for antibacterial or antifungal activity

Antibacterial agent	Fabrics	Antibacterial or antifungal tests
Silver-tricalcium phosphate NPs (Ag/TCP)	Polyamide	*E. coli* and *S. sanguinis*, 2011 [38]
Silver nanoparticles	Polyester	*E. coli* and *S. aureus*, 2010 [40]
Silver nanoparticles	Polyester and polyamide	*E. coli* and *S. aureus*; laundering durability pretreatment: corona, 2008 [41]
Silver nanoparticles	Polyester and polyamide	*C. albicans*; laundering durability pretreatment: corona, 2009 [42]
Silver nanoparticles	Polyester	*E. coli* and *S. aureus*; laundering durability pretreatment: radio frequency (RF) plasma, 2010 [43]
Silver nanoparticles	Polyester	*S. aureus*, *S. epidermidis*, *P. aeruginosa*, and *C. albicans*, 2008 [46]
Silver ammonia complex	Polyamide	*E. coli*; laundering durability pretreatment: glutaraldehyde (GDA), 2010 [47]
Silver ammonia complex	Polyamide	*E. coli* and *S. aureus*, 2012 [48]
Silver ammonia complex	Polyamide	*E. coli* and *S. aureus*; laundering durability, 2014 [49]
Silver nanoparticles	Polyamide	*E. coli* and *S. aureus*, 2015 [50]
Silver nanoparticles (in situ synthesis)	Polyester	*E. coli* and *S. aureus*, 2013 [51]
Copper nanoparticles (in situ synthesis)	Polyamide	*S. aureus*, 2013 [52]
TiO_2 nanoparticles	Polyester/wool	*E. coli*, 2011 [54]
TiO_2 nanoparticles	Polyester	*E. coli* pretreatment: oxygen and argon plasma, 2010 [58]
TiO_2 nanoparticles	Polyester	*E. coli* Pretreatment: corona/air RF plasma, 2011 [59]
SiO_2 nanoparticles	Polyester	*E. coli* and *S. aureus*; laundering durability binder: acrylate polymer, 2013 [55]
ZnO nanoparticles	Cotton/polyester	*E. coli* and *M. luteus*, 2014 [56]
Ag/ZnO composite nanoparticles	Cotton/polyester	*E. coli* and *M. luteus*, 2014 [61]
Silver-doped silica complex nanoparticles	Polyester	*E. coli* and *S. aureus*, 2014 [62]
Chitosan and silver-loaded chitosan nanoparticles	Polyester	*S. aureus*, 2011 [63]
Mixture of silver and TiO_2 nanoparticles	Polyester	*E. coli*, *S. aureus* and *C. albicans*, 2011 [64]
Silica sols with silver nanoparticles	Polyamide	*E. coli*; laundering durability, 2010 [65]
Gold nanoparticles	Polypropylene	*E. coli* and *S. aureus*; laundering durability pretreatment: air plasma treatment, 2013 [66]
Silver nanoparticles	Polypropylene/ polyethylene	*E. coli*, *Klebsiella pneumoniae*, and *S. aureus*, 2005 [67]

5. SUPERHYDROPHOBIC SURFACES

The fabrication of superhydrophobic materials with oil/water repellent properties has been receiving great attention in the textile industry [68]. The concept of superhydrophobicity can be defined as how those surfaces which present water contact angles higher than 150°, which is also commonly known as the artificial lotus effect. If the same contact angle (always higher than 150°) is obtained with oil, the surface is said to be superoleophobic [69]. Nowadays, there is a wide variety of methodologies which are based on the combination of a specific topography and an adequate chemical compound to produce a high contact angles [70, 71]. Theoretical models are based on both low surface free energy and high degree of roughness. Due to this, the superhydrophobicity depends strongly on the surface chemistry and the surface morphology [72]. Many approaches for building a superhydrophobic surface consist of a nanostructured surface and followed by a treatment with a fluoro-containing polymer or silane [73, 74]. A clear example onto polyester textile fabrics is presented by [75]. In this case, an alkaline hydrolysis plays a key role in the possible surface roughening effect and is applied to manipulate the surface morphology, whilst a further fluorocarbon layer is deposited in order to generate low surface energy. Other approaches based on the same methodology are presented by [69, 76]. Saraf et al. explored different techniques such as pulse plasma polymerization of perfluorodecyl acrylate (PFAC8), microwave-assisted fluorosioxane condensation, or fluorosiloxane condensation via wet processing in order to achieve superhydrophobicity and superoleophobicity on nylon nonwoven fabrics [69]. The treated fabrics showed very high contact angles for both water (168°–174°) and dodecane (153°–160°). Zhu et al. designed a superhydrophobic polyester fabric with a good mechanical stability and an easy repairability by coating the fabric with metallic Ag NPs and a further surface fluorination [76]. The superhydrophobic properties were maintained after immersion test, finger pressing, or abrasion with sandpaper, showing a high superhydrophobic permanency. However, when a loss of superhydrophobicity occurs, it is possible to regenerate the fabric surface by an easy repair process which consists of repeating silver deposition and surface fluorination process.

Alternative new approaches for the design of superhydrophobic surfaces in the textile industry are focused on the surface functionalization of SiO_2 nanoparticles with non-fluorinated alkylsilanes [77] or fluorosilanes [78–80]. As an example of the use of these functionalized SiO_2 nanoparticles, an interesting approach for imparting extremely high hydrophobicity to cotton and polyester fabrics is presented by Gao et al. [77]. For this case, a first step is based on the fabrication of a silica sol prepared by the respective hydrolysis and condensation of tetraethyl ortosilicate (TEOS) under alkaline conditions. Then, a second hydrophobization step was performed using hydrolized hexadecyltrimethoxysilane (HDMTS). The treated cotton and polyester textiles show excellent water repellent properties with a water contact angle of 155° on cotton and 143° on polyester. This high hydrophobicity is associated with the presence of a non-fluorinated compound such as HDMTS as well as an increase in roughness by the silica sol. An important aspect was the high durability of the treated textile fabrics because they were able to retain the most of the hydrophobic properties after 30 laundering cycles.

Xu and coworkers designed fluoropolymer/silica (FP/SiO_2) organic/inorganic nanocomposites coatings on polyester fabrics [78, 79]. For the preparation of these nanocomposites, firstly, vinyl silica ($V\text{-}SiO_2$) hydrosols were prepared by one-step water-based sol-gel method using the vinyl trimethoxy silane (VTMS) as the silane precursor and ammonium hydroxide as the catalyst. Secondly, a fluorinated acrylic polymer/silica nanocomposite was prepared by emulsion polymerization based on $V\text{-}SiO_2$ hydrosol. FTIR measurements corroborate that the $V\text{-}SiO_2$ nanoparticles had been successfully incorporated into the nanocomposite and the fluoropolymer had been chemically bonded with nano-SiO_2 by the vinyl bridge. A dip-pad-cure process was applied to deposit the nanocomposite onto the polyester fabrics. It has been demonstrated that the treated fabrics with these fluoropolymer/SiO_2 nanocomposites showed higher contact angle values than the fabrics which were only treated with a fluoropolymer. The nanocomposites exhibited excellent superhydrophobicity with a contact angle of 151.5° and a water shedding angle of 12°, whereas the polyester fabric treated by fluoropolymer without the presence of V-SiO_2 nanoparticles showed a contact angle of 132.3° and a water shedding angle of 29°. These results corroborate the importance of a rough surface structure on PET fabrics which is provided by the SiO_2 nanoparticles to improve the hydrophobic properties of the fluorinated polymers which provide low surface energy.

A novel approach for obtaining transparent and superhydrophobic surfaces by spin-coating technique is reported by Xu et al. using fluorosilane-functionalized SiO_2 nanoparticles (F-SiO_2 NP) [80]. In this case, firstly, commercial SiO_2 nanoparticle (100 nm in diameter) surfaces were functionalized with a fluorosilane in the presence of triethyl amine (TEA) and toluene (Fig. 5a). And secondly, these hydrophobic nanoparticles were applied onto the polyester surface using the spin-coating technique in a fluorinated solvent (Fig. 5b). The treated polyester fabrics exhibited superhydrophobicity with an advancing water contact angle greater than 150° and a water droplet roll-off angle less than 5°. In addition, it is important to remark the high optical transparency of the spin-coated films, showing greater than 95 % transmittance in the visible region. An important consideration is that SEM images reveal that by increasing the F-SiO_2 NP concentration, the assembly of the nanoparticles onto the substrate surface changed from random and nonclose packed (0.1 or 0.4 wt.%) to nearly close-packed (≥0.8 wt%) (Fig. 6). Due to this close-packed configuration, the spin-coated surface became superhydrophobic with an advancing water contact angle higher than 150°.

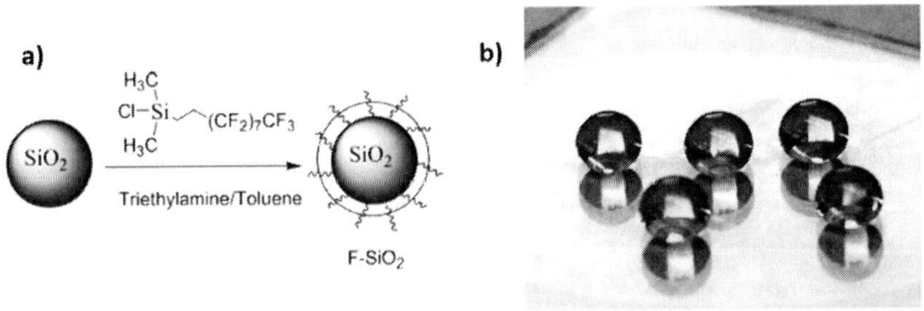

Figure 5 **a** Schematic illustration of the synthesis of fluorosilane-coated silica nanoparticles, F-SiO_2 NP; **b** optical photograph of water droplets (>10 μL) on F-SiO2 NP-coated polyester fabric. A small amount of dimethyl methylene blue dye was dissolved in water for illustration purpose. Reprinted with permission from [80]. Copyright (2012) American Chemical Society

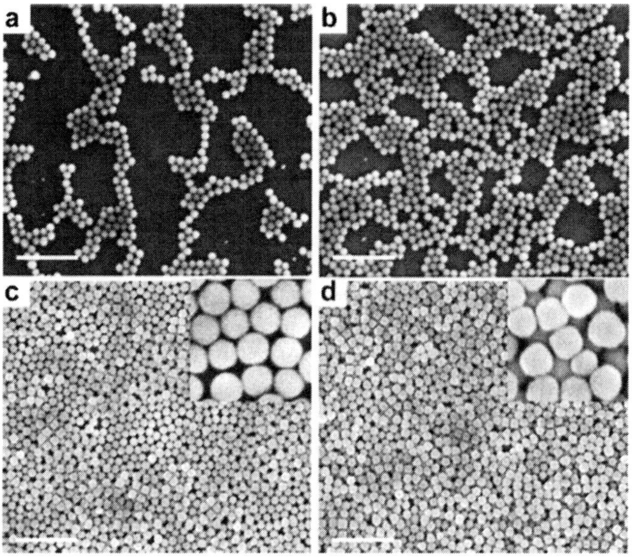

Figure 6 . SEM images of spin-coated 100-nm F-SiO2 NPs with different concentrations on TESPSA-functionalized Si wafers: **a** 0.1, **b** 0.4, **c** 0.8, and **d** 1.2 wt.%. The *insets* in **c** and **d** are high-magnification images. Scale bars: 1 μm. Reprinted with permission from [80]. Copyright (2012) American Chemical Society

Other different approaches for obtaining superhydrophobic synthetic fabrics are based on the incorporation of ZnO particles using different techniques [81–83]. For example, Fruza et al. reported that polyester fabrics were successfully coated with crystalline particles of ZnO using an electroless deposition [81]. Morphological analysis obtained by XRD and SEM indicated the formation of hexagonal prism crystallites (wurtzite type), having a variable size between 20 and 500 nm in diameter and up to 1 μm in length. The most important aspect of the incorporation of these ZnO prisms onto surface fabrics is the change from a hydrophobic surface (raw polyester fabric) to superhydrophobic surface because the water contact angles exceeded 150°.

Other alternative method to obtain superhydrophobic polyester fabrics is presented by Ashraf et al. [82]. In this specific work, ZnO particles with a nanorod shape were grown by seeding method with the aim to develop hierarchical roughness structure. The presence of the nanorods onto surface fabrics was corroborated by XRD analysis where three sharp peaks indicated a good crystallinity of the ZnO particles. Once the nanorods have been successfully incorporated on polyester fabrics, a second step is based on a further modification of the fabrics with octadecyltrimethoxysilane (ODS). It is important to remark that a change of the surface properties is observed in the PET fabrics from a previous treatment with ODS to PET treated with ODS. Before modification with ODS, the fabrics showed superhydrophilicity and water contact angle decreased as the seed concentration growth. However, after ODS treatment, the fabrics drastically changed their behavior, showing a lotus effect with very high values of water contact angle.

This same change in the surface properties is reported by Popescu et al. [83]. In this work, they proposed the functionalization of cotton/polyester woven fabrics with ZnO thin films or nanoparticles by pulsed laser deposition. It has been demonstrated that by changing the number of laser pulses, well-separated nanoparticles (ten pulses) or compact thin films (100 pulses) were obtained. In addition, the ZnO nanoparticles deposited onto fabrics can switch the wetting behavior of the textile from hydrophilic to hydrophobic as a function of ambient gas nature and the pressure in the deposition chamber. The best results were observed when it was deposited a thin film of ZnO in vacuum, showing superhydrophobicity with a water contact angle of 157°.

Finally, superhydrophobic surfaces onto polypropylene textiles can be also obtained [84, 85]. As an example, Zhu et al. proposed a simple, rapid, and scalable method to fabricate superhydrophobic polypropylene fabrics by solvent swelling method [85]. The key role of this work is that the fabrics surface is swelled by heated solvent (cyclohexane/heptane mixture) and then a shrinking of the polymeric chains is observed during the solvent evaporation process, leading the formation of submicron protuberances. As a result, the treated fabrics showed excellent repellence to various liquids such as blood, urine, milk, coffee, soy sauce, or vinegar (Fig. 7), showing a great durability and robustness.

Figure 7. Digital photographs of various liquid droplets of about 80 μL in volume testing on PP nonwoven fabrics. Reprinted with permission from [85]. Copyright (2015) American Chemical Society

Finally, the different methodologies for obtaining superhydrophobic surfaces as well as the corresponding nanoparticles or nanomaterial used for this purpose onto specific fabrics are summarized in Table 3.

Table 3 . Summary of the type of nanoparticles or nanomaterials used for superhydrophobic surfaces

Type of fabrics	Deposition process	Superhydrophobic surface
Polyester	Pad-dry-cure method	Alkaline hydrolysis and fluorocarbon layer, 2011 [75]
Polyester	Dip-coating	Silver nanoparticles and fluorination, 2012 [76]
Polyester	Dip-pad-cure process	Silica (SiO_2) nanoparticles and fluropolymer, 2015 and 2014 [78, 79]
Polyester	Spin-coating	Silica (SiO_2) nanoparticles and fluorosilanization, 2012 [80]
Polyester	Electroless deposition	Zinc oxide (ZnO) nanoparticles, 2013 [81]
Polyester	Solution or vapor deposition	Zinc oxide (ZnO) nanoparticles and octadecyltrimethoxysilane (ODS), 2013 [82]
Cotton/polyester	Pulse laser deposition	Zinc oxide (ZnO) nanoparticles, 2011 [83]
Polypropylene	Solvent swelling method	Swollen of the polymeric chains, 2015 [85]

6. NANOFIBERS AND NANOWEBS

A recent approach to create nanofibers is by the electrospinning method. This technique allows the production of synthetic nanofibers (polymeric, inorganic), denoted as electrospun nanofibers (ENFs). In 1995, Doshi and Reneker [86] reported for the first time the electrospinning process in a research article. At the beginning, this technique only consisted of a polymeric solution in a syringe with a needle, a high voltage source, and a metallic board as electrode, as shown in Fig. 8. The high voltage was applied to the polymeric solution with respect to the metallic board. This electrical field caused that an electrically charged polymeric jet was launched from the needle, forming the so-called Taylor cone (see Fig. 9a, b). The jet traveled in air, spinning into the Taylor cone, to the metallic board. The solvent of the solution was evaporated along the way, and ENFs were collected onto the board or the substrate to be coated [86–88].

In the last 15 years, the number of published electrospinning related works has experimented an almost exponential growth. This significant increase is due to its numerous advantages as simplicity, effectiveness, low-cost, quickness which allow the implementation of ENFs in a wide range of applications [89, 90]. Moreover, the ENFs structures present other strengths such as high surface area, small diameter, and specific physicochemical properties [91].

In this section, most used materials for the ENFs fabrication and novel ENFs with additional composites will be reported. Furthermore, a short list of the relevant applications of these types of nanofibers will be detailed, including some recent advances in diverse fields where ENFs are applied, as antibacterial coatings, water treatment, wound healing, and others.

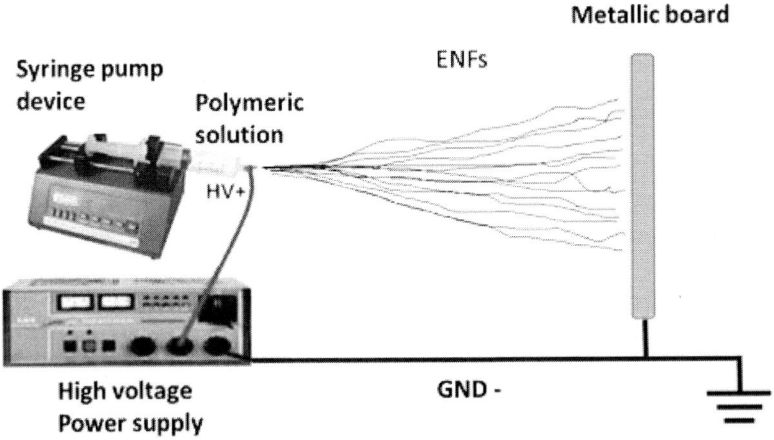

Figure 8 Setup scheme of the simple electrospinning technique to produce electrospun nanofibers (ENFs)

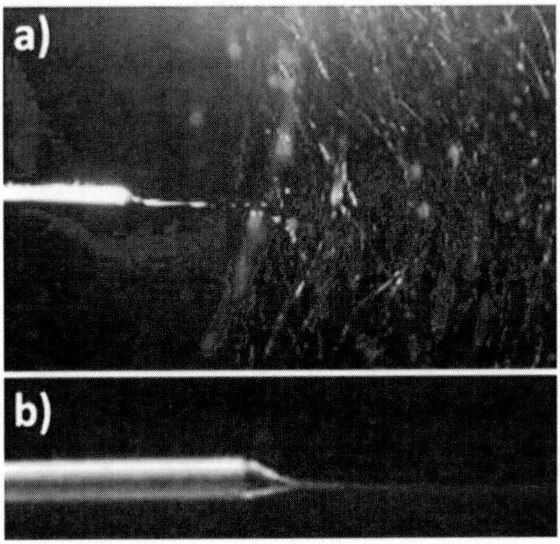

Figure 9 ENFs during the electrospinning process. **a** It shows the needle (*left*) and the fibers being electrospun towards the cathode (*right*). **b** Detailed view of the Taylor cone formed at the tip of the needle. Reprinted with permission from [88]. Copyright (2012) Wiley Periodicals Inc

7. ELECTROSPUN NANOFIBERS: COMPOSITION AND TYPES

The composition of ENFs is very diverse, and their diameter can vary from 10 to 1000 nm or even more. Natural or synthetic polymers are used depending on the target or final application. Electrospun natural nanofibers of chitin, chitosan, hyaluronic acid, silk, cellulose, collagen, or others can be found in literature [92]. Regarding to the synthetic ones, a wide list of chemical compounds have been used by electrospinning: PA, PAA, poly(vinyl alcohol) (PVA), poly(ethylene glycol) (PEG), poly(ethylene oxide) (PEO), poly(ethylene-co-vinyl acetate) (PEVA), poly(acrylonitril) (PAN), poly(vinyl acetate) (PVAc), poly(lactic acid) (PLA), and more. A complete list of polymers which have been electrospun can be also found in [93].

Generally, these polymers are dissolved in an adequate solvent to obtain the electrospinning solution. The most used solvents are water, ethanol, n,n-dimethylformamide (DMF), and isopropanol, among others. Other important factors in the tailoring of the ENFs are the viscosity of the solution, flow rate, voltage level applied, and distance between syringe and collector. The precision of these parameters plays a key role in the ENFs fabrication [94].

ENFs are commonly classified in two types: randomly oriented ENFs and aligned ENFs. On one hand, the use of a simple electrospinning setup, as in Fig. 8, produces the first type of ENFs. The result is an electrospun nanoweb or nanofiber mat deposited onto the substrate (see Fig. 10a). On the other hand, aligned ENFs are mainly created by collecting the fibers over a rotating collector [95], generally a drum or a rotating disc. The alignment of the ENFs depends on the final applications (see Fig. 10b) [96, 97].

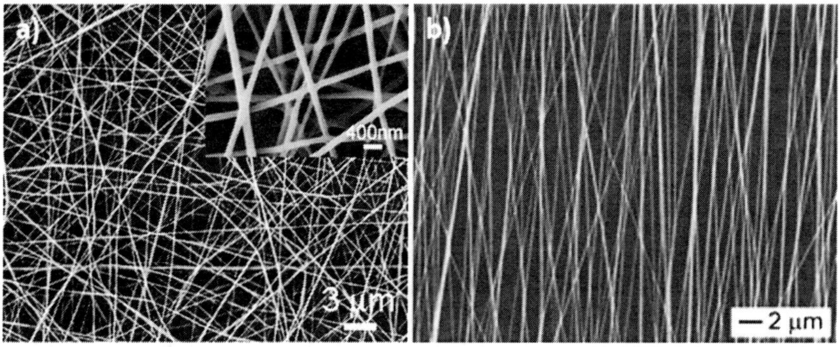

Figure 10 SEM images of electrospun nanofibers produced by electrospinning. **a** Nanofiber mat from random ENFs (*inset* shows the high-magnified SEM image); **b** aligned electrospun composite nanofibers. Reprinted (adapted) with permission from [96] and from [97], respectively. Copyright (2009) and Copyright (2003), American Chemical Society

Nowadays, the polymeric solutions can be combined with the incorporation and use of organic and inorganic nanostructured materials or nanoparticles, obtaining novel electrospun composite nanofibers (ECNFs) and providing additional properties as reusability, high porosity, antibacterial activity, corrosion or chemical resistance, biomolecular affinity, etc [98]. Thus, the synthesis of ECNFs is performed by various methods. The simplest one consists of the dispersion of NPs or other nanostructures into the polymeric solution [99] before the fabrication process. A different approach is the coaxial electrospinning [100, 101], which combines two different stock solutions and produces ENFs with a core/sheath nanostructure. Other alternatives in the ECNFs fabrication are some specific post-treatments, such as thermal [102] or UV [103] curing, chemical functionalization [104], or binding [105], among others.

8. ELECTROSPUN NANOFIBERS: APPLICATIONS

In the ENFs and ECNFs case, the applications depend strongly on the composition of the fibers and differ somewhat to the textile fiber applications. Thus, some of the most relevant applications of the ENFs and ECNFs are water treatment [106], antibacterial purposes [107], wound dressing [108], drug delivery [109], sensing [110, 111] and biosensing [112], tissue engineering [113], catalyst [114], microwave absorption [115], flame retardants [116], and other biomedical issues [117, 118].

As in the previous section, the recent advances in antibacterial coatings will be described. Furthermore, some relevant works for water treatment, wound healing and dressing, and other applications will be also presented because of their growing importance in the last years.

9. ANTIBACTERIAL

The fabrication of antibacterial ENFs can be classified in several groups, depending on the used antibacterial agent. These active agents are incorporated in the fibers by mixing them with the polymeric solution before the electrospinning process, by coaxial electrospinning, by other chemical treatments or binding onto the ENFs surface. The most used active agents for the development of antibacterial ENFs are Ag NPs, antibiotics, triclosan, chlorhexidine, quaternary ammonium salts, chitosan, and metal oxide NPs.

The use of the silver as an antibacterial agent is well-known and has been applied to biomedicine and other fields. In the last years, Ag NPs have experimented a growing interest for their large surface area and better release of metal cations [119]. There are three main methods to incorporate Ag NPs into ENFs: by simple mixing with the polymer before electrospinning, by the in situ Ag NPs synthesis into the polymeric solution, and by the in situ synthesis of the Ag NPs after electrospinning in a post-treatment.

The first incorporation method of Ag NPs into the polymeric solution is the simplest, and it has been widely used. Recent works have reported the development of antibacterial ENFs thanks to this method, using polymers such as PVA [120], PVP [121], or poly(vinylidene fluoride) (PVDF) [122].

The second method is based on the in situ Ag NPs synthesis into the electrospinning solution. For this purpose, silver nitrate ($AgNO_3$) is added to the polymeric solution. Here, the polymer acts as encapsulating agent for the silver cations. Next, a reducing agent is incorporated to the solution, thus forming Ag NPs. In some cases, the own solvent acts as a reducing agent. Examples of ECNFs fabricated with this method have been published by Le et al. [123] and Yang et al. [124] using PAN/$AgNO_3$/DMF mixing solution. In addition, other studies have successfully developed ECNFs from nylon/$AgNO_3$/formic acid [125] and PVA/$AgNO_3$/chitosan [126]. The main advantage of this method is the fact that it does not need to pre-synthetize the Ag NPs, and moreover, the ECNFS do not require further treatments.

The third method to obtain ENFs with Ag NPs is based on diverse treatments after the electrospinning process. In a first step, the ECNFs are fabricated. Then, a further thermal, UV, or other chemical treatments are performed. Thus, PVA/regenerated silk fibroin ENFs have been heated at 155 °C for 5 min or treated with UV for 3 h [127], PLA fibers at 80 °C for 48 h in a hydrogen atmosphere [128], or PAN ENFs at 160 °C [129] to allow the transition of Ag cations to metallic silver. Nevertheless, such post-treatments are not as effective as the Ag NPs synthesis in the polymeric solution [129]. However, in a recent separate study reported by Rivero et al. [88], PAA/cyclodextrin nanofibers were thermally treated and then processed by dip-coating as follows. In a first step, nanofibers were immersed in an $AgNO_3$ solution, charging the ENFs surfaces with Ag cations. Next, ENFs were immersed in dimethylamine borane, leading to the reduction of Ag cations to Ag NPs onto the ENF mat surface. A TEM image of the Ag NP-loaded ENFs and antibacterial test results from this work are shown in Figs. 11 and 12, respectively.

Not only Ag NPs exhibit antibacterial properties but also ZnO and TiO_2 NPs, as it was previously commented. These NPs present excellent antibacterial behavior after or during UV illumination [130, 131]. However, the necessity of a UV radiation makes them less attractive from the antibacterial point of view.

Figure 11 . TEM image of the Ag-loaded electrospun nanofibers. Ag NPs can be observed onto the ENFs surface. Reprinted with permission from [88]. Copyright (2012) Wiley Periodicals Inc

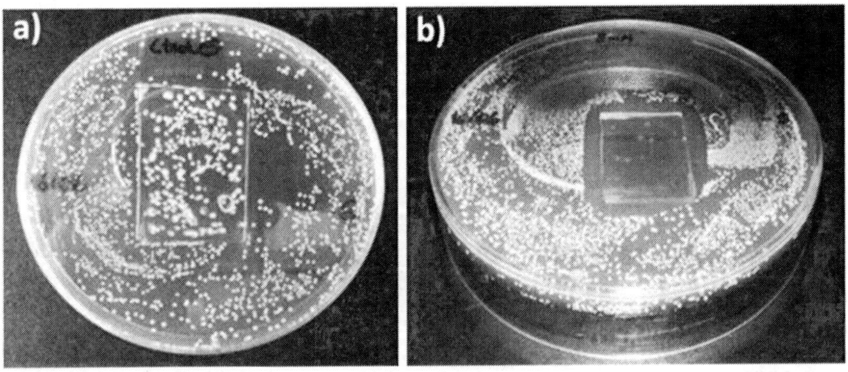

Figure 12 . Bacteriological cultures of *Lactobacillus plantarum* after 24 h with two different samples. **a** Reference sample of PAA electrospun nanofiber mat with no silver nanoparticles; **b** silver-loaded PAA electrospun fiber mat. Substrate: glass slides. Reprinted with permission from [88]. Copyright (2012) Wiley Periodicals Inc

In the antibiotics case, the amount of drug is adjusted in the electrospinning solution. The concentration of each antibiotic is different for each solution. Thus, Kenawy et al. [132] used blends of PLA and PEVA containing 5 wt.% tetracycline hydrochloride. In other works, the quantities are 30 wt.% of Moxi in coPLA [133] and 7.5 wt.% mupirocin in PLA fibers [134], respectively. Although the antibiotic-polymer mixture has advantages as simplicity and flexibility in the amount of the drug loaded, the ENFs tend to leach out the active agent too fast into the solution, which is known as burst release phenomenon. To avoid this and provide a well-controlled release, some approaches have been developed using coaxial electrospinning, fabricating core/sheath structures. For example, antibiotics as ampicillin [135], tetracycline hydrochloride [136], or gentamycin [137] have been successfully encapsulated in diverse ENFs.

Other biocide compounds have been also mixed into electrospinning solution as chlorhexidine [138] in cellulose acetate or amine groups [139] in PAN. In another approach reported by Kayaci et al. [140], triclosan (TR) have been complexed with cyclodextrin (CD) and mixed with PLA. Thanks to the complexation, the ECNFs showed an increase in their antibacterial activity for *E. coli* and *S. aureus*. A schematic of the ECNFs fabrication and their antibacterial results are shown in Figs. 13 and 14, respectively.

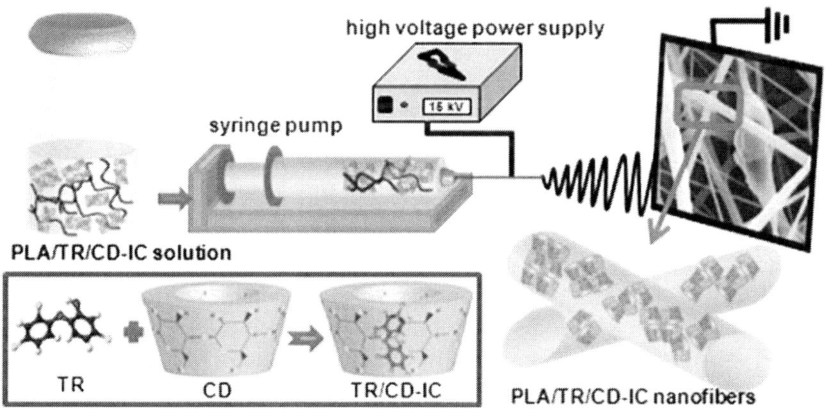

Figure 13. Schematic of the fabrication ECNFs process of triclosan/cyclodextrin-induced complexes (TR/CD-IC) with PLA performed by Kayazi et al. Reprinted with permission from [140]. Copyright (2013) American Chemical Society

Chitosan is another natural substance which has an inherent antibacterial ability, and therefore, it has been also used to fabricate antibacterial ENFs. Due to its weak mechanical properties, chitosan has been mixed with PVA,

PLA, PEO, or PA to develop stronger ECNFs. More information about the use of chitosan in antibacterial ENFs can be found in literature [141].

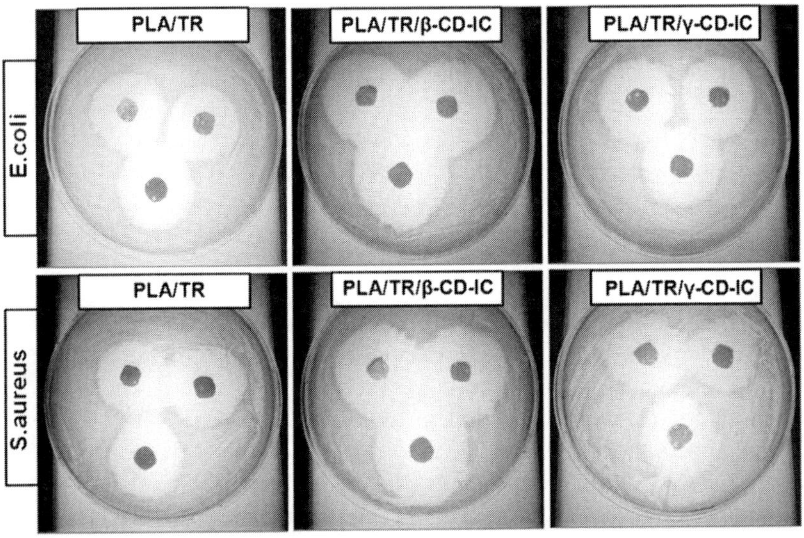

Figure14 . Antibacterial results of the ECNFs with and without different types of CD. The bacterial inhibition growth is higher for the PLA/TR/β-CD-IC nanofibers. Reprinted with permission from [140]. Copyright (2013) American Chemical Society

10. WOUND DRESSING

Selected wound dressing materials are required with special attention for the wound healing process. Among the overall compounds of great importance, Ag NPs are used as antimicrobial in hydrocolloids, alginates, and hydrofibrous materials fabricated by electrospinning [142]. For example, ECNFs bandages have been developed using iodines encapsulated into PVP and PEO/PVP ENFs [143]. Other relevant works focused on the preparation of wound dressing mats by electrospinning have been presented by Hong [144] and Rujitanaroj et al. [145].

Hong reported a novel wound dressing material based on PVA/$AgNO_3$ nanofiber mats. In a first stage, PVA and $AgNO_3$ were mixed and launched by electrospinning. After that, the resultant nanofiber webs were treated by heat or UV radiation. The study concluded that the only heat-treated electrospun PVA/$AgNO_3$ fiber web was a good material as wound dressings because it had structural stability in moisture environment as well as excellent antimicrobial ability and quick and continuous release of the effectiveness [144].

Rujinaroj et al. developed successful gelatin fiber mats containing Ag NPs. The stability of the fiber mats were tested in an aqueous solution, the release of the Ag cations was studied in a simulated body fluid (pH 7.4) at skin temperature (32 °C) and physiologic temperature (37 °C), and the antibacterial activity was analyzed [145].

11. WATER TREATMENT

Water pollution is a high problem nowadays, and the wastewater and pollutants may impact dramatically in the aquatic life and humans. To overcome this, there are some techniques which are developing interesting solutions. One of these processes is the photocatalysis, which allows the photodegradation of some pollutants present in water. Recently, new advances in ENFs with photocatalytic nanostructures have been reported. In a study, Panthi et al. [114] developed ECNFs using PAN and in situ synthesized CoS NPs. Their photocatalytic activity was analyzed, and results showed excellent behavior for photodegradation of methylene blue and methyl red dyes under solar light radiation. In other example, Yu et al. [146] fabricated ECNFs based on PAN and Ag_3PO_4 and analyzed their photocatalytic activity under visible radiation. They demonstrated that the ECNFs exhibit higher photocatalytic performances than pure Ag_3PO_4 powder. They indicated that the combination of particle-polymer for the ECNFs preparation may be adopted for more applications. Furthermore, TiO_2 and ZnO have well-known photocatalytic properties. Both metal oxides have been also used to fabricate ECNFs for wastewater treatment [147, 148].

12. OTHER APPLICATIONS

Between other important applications are tissue engineering and drug delivery. In tissue engineering, the ECNF material must be biocompatible. For example, poly(l-lactide-co-ε-caprolactone) has been used to fabricate electrospun tubular scaffolds for tissue engineered vascular grafts [149]. Another vascular application was reported by He et al. [150]. They demonstrated in an in vitro study that the incorporation of electrospun collagen-blended poly(l-lactic acid)-co-poly(ε-caprolactone) scaffolds enhances human coronary artery endothelial cell viability, spreading, and attachment, while also preserving the endothelial cell phenotype . Other emergent developments in tissue engineering based on electrospinning are focused on potential scaffolds for ligaments [151], tendons [152], and bones [153], thanks to the alignment and the mechanical properties of the ECNFs.

Regarding drug delivery applications, ENFs with high porosity allow drug loadings and have the ability to overcome mass transfer limitations associated with other polymeric systems. Thus, in some works, drug is located by the absorption onto the surface of the ENFs and then their efficacy was analyzed in in vivo studies [154]. Also in the drug delivery field, the targeted and controlled delivery of anticancer drugs is being one of the widest areas of research now [155, 156].

Another recent application based on ECNFs is their use as flame retardants. Wu et al. [157] reported novel ECNFs composed of nylon 6 with two additives: montmorillorite nanoclay and a non-halogenated flame retardant based on organic phosphinates. The three components were mixed, and the ECNFs were performed and analyzed. Although the termogravimetric analysis presented worse results, the loss mass rate of the samples was overall decreased, and the additives played a key role in reducing flammability. In other study, Selvakumar et al. [116] developed ENFs of PA with boronic acid NPs onto a cotton substrate. With a very low concentration of these resultant ECNFs, they increased the flame-spreading time in 80 %, obtaining excellent fire protection results.

Finally, a summary of some of the most relevant works and their applications are summarized in Table 4, showing in each case, the polymeric precursors and the active agents used to the ECNFs fabrication.

13. CONCLUSIONS

Nanoparticles are a new promising kind of materials for functionalization of fibers and textiles. Their outstanding properties derived from their size and extremely high specific surface make them one of the most useful additives to provide of additional functionality to fibers and textiles. Their low cost, lower than other nanostructures, make them suitable for industrial applications. Although there are still some problems with their incorporation to commercial applications, there are numerous significant scientific research works that report their suitability for high-value functional textile applications.

Table 4 . Summary of some of the most relevant works classified by applications, indicating the polymer precursors and the loaded active agents for the electrospun nanofibers fabrication

Antibacterial applications		
Antibacterial agent	Polymeric precursor	Antibacterial tests
Ag NPs (from $AgNO_3$ reduction)	Poly(acrylic acid) (PAA)/cyclodextrin	*L. plantarum*, 2012 [88]
Ag NPs (from aqueous solution)	Poly(vinyl alcohol) (PVA)	*S. aureus* and *E. coli*, 2010 [107]; 2011 [120]
Ag NPs (by seed mediate growth method)	Poly(vinyl pyrrolidone) (PVP)	*S. aureus*, *K. pneumoniae*, and *E. coli*, 2011 [121]
Ag NPs (in ethanol solution)	Poly(vinylidene fluoride) (PVDF)	*S. aureus* and *K. pneumoniae*, 2010 [122]
Ag NPs (from $AgNO_3$ reduction)	Poly(vinyl alcohol) (PVA) and chitosan	*E. coli*, 2012 [126]
Ag NPs (from $AgNO_3$ reduction)	Polyvinyl alcohol (PVA)/regenerated silk fibroin	*S. aureus* and *E. coli*, 2011 [127]
Ag NPs (from $AgNO_3$ reduction)	Poly(acrylonitrile) (PAN)	Not tested, 2005, 2003 [123]; [124]. *E. coli*, 2012 [129]
Ag NPs (from $AgNO_3$ reduction)	Nylon 6	*B. cereus* and *E. coli*, 2011 [125]
Quaternary ammonium salts	Diblock copolymers with polyacrylates	*S. aureus* and *E. coli*, 2008 [139]
Triclosan	Poly(lactic acid) (PLA), cyclodextrin	*S. aureus* and *E. coli*, 2013 [140]
Chlorhexidine	Cellulose acetate	*E. coli* and *S. epidermidis*, 2008 [138]
Other Applications		
Active agent	Polymeric precursor	Application
Mupirocin (antibiotic)	Poly-L-lactic acid	Drug release, 2008 [134]
Tetracycline hydrochloride (antibiotic)	Poly(lactic acid) (PLA), poly(ethylene-co-vinyl acetate) (PEVA); poly(l-lactid-co-ε-caprolactone)	Drug release, 2002 [132], 2009 [136]
Fluoroquinolone antibiotics	Poly(l-lactide-co-d,l-lactide) and coPLA/poly (ethylene glycol)	Drug release, 2012 [133]
Ampicillin (antibiotic)	Poly(methyl methacrylate)-nylon 6	Drug release, 2013 [135]
Gentamycin sulfate (antibiotic)	Polycaprolactone	Drug release, 2006 [137]
Iodines	Poly(vinyl pyrrolidone) (PVP)	Wound dressing, 2007 [143]
Ag NPs	Gelatin; poly(vinyl alcohol)	Wound dressing mats, 2007, 2008 [144, 145]
CoS NPs, Ag_3PO_4	Poly(acrylonitrile) (PAN)	Water treatment, photocatalyst, 2013, [114, 146]
CdO, ZnO, TiO_2 (photocatalytic)	Poly(vinyl alcohol)	Water treatment, 2012 [147], 2010 [148]
Collagen; Cell adhesive peptides	Poly(L-lactic acid)-co-poly(ε-caprolactone); poly(B,L-lactic-co-glycolic acid) (PLGA)	Tissue engineering, 2006 [149], 2005 [150]
Hydroxyapatite, PLGA	Poly-L-lactic acid; PLGA	Tendons/ligaments/bones tissue engineering, 2006 [152], 2007 [153]
Boronic acid NPs	Polyamide 6	Flame retardant, 2012 [116], 2014 [157]

The addition of nanoparticles to traditional synthetic fibers is usually performed using well-known techniques such as pad-dry-cure processes; nevertheless, the appropriate surface chemistry of nanoparticles to achieve a good adhesion or even chemical bonding to the fibers is where the technological challenge is. Other techniques such as corona treatments and in situ synthesis of the nanoparticles have been successfully used as alternative functionalization techniques with very good results. Regarding to the scientific bibliography, the antibacterial functionality is the most demanded application for nanoparticles into textile fibers, due to their use in the elaboration of clothes, but there are other properties such as flammability, light absorption, and wettability that can be dramatically affected by the presence of the appropriate nanoparticles.

Furthermore, there are special nonwoven micro and nano textiles which can be obtained using the electrospinning technique. Their ultrathin diameter and the good control of their internal structure, density, and porosity make them ideal for very specific high technology applications. Their controllable composition and hosting properties for different nanoparticles allow to create breathable fiber mats with antibacterial properties which can help in the healing of wounded skin. But such porous membranes have been also investigated for filters and water treatment applications and other kinds of tunable membranes. The apparition of all these functional artificial fibers and smart fabrics mean undoubtedly the new revolution for the textile industry.

ACKNOWLEDGEMENTS

This work was supported in part by the Spanish Ministry of Economy and Competitiveness—CICYT-FEDER TEC2013-43679-R research grant, and an UPNA pre-doctoral research grant.

COMPETING INTERESTS

The authors declare that they have no competing interests.

AUTHORS' CONTRIBUTIONS

PJR carried out the work related to traditional synthetic fibers. AU carried out the work related to nanofibers and nanowebs. JG participated in writing the "Introduction" and "Conclusions" sections obtained by this work. FJA participated in the design of the study and helped to draft the manuscript. All authors read and approved the final manuscript.

REFERENCES

1. Balter M (2009) Clothes make the (Hu) man. Science 325:1329

2. Tunzelmann GNV (1995) Time-saving technical change: the cotton industry in the English Industrial Revolution. Explor Econ Hist 32:1–27

3. Kauffman GB (1993) Rayon: the first semi-synthetic fiber product. J Chem Educ 62:887–893

4. Tyler DJ (2005) Textile digital printing technologies. Text Prog 37:1-65.

5. Castano LMm, Flatau AB (2014) Smart fabric sensors and e-textile technologies: a review. Smart Mater Struct 23:1-27.

6. Holme I (2007) Innovative technologies for high performance textiles. Color Technol 123:59–73

7. Perepelkin KE (2005) Principles and methods of modification of fibres and fibre materials. A review. Fibre Chem 37:123–140

8. Perepelkin KE (2003) Trends and changes in world chemical fibre production. An analytical review. Part 2. Trends in the development of traditional and promising kinds of chemical fibres. Fibre Chem 35:241–249.

9. Coyle S, Wu Y, Lau K, De Rossi D, Wallace G, Diamond D (2007) Smart Nanotextiles: A Review of Materials and Applications. MRS Bull 32:434-442.

10. Da Rocha A (2005) Development of textile-based high-tech products: The new challenge. Studies in Health Technology and Informatics 108:330-334.

11. Tang SLP, Stylios GK (2006) An overview of smart technologies for clothing design and engineering. Int J Cloth Sci Technol 18:108–128

12. Levchik SV, Weil ED (2005) Flame retardancy of thermoplastic polyester—a review of the recent literature. Polym Int 54:11–35

13. Alongi J, Carosio F, Malucelli G (2014) Current emerging techniques to impart flame retardancy to fabrics: an overview. Polym Degrad Stab 106:138–149

14. Didane N, Giraud S, Devaux E, Lemort G, Capon G (2012) Thermal and fire resistance of fibrous materials made by PET containing flame retardant agents. Polym Degrad Stab 97:2545–2551

15. van der Veen I, de Boer J (2012) Phosphorus flame retardants: properties, production, environmental occurrence, toxicity and analysis. Chemosphere 88:1119–1153

16. Alongi J, Tata J, Carosio F, Rosace G, Frache A, Camino G (2015) A comparative analysis of nanoparticle adsorption as fire-protection approach for fabrics. Polymers 7:47–68

17. Carosio F, Alongi J, Frache A (2011) Influence of surface activation by plasma and nanoparticle adsorption on the morphology, thermal

stability and combustion behavior of PET fabrics. Eur Polym J 47:893–902

18. El-Hady MMA, Farouk A, Sharaf S (2013) Flame retardancy and UV protection of cotton based fabrics using nano ZnO and polycarboxylic acids. Carbohydr Polym 92:400–406

19. Nazari A, Montazer M, Rashidi A, Yazdanshenas M, Anary-Abbasinejad M (2009) Nano TiO2 photo-catalyst and sodium hypophosphite for cross-linking cotton with poly carboxylic acids under UV and high temperature. Appl Catal A Gen 371:10–16

20. Carosio F, Laufer G, Alongi J, Camino G, Grunlan JC (2011) Layer-by-layer assembly of silica-based flame retardant thin film on PET fabric. Polym Degrad Stab 96:745–750

21. Tata J, Alongi J, Frache A (2012) Optimization of the procedure to burn textile fabrics by cone calorimeter: part II. Results on nanoparticle-finished polyester. Fire Mater 36:527–536

22. Decher G (1997) Fuzzy nanoassemblies: toward layered polymeric multicomposites. Science 277:1232–1237

23. Apaydin K, Laachachi A, Ball V, Jimenez M, Bourbigot S, Toniazzo V, Ruch D (2014) Intumescent coating of (polyallylamine-polyphosphates) deposited on polyamide fabrics via layer-by-layer technique. Polym Degrad Stab 106:158–164

24. Carosio F, Alongi J, Malucelli G (2012) Layer by layer ammonium polyphosphate-based coatings for flame retardancy of polyester-cotton blends. Carbohydr Polym 88:1460–1469

25. Apaydin K, Laachachi A, Ball V, Jimenez M, Bourbigot S, Ruch D (2015) Layer-by-layer deposition of a TiO2-filled intumescent coating and its effect on the flame retardancy of polyamide and polyester fabrics. Colloids Surf Physicochem Eng Aspects 469:1–10

26. Gonçalves AG, Jarrais B, Pereira C, Morgado J, Freire C, Pereira MFR (2012) Functionalization of textiles with multi-walled carbon nanotubes by a novel dyeing-like process. J Mater Sci 47:5263–5275

27. Morones JR, Elechiguerra JL, Camacho A, Holt K, Kouri JB, Ramírez JT, Yacaman MJ (2005) The bactericidal effect of silver nanoparticles. Nanotechnology 16:2346–2353

28. Panácek A, Kvítek L, Prucek R, Kolár M, Vecerová R, Pizúrová N, Sharma VK, Nevecná T, Zboril R (2006) Silver colloid nanoparticles: synthesis, characterization, and their antibacterial activity. J Phys Chem B 110:16248–16253

29. Mohammed Fayaz A, Balaji K, Girilal M, Kalaichelvan PT, Venkatesan R (2009) Mycobased synthesis of silver nanoparticles and their incorporation into sodium alginate films for vegetable and fruit preservation. J Agric Food Chem 57:6246–6252

30. Jain J, Arora S, Rajwade JM, Omray P, Khandelwal S, Paknikar KM (2009) Silver nanoparticles in therapeutics: development of an antimicrobial gel formulation for topical use. Mol Pharm 6:1388–1401

31. Chaloupka K, Malam Y, Seifalian AM (2010) Nanosilver as a new generation of nanoproduct in biomedical applications. Trends Biotechnol 28:580–588

32. Furno F, Morley KS, Wong B, Sharp BL, Arnold PL, Howdle SM, Bayston R, Brown PD, Winship PD, Reid HJ (2004) Silver nanoparticles and polymeric medical devices: a new approach to prevention of infection? J Antimicrob Chemother 54:1019–1024

33. Li Q, Mahendra S, Lyon DY, Brunet L, Liga MV, Li D, Alvarez PJJ (2008) Antimicrobial nanomaterials for water disinfection and microbial control: potential applications and implications. Water Res 42:4591–4602

34. Gao Y, Cranston R (2008) Recent advances in antimicrobial treatments of textiles. Text Res J 78:60–72

35. Lee HJ, Yeo SY, Jeong SH (2003) Antibacterial effect of nanosized silver colloidal solution on textile fabrics. J Mater Sci 38:2199–2204

36. Marambio-Jones C, Hoek EMV (2010) A review of the antibacterial effects of silver nanomaterials and potential implications for human health and the environment. J Nanoparticle Res 12:1531–1551

37. Radetic M (2012) Functionalization of textile materials with silver nanoparticles. J Mater Sci 48:95–107

38. Gerber LC, Mohn D, Fortunato G, Astasov-Frauenhoffer M, Imfeld T, Waltimo T, Zehnder M, Stark WJ (2011) Incorporation of reactive silver-tricalcium phosphate nanoparticles into polyamide 6 allows preparation of self-disinfecting fibers. Polym Eng Sci 51:71–77

39. Shastri JP, Rupani MG, Jain RL (2012) Antimicrobial activity of nanosilver-coated socks fabrics against foot pathogens. J Text Inst 103:1234–1243

40. Jiang SX, Qin WF, Guo RH, Zhang L (2010) Surface functionalization of nanostructured silver-coated polyester fabric by magnetron sputtering. Surf Coat Technol 204:3662–3667

41. Radetic M, Ilic V, Vodnik V, Dimitrijevic S, Jovancic P, Šaponjic Z, Nedeljkovic JM (2008) Antibacterial effect of silver nanoparticles deposited on corona-treated polyester and polyamide fabrics. Polym Adv Technol 19:1816–1821

42. Ilic V, Šaponjic Z, Vodnik V, Molina R, Dimitrijevic S, Jovancic P, Nedeljkovic J, Radetic M (2009) Antifungal efficiency of corona pretreated polyester and polyamide fabrics loaded with Ag nanoparticles. J Mater Sci 44:3983–3990

43. Ilic V, Šaponjic Z, Vodnik V, Lazovic S, Dimitrijevic S, Jovancic P, Nedeljkovic JM, Radetic M (2010) Bactericidal efficiency of silver nanoparticles deposited onto radio frequency plasma pretreated polyester fabrics. Ind Eng Chem Res 49:7287–7293

44. Liz-Marzán LM (2006) Tailoring surface plasmons through the morphology and assembly of metal nanoparticles. Langmuir 22:32–41

45. Rivero PJ, Goicoechea J, Urrutia A, Arregui FJ (2013) Effect of both protective and reducing agents in the synthesis of multicolor silver nanoparticles. Nanoscale Res Lett 8:1–9

46. Falletta E, Bonini M, Fratini E, Lo Nostro A, Pesavento G, Becheri A, Lo Nostro P, Canton P, Baglioni P (2008) Clusters of poly(acrylates) and silver nanoparticles: structure and applications for antimicrobial fabrics. J Phys Chem C 112:11758–11766

47. Textor T, Fouda MMG, Mahltig B (2010) Deposition of durable thin silver layers onto polyamides employing a heterogeneous Tollens' reaction. Appl Surf Sci 256:2337–2342

48. Montazer M, Shamei A, Alimohammadi F (2012) Synthesizing and stabilizing silver nanoparticles on polyamide fabric using silver-ammonia/PVP/UVC. Prog Org Coat 75:379–385

49. Montazer M, Shamei A, Alimohammadi F (2014) Synthesis of nanosilver on polyamide fabric using silver/ammonia complex. Mater Sci Eng C 38:170–176

50. Babaahmadi V, Montazer M (2015) A new route to synthesis silver nanoparticles on polyamide fabric using stannous chloride. J Text Inst 106:970–977

51. Allahyarzadeh V, Montazer M, Nejad NH, Samadi N (2013) In situ synthesis of nano silver on polyester using NaOH/Nano TiO 2. J Appl Polym Sci 129:892–900

52. Komeily-Nia Z, Montazer M, Latifi M (2013) Synthesis of nano copper/nylon composite using ascorbic acid and CTAB. Colloids Surf Physicochem Eng Aspects 439:167–175

53. Abou-Elela SI, Ibrahim HS, Kamel MM, Gouda M (2014) Application of nanometal oxides in situ in nonwoven polyester fabric for the removal of bacterial indicators of pollution from wastewater. Sci World J 2014

54. Montazer M, Seifollahzadeh S (2011) Enhanced self-cleaning, antibacterial and UV protection properties of nano TiO 2 treated textile through enzymatic pretreatment. Photochem Photobiol 87:877–883

55. El-Gabry LK, Allam OG, Hakeim OA (2013) Surface functionalization of viscose and polyester fabrics toward antibacterial and coloration properties. Carbohydr Polym 92:353–359

56. Farouk A, Moussa S, Ulbricht M, Schollmeyer E, Textor T (2014) ZnO-modified hybrid polymers as an antibacterial finish for textiles. Text Res J 84:40–51

57. Radetic M (2013) Functionalization of textile materials with TiO2 nanoparticles. J Photochem Photobiol C 16:62–76

58. Mihailovic D, Šaponjic Z, Molina R, Puac N, Jovancic P, Nedeljkovic J, Radetic M (2010) Improved properties of oxygen and argon RF plasma-activated polyester fabrics loaded with TiO2 nanoparticles. ACS Appl Mater Interfaces 2:1700–1706

59. Mihailovic D, Šaponjic Z, Molina R, Radoicic M, Esquena J, Jovancic P, Nedeljkovic J, Radetic M (2011) Multifunctional properties of polyester fabrics modified by corona discharge/air RF plasma and colloidal TiO2 nanoparticles. Polym Compos 32:390–397

60. Mihailovic D, Šaponjic Z, Radoicic M, Molina R, Radetic T, Jovancic P, Nedeljkovic J, Radetic M (2011) Novel properties of PES fabrics modified by corona discharge and colloidal TiO2 nanoparticles. Polym Adv Technol 22:703–709

61. Ibanescu M, Musat V, Textor T, Badilita V, Mahltig B (2014) Photocatalytic and antimicrobial Ag/ZnO nanocomposites for functionalization of textile fabrics. J Alloys Compounds 610:244–249

62. Shin Y, Park M, Kim H, Jin F, Park S (2014) Synthesis of silver-doped silica-complex nanoparticles for antibacterial materials. Bull Kor Chem Soc 35:2979–2984

63. Ali SW, Rajendran S, Joshi M (2011) Synthesis and characterization of chitosan and silver loaded chitosan nanoparticles for bioactive polyester. Carbohydr Polym 83:438–446

64. Mihailovic D, Šaponjic Z, Vodnik V, Potkonjak B, Jovancic P, Nedeljkovic JM, Radetic M (2011) Multifunctional PES fabrics modified with colloidal Ag and TiO2 nanoparticles. Polym Adv Technol 22:2244–2249

65. Mahltig B, Textor T (2010) Silver containing sol-gel coatings on polyamide fabrics as antimicrobial finish-description of a technical application process for wash permanent antimicrobial effect. Fibers Polymers 11:1152–1158

66. Radic N, Obradovic BM, Kostic M, Dojcinovic B, Hudcová M, Kuraica MM, Cernák M (2013) Deposition of gold nanoparticles on polypropylene nonwoven pretreated by dielectric barrier discharge and diffuse coplanar surface barrier discharge. Plasma Chem Plasma Process 33:201–218

67. Jeong SH, Hwang YH, Yi SC (2005) Antibacterial properties of padded PP/PE nonwovens incorporating nano-sized silver colloids. J Mater Sci 40:5413–5418

68. Guo P, Zhai S, Xiao Z, Zhang F, An Q, Song X (2014) Preparation of superhydrophobic materials for oil/water separation and oil absorption using PMHS–TEOS-derived xerogel and polystyrene. J Sol Gel Sci Technol 72:385–393

69. Saraf R, Lee HJ, Michielsen S, Owens J, Willis C, Stone C, Wilusz E (2011) Comparison of three methods for generating superhydrophobic, superoleophobic nylon nonwoven surfaces. J Mater Sci 46:5751–5760

70. Gao L, McCarthy TJ (2006) "Artificial lotus leaf" prepared using a 1945 patent and a commercial textile. Langmuir 22:5998–6000

71. Wang H, Ding J, Xue Y, Wang X, Lin T (2010) Superhydrophobic fabrics from hybrid silica sol-gel coatings: structural effect of precursors on wettability and washing durability. J Mater Res 25:1336–1343

72. Cassie ABD, Baxter S (1944) Wettability of porous surfaces. Trans Faraday Soc 40:546–551

73. Bravo J, Zhai L, Wu Z, Cohen RE, Rubner MF (2007) Transparent superhydrophobic films based on silica nanoparticles. Langmuir 23:7293–7298

74. Zhang X, Shi F, Niu J, Jiang Y, Wang Z (2008) Superhydrophobic surfaces: from structural control to functional application. J Mater Chem 18:621–633

75. Mazrouei-Sebdani Z, Khoddami A (2011) Alkaline hydrolysis: a facile method to manufacture superhydrophobic polyester fabric by fluorocarbon coating. Prog Org Coat 72:638–646

76. Zhu X, Zhang Z, Yang J, Xu X, Men X, Zhou X (2012) Facile fabrication of a superhydrophobic fabric with mechanical stability and easy-repairability. J Colloid Interface Sci 380:182–186

77. Gao Q, Zhu Q, Guo Y, Yang CQ (2009) Formation of highly hydrophobic surfaces on cotton and polyester fabrics using silica sol nanoparticles and nonfluorinated alkylsilane. Ind Eng Chem Res 48:9797–9803

78. Xu L, Shen Y, Wang L, Ding Y, Cai Z (2015) Preparation of vinyl silica-based organic/inorganic nanocomposites and superhydrophobic polyester surfaces from it., Colloid Polym. Sci

79. Xu L, Cai Z, Shen Y, Wang L, Ding Y (2014) Facile preparation of superhydrophobic polyester surfaces with fluoropolymer/SiO2 nanocomposites based on vinyl nanosilica hydrosols. J Appl Polym Sci 131

80. Xu L, Karunakaran RG, Guo J, Yang S (2012) Transparent, superhydrophobic surfaces from one-step spin coating of hydrophobic nanoparticles. ACS Appl Mater Interfaces 4:1118–1125

81. Frunza L, Preda N, Matei E, Frunza S, Ganea CP, Vlaicu AM, Diamandescu L, Dorogan A (2013) Synthetic fabrics coated with zinc oxide nanoparticles by electroless deposition: structural characterization and wetting properties. J Polym Sci Part B 51:1427–1437

82. Ashraf M, Campagne C, Perwuelz A, Champagne P, Leriche A, Courtois C (2013) Development of superhydrophilic and superhydrophobic polyester fabric by growing zinc oxide nanorods. J Colloid Interface Sci 394:545–553

83. Popescu AC, Duta L, Dorcioman G, Mihailescu IN, Stan GE, Pasuk I, Zgura I, Beica T, Enculescu I, Ianculescu A, Dumitrescu I (2011) Radical modification of the wetting behavior of textiles coated with ZnO thin films and nanoparticles when changing the ambient pressure in the pulsed laser deposition process. J Appl Phys 110

84. Huang F, Zhang H, Hui Q, Wang T, Wei Q (2012) Fabrication of switchable superhydrophobic nonwoven fabrics via cosputtering. Adv Sci Lett 10:599–601

85. Zhu T, Cai C, Duan C, Zhai S, Liang S, Jin Y, Zhao N, Xu J (2015) Robust polypropylene fabrics super-repelling various liquids: a simple, rapid and scalable fabrication method by solvent swelling. ACS Appl Mater Interfaces 7:13996–14003

86. Doshi J, Reneker DH (1995) Electrospinning process and applications of electrospun fibers. J Electrostatics 35:151–160

87. Reneker DH, Chun I (1996) Nanometre diameter fibres of polymer, produced by electrospinning. Nanotechnology 7:216–223

88. Rivero PJ, Urrutia A, Goicoechea J, Rodríguez Y, Corres JM, Arregui FJ, Matías IR (2012) An antibacterial submicron fiber mat with in situ synthesized silver nanoparticles., J Appl Polym Sci

89. Bhardwaj N, Kundu SC (2010) Electrospinning: a fascinating fiber fabrication technique. Biotechnol Adv 28:325–347

90. Reneker DH, Yarin AL (2008) Electrospinning jets and polymer nanofibers. Polymer 49:2387–2425

91. Li D, Xia Y (2004) Direct fabrication of composite and ceramic hollow nanofibers by electrospinning. Nano Lett 4:933–938

92. Venugopal J, Ramakrishna S (2005) Applications of polymer nanofibers in biomedicine and biotechnology. Appl Biochem Biotechnol Part A Enzym Eng Biotechnol 125:147–157

93. Huang Z, Zhang Y, Kotaki M, Ramakrishna S (2003) A review on polymer nanofibers by electrospinning and their applications in nanocomposites. Composites Sci Technol 63:2223–2253

94. Li D, Xia Y (2004) Electrospinning of nanofibers: reinventing the wheel? Adv Mater 16:1151–1170

95. Katta P, Alessandro M, Ramsier RD, Chase GG (2004) Continuous electrospinning of aligned polymer nanofibers onto a wire drum collector. Nano Lett 4:2215–2218

96. He D, Hu B, Yao Q, Wang K, Yu S (2009) Large-scale synthesis of flexible free-standing SERS substrates with high sensitivity: electrospun PVA nanofibers embedded with controlled alignment of silver nanoparticles. ACS Nano 3:3993–4002.

97. Li D, Wang Y, Xia Y (2003) Electrospinning of polymeric and ceramic nanofibers as uniaxially aligned arrays. Nano Lett 3:1167–1171.

98. Panthi G, Park M, Kim H, Park S (2015) Electrospun polymeric nanofibers encapsulated with nanostructured materials and their applications: a review. J Ind Eng Chem 24:1–13

99. Reneker DH, Yarin AL, Zussman E, Xu H (2007) Electrospinning of nanofibers from polymer solutions and melts. Adv Appl Mech 41:43-"195,345-346"

100. Jiang H, Hu Y, Li Y, Zhao P, Zhu K, Chen W (2005) A facile technique to prepare biodegradable coaxial electrospun nanofibers for controlled release of bioactive agents. J Control Release 108:237–243

101. Zhang YZ, Wang X, Feng Y, Li J, Lim CT, Ramakrishna S (2006) Coaxial electrospinning of (fluorescein isothiocyanate-conjugated bovine serum albumin)-encapsulated poly(e-caprolactone) nanofibers for sustained release. Biomacromolecules 7:1049–1057

102. Shalumon KT, Binulal NS, Selvamurugan N, Nair SV, Menon D, Furuike T, Tamura H, Jayakumar R (2009) Electrospinning of carboxymethyl chitin/poly(vinyl alcohol) nanofibrous scaffolds for tissue engineering applications. Carbohydr Polym 77:863–869

103. Xu X, Zhang J, Fan Y (2010) Fabrication of cross-linked polyethyleneimine microfibers by reactive electrospinning with in situ photo-cross-linking by UV radiation. Biomacromolecules 11:2283–2289

104. Formo E, Lee E, Campbell D, Xia Y (2008) Functionalization of electrospun TiO 2 nanofibers with Pt nanoparticles and nanowires for catalytic applications. Nano Lett 8:668–672

105. Elzoghby AO, Samy WM, Elgindy NA (2012) Albumin-based nanoparticles as potential controlled release drug delivery systems. J Control Release 157:168–182

106. Botes M, Cloete TE (2010) The potential of nanofibers and nanobiocides in water purification. Crit Rev Microbiol 36:68–81

107. Park JH, Karim MR, Kim IK, Cheong IW, Kim JW, Bae DG, Cho JW, Yeum JH (2010) Electrospinning fabrication and characterization of poly(vinyl alcohol)/montmorillonite/silver hybrid nanofibers for antibacterial applications. Colloid Polym Sci 288:115–121

108. Zahedi P, Rezaeian I, Ranaei-Siadat S, Jafari S, Supaphol P (2010) A review on wound dressings with an emphasis on electrospun nanofibrous polymeric bandages. Polym Adv Technol 21:77–95

109. Zeng J, Xu X, Chen X, Liang Q, Bian X, Yang L, Jing X (2003) Biodegradable electrospun fibers for drug delivery. J Control Release 92:227–231

110. Urrutia A, Goicoechea J, Rivero PJ, Matías IR, Arregui FJ (2013) Electrospun nanofiber mats for evanescent optical fiber sensors. Sensors Actuators B Chem 176:569–576

111. Katoch A, Kim J, Kim SS (2015) Significance of the nanograin size on the H2S-sensing ability of CuO-SnO2 composite nanofibers. J Sens 2015:1–7

112. Ren G, Xu X, Liu Q, Cheng J, Yuan X, Wu L, Wan Y (2006) Electrospun poly(vinyl alcohol)/glucose oxidase biocomposite membranes for biosensor applications. React Funct Polym 66:1559–1564

113. Oh B, Lee CH (2013) Nanofiber for cardiovascular tissue engineering. Expert Opin Drug Deliv 10:1565–1582

114. Panthi G, Barakat NAM, Abdelrazek Khalil K, Yousef A, Jeon K, Kim HY (2013) Encapsulation of CoS nanoparticles in PAN electrospun nanofibers: effective and reusable catalyst for ammonia borane hydrolysis and dyes photodegradation. Ceram Int 39:1469–1476

115. Chiscan O, Dumitru I, Postolache P, Tura V, Stancu A (2012) Electrospun PVC/Fe 3O 4 composite nanofibers for microwave absorption applications. Mater Lett 68:251–254

116. Selvakumar N, Azhagurajan A, Natarajan TS, Mohideen Abdul Khadir M (2012) Flame-retardant fabric systems based on electrospun polyamide/boric acid nanocomposite fibers. J Appl Polym Sci 126:614–619

117. Ignatova M, Rashkov I, Manolova N (2013) Drug-loaded electrospun materials in wound-dressing applications and in local cancer treatment. Expert Opin Drug Deliv 10:469–483

118. Muzzarelli RAA, El Mehtedi M, Mattioli-Belmonte M (2014) Emerging biomedical applications of nano-chitins and nano-chitosans obtained via advanced eco-friendly technologies from marine resources. Mar Drugs 12:5468–5502

119. Rai MK, Deshmukh SD, Ingle AP, Gade AK (2012) Silver nanoparticles: the powerful nanoweapon against multidrug-resistant bacteria. J Appl Microbiol 112:841–852

120. Park JH, Kim IK, Choi JY, Karim MR, Cheong IW, Oh W, Yeum JH (2011) Electrospinning fabrication of poly(vinyl alcohol)/waterborne polyurethane/silver composite nanofbre mats in aqueous solution for anti-bacterial exploits. Polym Polym Compos 19:753–761

121. Hwang S, Jeong S (2011) Electrospun nano composites of poly(vinyl pyrrolidone)/nano-silver for antibacterial materials. J Nanosci Nanotechnol 11:610–613

122. Yuan J, Geng J, Xing Z, Shen J, Kang I, Byun H (2010) Electrospinning of antibacterial poly(vinylidene fluoride) nanofibers containing silver nanoparticles. J Appl Polym Sci 116:668–672

123. Lee HK, Jeong EH, Baek CK, Youk JH (2005) One-step preparation of ultrafine poly(acrylonitrile) fibers containing silver nanoparticles. Mater Lett 59:2977–2980

124. Yang QB, Li DM, Hong YL, Li ZY, Wang C, Qiu SL, Wei Y (2003) Preparation and characterization of a PAN nanofibre containing Ag nanoparticles via electrospinning. Synth Met 137:973–974

125. Shi Q, Vitchuli N, Nowak J, Noar J, Caldwell JM, Breidt F, Bourham M, McCord M, Zhang X (2011) One-step synthesis of silver nanoparticle-filled nylon 6 nanofibers and their antibacterial properties. J Mater Chem 21:10330–10335

126. Zhao Y, Zhou Y, Wu X, Wang L, Xu L, Wei S (2012) A facile method for electrospinning of Ag nanoparticles/poly (vinyl alcohol)/carboxymethyl-chitosan nanofibers. Appl Surf Sci 258:8867–8873

127. Li W, Wang J, Chi H, Wei G, Zhang J, Dai L (2012) Preparation and antibacterial activity of polyvinyl alcohol/regenerated silk fibroin composite fibers containing Ag nanoparticles. J Appl Polym Sci 123:20–25

128. Xu X, Yang Q, Wang Y, Yu H, Chen X, Jing X (2006) Biodegradable electrospun poly(l-lactide) fibers containing antibacterial silver nanoparticles. Eur Polym J 42:2081–2087

129. Mahapatra A, Garg N, Nayak BP, Mishra BG, Hota G (2012) Studies on the synthesis of electrospun PAN-Ag composite nanofibers for antibacterial application. J Appl Polym Sci 124:1178–1185

130. Kong H, Song J, Jang J (2010) Photocatalytic antibacterial capabilities of TiO2-biocidal polymer nanocomposites synthesized by a surface-initiated photopolymerization. Environ Sci Technol 44:5672–5676

131. Kim SW, An Y (2012) Effect of ZnO and TiO 2 nanoparticles preilluminated with UVA and UVB light on Escherichia coli and Bacillus subtilis. Appl Microbiol Biotechnol 95:243–253

132. Kenawy E, Bowlin GL, Mansfield K, Layman J, Simpson DG, Sanders EH, Wnek GE (2002) Release of tetracycline hydrochloride

from electrospun poly(ethylene-co-vinylacetate), poly(lactic acid), and a blend. J Control Release 81:57–64

133. Toncheva A, Paneva D, Maximova V, Manolova N, Rashkov I (2012) Antibacterial fluoroquinolone antibiotic-containing fibrous materials from poly(l-lactide-co-d,l-lactide) prepared by electrospinning. Eur J Pharm Sci 47:642–651

134. Thakur RA, Florek CA, Kohn J, Michniak BB (2008) Electrospun nanofibrous polymeric scaffold with targeted drug release profiles for potential application as wound dressing. Int J Pharm 364:87–93

135. Sohrabi A, Shaibani PM, Etayash H, Kaur K, Thundat T (2013) Sustained drug release and antibacterial activity of ampicillin incorporated poly(methyl methacrylate)–nylon6 core/shell nanofibers. Polymer 54:2699–2705

136. Su Y, Li X, Wang H, He C, Mo X (2009) Fabrication and characterization of biodegradable nanofibrous mats by mix and coaxial electrospinning. J Mater Sci Mater Med 20:2285–2294

137. Huang Z, He C, Yang A, Zhang Y, Han X, Yin J, Wu Q (2006) Encapsulating drugs in biodegradable ultrafine fibers through co-axial electrospinning. J Biomed Mat Res Part A 77:169–179

138. Chen L, Bromberg L, Hatton TA, Rutledge GC (2008) Electrospun cellulose acetate fibers containing chlorhexidine as a bactericide. Polymer 49:1266–1275

139. Guo-Dong F, Fang Y, Zhigang L, Xinsong L (2008) Solvent-resistant antibacterial microfibers of self-quaternized block copolymers from atom transfer radical polymerization and electrospinning. J Mater Chem 18:859–867

140. Kayaci F, Umu OCO, Tekinay T, Uyar T (2013) Antibacterial electrospun poly(lactic acid) (PLA) nanofibrous webs incorporating triclosan/cyclodextrin inclusion complexes. J Agric Food Chem 61:3901–3908

141. Ignatova M, Manolova N, Rashkov I (2013) Electrospun antibacterial chitosan-based fibers. Macromol Biosci 13:860–872

142. Meaume S, Vallet D, Morere MN, Téot L (2005) Evaluation of a silver-releasing hydroalginate dressing in chronic wounds with signs of local infection. J Wound Care 14:411–419

143. Ignatova M, Manolova N, Rashkov I (2007) Electrospinning of poly(vinyl pyrrolidone)–iodine complex and poly(ethylene

oxide)/poly(vinyl pyrrolidone)–iodine complex—a prospective route to antimicrobial wound dressing materials. Eur Polym J 43:1609–1623

144. Hong KH (2007) Preparation and properties of electrospun poly (vinyl alcohol)/silver fiber web as wound dressings. Polym Eng Sci 47:43–49

145. Rujitanaroj P, Pimpha N, Supaphol P (2008) Wound-dressing materials with antibacterial activity from electrospun gelatin fiber mats containing silver nanoparticles. Polymer 49:4723–4732

146. Yu H, Jiao Z, Hu H, Lu G, Ye J, Bi Y (2013) Fabrication of Ag3PO4-PAN composite nanofibers for photocatalytic applications. Crystengcomm 15:4802–4805

147. Yousef A, Barakat NAM, Amna T, Unnithan AR, Al-Deyab SS, Yong Kim H (2012) Influence of CdO-doping on the photoluminescence properties of ZnO nanofibers: effective visible light photocatalyst for waste water treatment. J Lumin 132:1668–1677

148. Yun J, Jin D, Lee Y, Kim H (2010) Photocatalytic treatment of acidic waste water by electrospun composite nanofibers of pH-sensitive hydrogel and TiO2. Mater Lett 64:2431–2434

149. Kim TG, Park TG (2006) Biomimicking extracellular matrix: cell adhesive RGD peptide modified electrospun poly(D, L-lactic-co-glycolic acid) nanofiber mesh. Tissue Eng 12:221–233

150. He W, Yong T, Teo WE, Ma Z, Ramakrishna S (2005) Fabrication and endothelialization of collagen-blended biodegradable polymer nanofibers: potential vascular graft for blood vessel tissue engineering. Tissue Eng 11:1574–1588

151. Lee CH, Shin HJ, Cho IH, Kang Y, Kim IA, Park K, Shin J (2005) Nanofiber alignment and direction of mechanical strain affect the ECM production of human ACL fibroblast. Biomaterials 26:1261–1270

152. Sahoo S, Ouyang H, James C, Tay TE, Toh SL (2006) Characterization of a novel polymeric scaffold for potential application in tendon/ligament tissue engineering. Tissue Eng 12:91–99

153. Sui G, Yang X, Mei F, Hu X, Chen G, Deng X, Ryu S (2007) Poly-L-lactic acid/hydroxyapatite hybrid membrane for bone tissue regeneration. J Biomed Mater Res A 82:445–454

154. Bölgen N, Vargel I, Korkusuz P, Menceloglu YZ, Piskin E (2007) In vivo performance of antibiotic embedded electrospun PCL membranes for prevention of abdominal adhesions. J Biomed Mater Res B Appl Biomater 81:530–543

155. Xu X, Chen X, Xu X, Lu T, Wang X, Yang L, Jing X (2006) BCNU-loaded PEG-PLLA ultrafine fibers and their in vitro antitumor activity against Glioma C6 cells. J Control Release 114:307–316

156. Xie J, Wang C (2006) Electrospun micro- and nanofibers for sustained delivery of paclitaxel to treat C6 glioma in vitro. Pharm Res 23:1817–1826

157. Wu H, Krifa M, Koo JH (2014) Flame retardant polyamide 6/nanoclay/intumescent nanocomposite fibers through electrospinning. Text Res J 84:1106–1118

CHAPTER 7

A Smart Shirt Made with Conductive Ink and Conductive Foam for the Measurement of Electrocardiogram Signals with Unipolar Precordial Leads

Yasunori Tada [1,*], Yusaku Amano [2,†], Tomonobu Sato [3,†], Shigeru Saito [3,†] and Masahiro Inoue [1,†]

[1] Advanced Scientific Research Leaders Development Unit, Gunma University, 1-5-1 Tenjin-cho, Kiryu, Gunma 376-8515, Japan

[2] Department of Mechanical Science and Technology, Gunma University, 1-5-1 Tenjin-cho, Kiryu, Gunma 376-8515, Japan

[3] Department of Anesthesiology, Gunma University, 3-39-22, Showa-machi, Maebashi, Gunma 371-8511, Japan

ABSTRACT

The Holter monitor is used to measure an electrocardiogram (ECG) signal while a subject moves. However, the Holter monitor is uncomfortable for the subject. Another method of measuring the ECG signal uses a smart shirt. We developed a smart shirt that has six electrodes on the chest and can measure a detailed ECG, obtained with unipolar precordial leads. The electrodes and wires of the shirt are made of conductive ink that is flexible and stretchable.

The smart shirt is stretchable and fits the body well. However, because of the gap between the smart shirt and the body, electrodes V1 and V2 do not touch the body consistently. We developed a conductive foam block that fills this gap. We investigated the characteristics of the conductive foam block, and measured ECG signals using the smart shirt. The electrical resistance of the conductive foam block was reduced by pressure. This characteristic could be utilized to measure the ECG signal because the block was pressed by the body and smart shirt. We could measure the ECG signal using the smart shirt and blocks while the subject walked and could detect peaks of the ECG signal while the subject jogged slowly.

KEYWORDS

smart shirt; conductive ink; conductive foam; electrocardiogram measurement

1. INTRODUCTION

In the fields of medicine and sports science, the Holter monitor is used to measure an electrocardiogram (ECG) signal while a subject moves. Electrodes must be pasted to the skin of the subject using electrical conductive adhesion gel, and cables are used to connect the electrodes to the Holter monitor. These electrodes and cables make ECG measurements uncomfortable for the subject. To eliminate the uncomfortable feeling of the electrodes and cables, other methods using dry electrodes have been proposed [1,2,3,4,5,6,7,8,9,10,11,12,13,14,15,16]. The electrodes and wires in such methods are made of conductive textile [1,2,3,5,6,7,8,9,10, 12,13,14,15,16] or conductive ink [4,11], and are embedded in a shirt or belt. Such a shirt is called a smart shirt and the biological signals are measured for a person wearing the shirt [1,2,3,4,5,6,7,8,9]. The belt is wound around a subject's body and measures biological signals [10,11,12,13,14,15,16]. Smart shirts and belts measure various biological signals, including ECG signals [1,2,3,4,5,6,7,8,9,10,11,12,13,14,15,16], electromyogram (EMG) signals [1,2,7,11], and signals of the respiration rate [1,7,14,15,16] and other body properties [1,6,7,11,12,14,15]. We consider an ECG measurement made using smart shirts and belts.

Smart shirts and belts for ECG measurement can be categorized by the number of electrodes. The number of the electrodes relates to the method of ECG measurement. Different methods of the ECG measurement employ bipolar leads, unipolar leads, standard limb leads, and unipolar precordial leads. Almost all existing smart shirts and belts adopt unipolar leads to measure the ECG signal, with the use of three electrodes being the simplest method [2,3,4,5,6,7,10,11,12,13,14,15,16]. Unipolar leads allow the measurement of the signal of a simple ECG signal but cannot be used to measure the signal of a detailed ECG. A few smart shirts have six electrodes on the chest with unipolar precordial leads for the measurement of the signal of a detailed ECG [1,8,9]. However, existing smart shirts that adopt unipolar precordial leads have not yet been discussed in terms of the subject's motion. There are two problems to overcome in conducting an ECG measurement using the smart shirt if the subject moves. The first problem is the size of the electrode on the smart shirt. Each electrode has to be small enough to avoid contact between neighboring electrodes, yet a stable measurement of the ECG signal is difficult if the electrode is small. The second problem relates to the positions of the V1 and V2 electrodes. These electrodes are placed on a concave area of the chest. Thus, these electrodes on the smart shirt are not in stable contact with the subject's skin. This problem is difficult to resolve because it originates from the structure of the human body. To resolve this problem, a novel electrode that can fill the gap between the smart shirt and body is required.

In this study, we developed a smart shirt that resolves the two problems above. The smart shirt has electrodes and wires made of conductive ink, and had 10 small electrodes and unipolar precordial leads for measuring the ECG signal. Six electrodes functioning as active electrodes were placed on the chest of the smart shirt. The remaining four electrodes were used as indifferent electrodes and a neutral electrode, and were placed at the scapulae and lumbar. Furthermore, we developed a conductive foam block that fills the gap between the smart shirt and body. The electrical resistance of the block was low enough for the ECG measurement and was reduced by compression pressure. We investigated whether the conductive foam block could fill the gap and be used for the ECG measurement. Finally, we made an ECG measurement while the subject not only rested but also walked.

2. EXPERIMENTAL MATERIALS AND METHODS

2.1. Manufacturing of the Smart Shirt

We used a compression shirt (UA Heat Gear Armour, manufactured by DOME Corporation, Tokyo, Japan) as a substrate of the smart shirt. The compression shirt was very stretchable and well fitted a subject's body. Electrodes and wires for the ECG measurement were conductive ink printed onto the compression shirt. Placement of the electrodes is shown in Figure 1. This placement is referred to as the placement of unipolar precordial leads

[17]. The conductive ink contained 35 wt% urethane elastomer (AR binder GS, manufactured by Matsui Shikiso Chemical Co., Ltd., Kyoto, Japan) and 65 wt% Ag flakes (AgC-A, manufactured by Fukuda Metal Foil & Powder Co., Ltd., Kyoto, Japan). The conductivity of this ink was 6.1×10^{-5} Ωcm when the ink was pasted on a glass slide and cured at 70 °C for 20 min. The conductive ink wire was stretchable and had elongation durability [18]. In a previous study [18], the conductive ink (80 wt% Ag flakes contained) was printed on polyester and polyurethane elastic webbing; the dimensions of the printed ink were 2 mm × 45 mm × 30 μm. The conductive ink was repeatedly stretched at 10% strain for 4500 cycles. The resistance of the conductive ink after stretching was approximately 1 kΩ. This elongation durability was expected for our smart shirt.

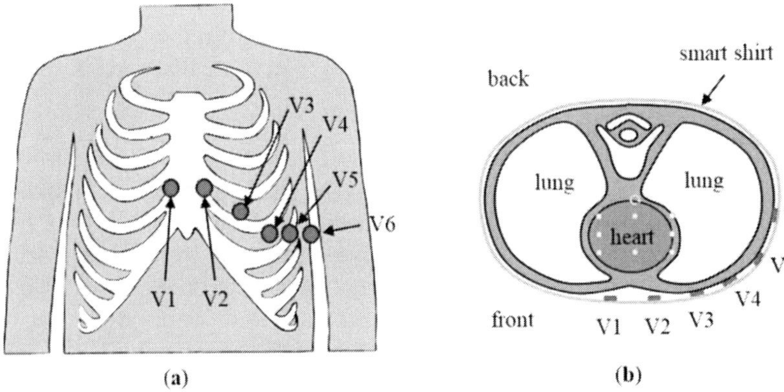

Figure 1. Placement of electrodes with unipolar precordial leads used to measure the ECG: (**a**) front view; (**b**) cross-sectional view of the chest.

The conductive ink electrode and wire were manufactured by procedures shown in Figure 2. In step (1), non-conductive ink made of urethane elastomer (AR binder GS, the same as that mentioned above) was printed on the compression shirt to create a bottom insulation layer. The ink was cured at 70 °C for 20 min. Curing was repeated after each subsequent step. In step (2), conductive ink was printed on the insulation layer to create an electrode and wire. In step (3), non-conductive ink was printed on the conductive ink, but the electrode and terminal were left unprinted. These procedures produced three layers of conductive ink wire that had a water resistant property. In step (4), a snap button was mounted on each terminal. A portable electrocardiograph was connected to the smart shirt through the snap buttons. To reinforce the fabric of the smart shirt, conductive fabric tape (E05R, manufactured by Seiwa Electric Mfg. Co., Ltd., Kyoto, Japan), which was not stretchable, was pasted on the terminal.

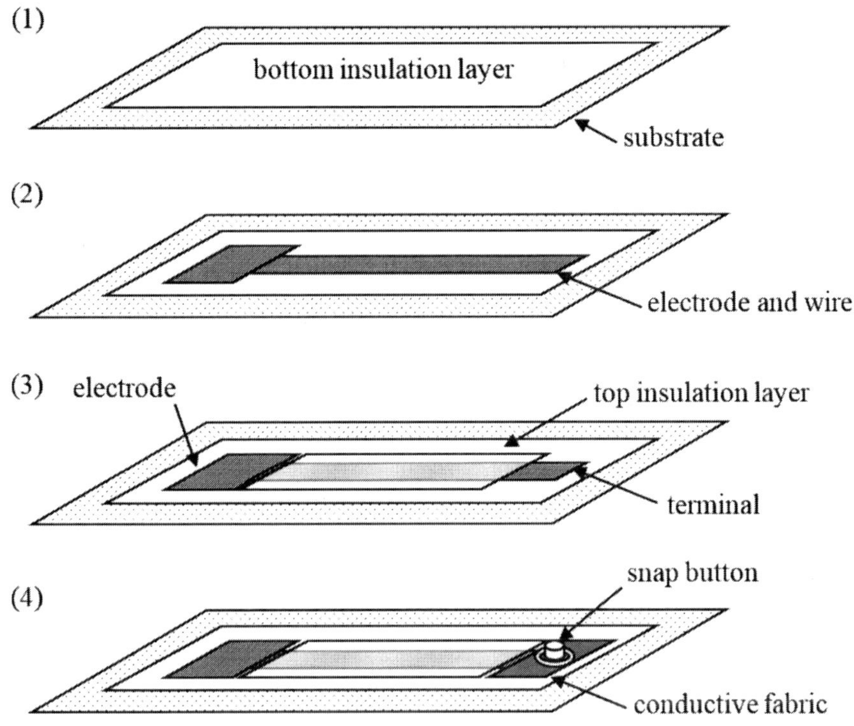

Figure 2. Procedure of the manufacture of the conductive ink electrode and wire.

2.2. Manufacturing of a Conductive Foam Block

Our smart shirt well fitted the subject's body. However, electrodes V1 and V2 were not in stable contact with the skin because of the body structure shown in Figure 1. To improve the contact condition, conductive foam blocks were inserted between the smart shirt and body. We manufactured the conductive foam block as follows.

The substrate of the conductive foam block was made of low-rebound urethane foam (EGR-6H, manufactured by Softpren Industry Corporation, Saitama, Japan). We added an electrical conduction property to the urethane foam by constructing nanostructured Ag in a sonoprocess [19]. We used reaction liquid that contained 1.0 wt% Ag_2O, 98.9 wt% ethanol, and 0.1 wt% ethylene glycol. The urethane foam block with dimensions of 20 mm × 20 mm × 15 mm was immersed in the reaction liquid. The urethane foam block with the reaction liquid was irradiated using an ultrasonic washing machine (nominal frequency of 38 kHz) for 21 h. Following this procedure, nanostructured Ag was constructed on the urethane foam block, and the conductivity was expressed.

2.3. Measuring the Contact Pressure at Each Electrode

The contact pressure of the smart shirt was different at each electrode. We investigated the contact pressure of the smart shirt to confirm the contact condition of the electrode. The contact pressure was measured by a pressure sensor (PDA-500KPA, manufactured by Tokyo Sokki Kenkyujo Co., Ltd., Tokyo, Japan). The pressure sensor was used only before measuring the ECG and was fixed at the placement of each electrode on the subject. Moreover, the increase in pressure due to the insertion of the pressure sensor could be ignored because of the small size of the sensor; the diameter and thickness were 6 and 2 mm, respectively.

2.4. Measuring the ECG

The subject wore the manufactured smart shirt. A portable electrocardiograph (BAQT-0001, manufactured by Bio Signal Co., Ltd., Osaka, Japan) was connected to the smart shirt. The portable electrocardiograph had four differential input channels for which the input impedance was 10 MΩ, and measured the ECG signals at 200 Hz. Measured ECG signals were transferred to a personal computer by a wireless connection using an XBee module (XB24-AWI-001, manufactured by Digi International Inc., MN, USA) in real time. In this experiment, the portable electrocardiograph was unable to measure the complete ECG signal with unipolar precordial leads owing to its simple design; i.e., the electrocardiograph had only four input channels. In this study, we measured signals from four of the six electrodes that are used in unipolar precordial lead measurements. Moreover, the skin moisture of the subject was measured by a moisture sensor (MY-808S, manufactured by Scalar Corporation, Tokyo, Japan). In the experiments, we measured the ECG signals while the subject rested, walked, and slowly jogged. Furthermore, we confirmed the usability of the conductive foam block.

3. RESULTS AND DISCUSSION

3.1. Manufactured Smart Shirt

A manufactured smart shirt is shown in Figure 3. The smart shirt in the figure is presented inside out to show the electrodes. Figure 3a shows the front of the smart shirt. The dimensions of electrodes V1 and V2 were 20 mm × 20 mm, while the dimensions of the other electrodes were 20 mm × 30 mm. Positions of electrodes V1 to V6 corresponded to active electrodes with unipolar precordial leads. In the standard measurement using unipolar precordial leads, indifferent and neutral electrodes are placed on both wrists and both ankles. In this study, however, these electrodes were printed on the back of the smart shirt shown in Figure 3b. The indifferent electrodes R, L, and F were printed on the scapulae and left lumbar part on the smart shirt. These electrodes were connected to one wire because the wiring was defined by the unipolar precordial lead configuration. The neutral electrode N was printed on the right lumbar part of the smart shirt. The indifferent and neutral electrodes were placed on the convex area of the scapulae and lumbar. These electrodes were therefore in stable contact with the subject's body. The longest length of wire was 480 mm for electrode R. The electrical resistance of this wire was 7.0 Ω before the smart shirt was worn. The resistance increased to 17.1 Ω when the subject wore the smart shirt, but the resistance was low enough for measurement of the ECG signal.

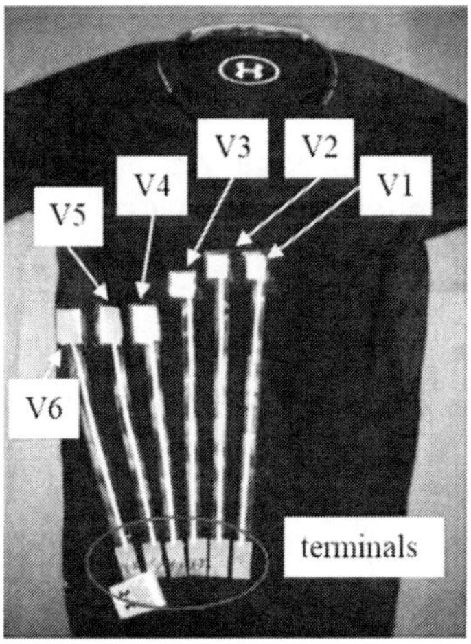

(a)

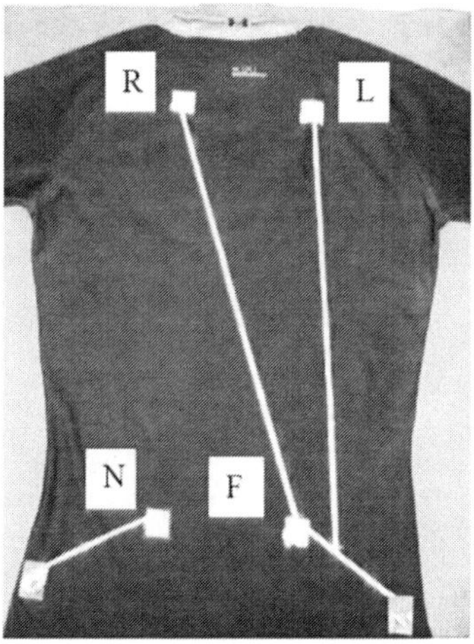

(b)

Figure 3. Manufactured smart shirt. The smart shirt is inside out in the photographs to show electrodes and wires: (**a**) front view; (**b**) back view.

3.2. Manufactured Conductive Foam Block

A block of low-rebound urethane foam in reaction liquid was irradiated with ultrasound for 21 h. After irradiation, the urethane foam block was washed with ethanol and dried at 60 °C for 60 min. Figure 4 shows the urethane foam block before and after irradiation. The irradiated urethane foam block changed color to dark gray because of the nanostructured Ag and became electrically conductive. A preliminary study of the nanostructured Ag showed that the nanostructured Ag was strongly attached to the urethane foam and did not peel off easily.

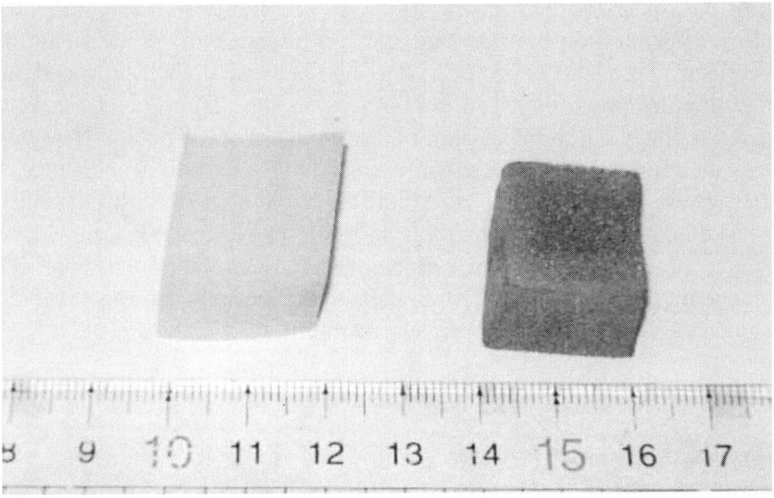

Figure 4. (Left) Low-rebound urethane foam block before irradiation with ultrasound; **(Right)** Conductive foam block produced by irradiating the urethane foam block with ultrasound in the reaction liquid.

The initial electrical resistance of the conductive foam block was different for each pair of opposing faces and ranged from 22.2 to 460.0 Ω. In the measurement of electrical resistance, the block was not compressed by probes of the resistance meter. However, the block received pressure from the smart shirt and the subject's body when the block was attached to the smart shirt. We investigated the resistance of the block when the block was pressed. Figure 5 shows relationships between the compression pressure and resistance. The resistance was measured for pairs 1, 2, and 3 of opposing faces. The figure indicates that the resistance was different for each pair of faces but was reduced by pressure in each case. The difference in resistance for each pair of opposing faces was due to the unbalanced ultrasound irradiation; the ultrasound was transmitted in one direction, and the block in the reaction liquid was not rotated. However, the resistance at pressure of 2.0 kPa was less than 30 Ω even though the initial resistance was 460.0 Ω (pair 1); this was sufficient for the measurement of the ECG signal. Moreover, we measured the resistance of the other blocks manufactured by employing the same method. The resistance and characteristics of those blocks were similar to those mentioned above.

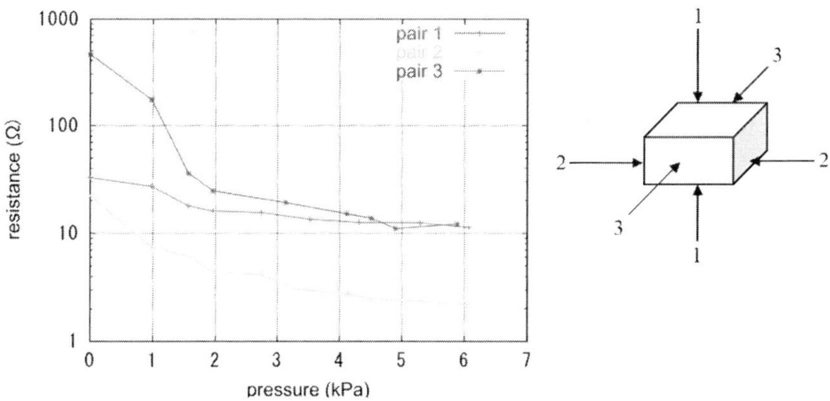

Figure 5. Relationships between compression pressure and electrical resistance. The schematic on the right labels pairs of opposing faces in the measurement.

We here discuss the reason why the resistance was reduced by pressure. We assumed that the decrease in resistance was due to the change in contact area, and investigated the contact area before and after compressing the conductive foam block. The block before compression was filled with resin and polished. A surface of the block filled with resin was observed under a microscope. Figure 6 shows the surface of the block. The figure is a composite of 16 images because the field of view of the microscope was narrow. The spiky long and thin features were fibers of conductive urethane foam. The broken-line rectangle (2.0 mm × 1.5 mm) was the observation area in this experiment. Moreover, we converted the image to the monochrome image shown in Figure 7a to improve visibility. The dimensions of each figure in Figure 7 are 1.5 mm × 2.0 mm; the images are rotated for ease of comparison.

Black areas in the figure indicate the conductive foam; i.e., the contact area. The combined area of contact was 0.16 mm^2. The pressed conductive foam blocks were observed employing the same method. Figure 7b,c show surfaces of the blocks pressed at 2.0 and 3.0 kPa, respectively; summations of the contact area were 0.35 and 0.75 mm², respectively. It is thus seen that the contact area of the block was increased by compression pressure, and the resistance of the conductive foam block was decreased by increasing the contact area of the block.

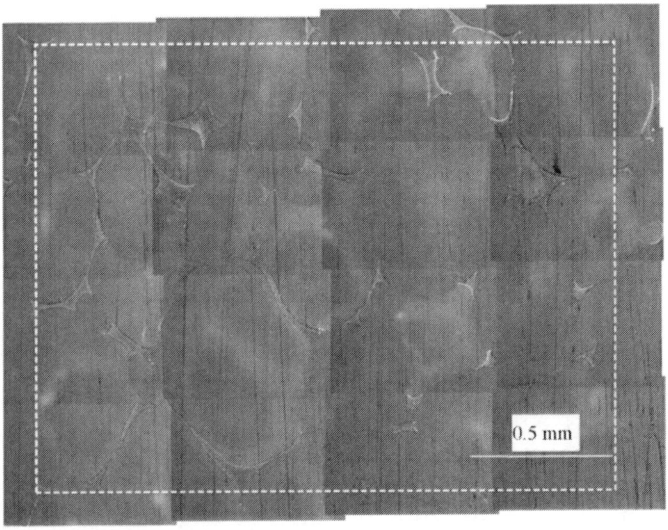

Figure 6. Surface of the conductive foam block. The compression pressure was 0.0 kPa. This figure is the combination of 16 images. Spiky long and thin features are fibers of the conductive foam. The broken-line rectangle shows the observation area (2.0 mm × 1.5 mm).

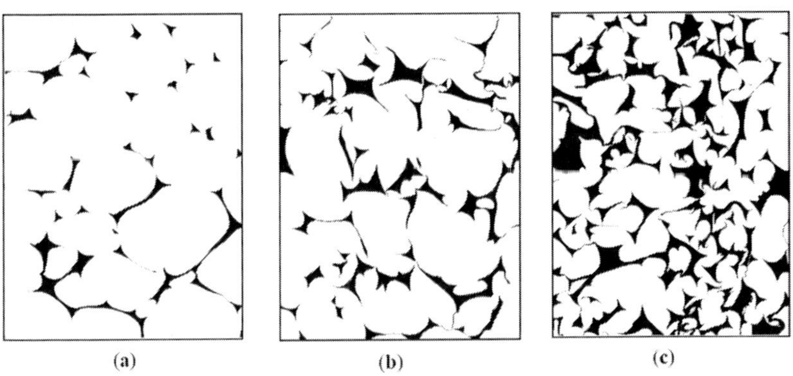

Figure 7. Comparison of contact areas of the conductive foam. Black areas indicate the surface of the conductive foam: (**a**) no compression (0.0 kPa); (**b**) compressed at 2.0 kPa; (**c**) compressed at 3.0 kPa.

3.3. Measured Contact Pressure at Each Electrode

We measured the contact pressure at each electrode when the subject wore the smart shirt. At first, the contact pressure was measured without the conductive foam blocks inserted between the shirt and body. We then measured the contact pressure when the conductive foam blocks were attached to the smart shirt (Figure 8).

Measured pressures at each electrode are given in Table 1. The upper row of the table gives the contact pressure when the blocks were not attached. The pressures at electrodes V1 and V2 have values of 0.0 kPa because these electrodes did not touch the body. The other pressures were approximately 2 to 3 kPa. In contrast, pressures were approximately 3 to 5 kPa when the blocks were attached to the shirt; i.e., all electrodes were in contact with the body because of the conductive foam blocks.

Table 1. Compression pressures at electrodes V1 to V6.

Electrode	V1	V2	V3	V4	V5	V6
without conductive foam blocks (kPa)	0.0	0.0	2.7	2.4	1.8	2.0
with conductive foam blocks (kPa)	3.7	3.5	4.5	4.9	3.5	3.3

Figure 8. Conductive foam blocks on the smart shirt.

3.4. Measured ECG Signals

We measured the subject's ECG signals using the smart shirt (Figure 9). In experiments, we measured the signals from four electrodes—V1, V2, V4, and V5—because the amplifier had only four input channels. This selection was decided by the contact conditions of the electrodes. The contact conditions (noncontact or contact) of electrodes V1 and V2 changed drastically if the conductive foam blocks were used. In contrast, electrodes V4 and V5 were in contact with the body irrespective of whether the blocks were used.

At first, the conductive foam blocks were not attached to the smart shirt. The subject was slightly sweaty; skin moisture was approximately 40%. The ECG measurement when the subject rested in a chair is shown in Figure 10a. There was a humming noise in the V2 signal. In contrast, the V4 and V5 signals were stable. V1 and V2 signals appeared like the ECG signals even though the contact pressure was 0.0 kPa as described in section 3.3. This was because the electrodes were in slight contact with the body. Figure 10b shows the ECG signals recorded while the subject walked. V1 and V2 signals were not discriminable as ECG signals because of swing noise and irregular peaks. In contrast, V4 and V5 signals were stable and discriminable as ECG signals. Moreover, we measured the ECG signals shown in Figure 10c while the subject jogged slowly. V1 and V2 signals had much swing noise and could not be distinguished as ECG signals. There were slight swings in V4 and V5 signals, but peaks of the ECG signal were detectable.

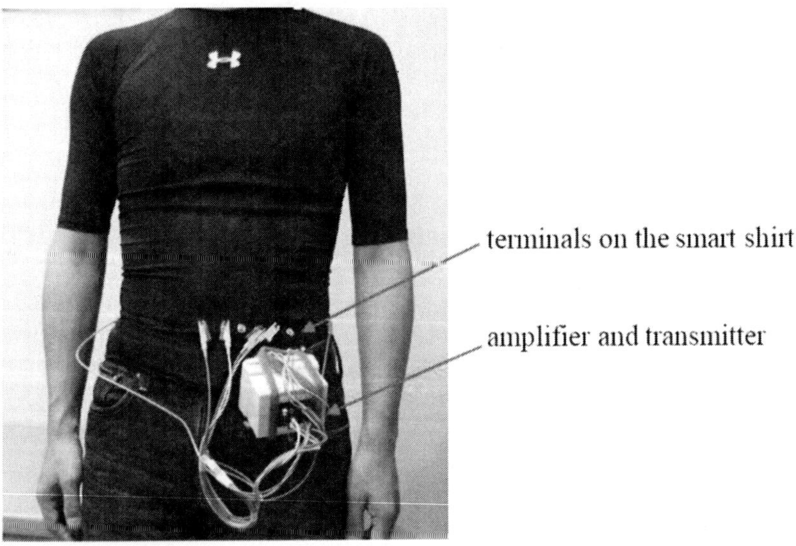

Figure 9. Subject wearing our smart shirt.

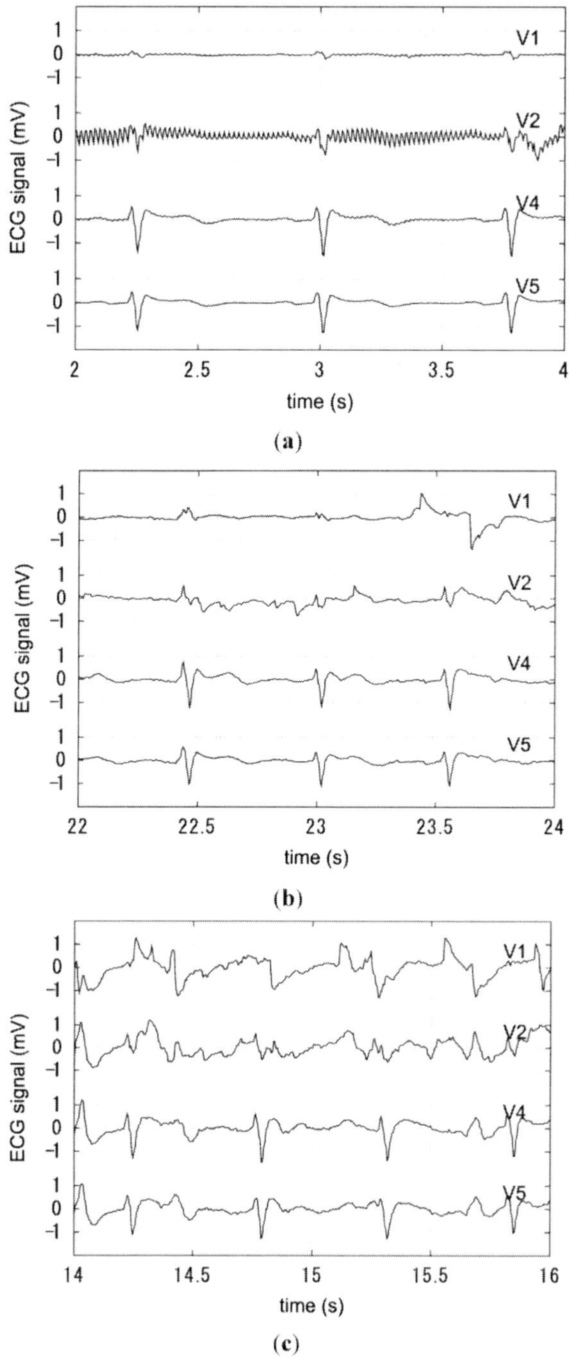

Figure 10. ECG signals measured by the smart shirt without the conductive foam blocks: (**a**) the subject resting in a chair; (**b**) the subject walking; (**c**) the subject slowly jogging.

In the next experiments, conductive foam blocks were attached at electrodes V1, V2, V4 and V5 on the smart shirt. We measured the ECG signal while the subject rested, walked, and slowly jogged. Figure 11a shows the ECG signals recorded while the subject rested. The ECG signals were very stable and easy to discriminate. The ECG signals recorded while the subject walked are shown in Figure 11b. The ECG signal was easy to discriminate, but there was small swing noise. Finally, Figure 11c shows the ECG signals recorded while the subject jogged slowly. V1 and V2 signals had much swing noise and could not be distinguished as ECG signals. V4 and V5 signals had small swing noise and distinguishable peaks of ECG signals.

The above experimental results are summarized as follows. The ECG signals from electrodes V1 and V2 that were difficult to measure stably without the conductive foam blocks could be measured with the conductive foam blocks. The ECG signals from all electrodes were measured stably even though the subject was walking. We thus considered that our smart shirt could measure ECG signals during light activities of daily life. However, measuring the ECG was difficult when the subject was exercising.

Figure 12 shows the ECG signals recorded when the subject jogged for 69 s and then rested in a standing position. The figure indicates that the measured ECG signals became stable immediately when the subject stopped exercising. Therefore, the smart shirt could measure the ECG signals intermittently when the subject was exercising.

The stability of signals from electrodes V1 and V2 has to be improved. The contact condition of these electrodes was greatly affected by the vibration of jogging. We considered that the change in the contact condition was a cause of the unstable signal. If the contact pressure increases, the electrodes are pushed close against the body and the ECG signal can be measured stably.

4. CONCLUSIONS

We developed a smart shirt that had 10 electrodes made of conductive ink and could measure the ECG with a unipolar precordial lead configuration. The smart shirt well fitted the subject's body. However, electrodes V1 and V2 were not in stable contact with the body because they were placed on a concave area of the chest. To fill the gap between the smart shirt and body, we developed conductive foam blocks. The substrate of each block was made of the urethane foam and was irradiated by ultrasound within a reaction liquid. The irradiated urethane foam block became conductive because of nanostructured Ag. Moreover, the resistance of the conductive foam block was reduced by compression of the block. This characteristic is suitable for the smart shirt because the block attached to the shirt is compressed when the shirt is worn.

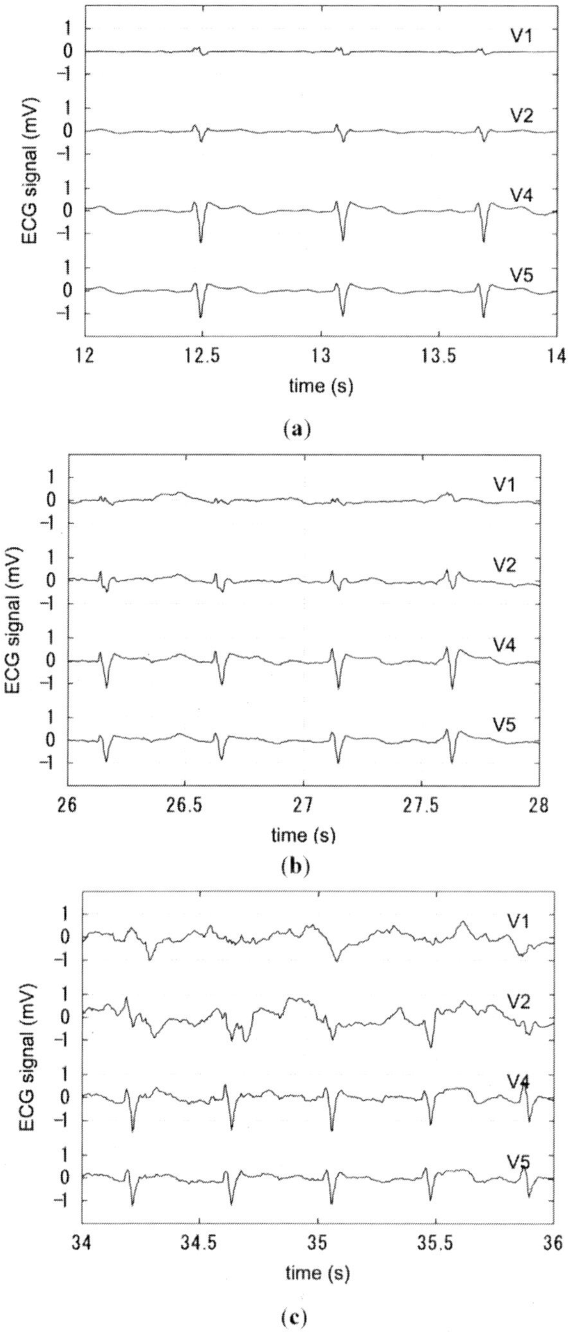

Figure 11. ECG signals measured by the smart shirt with the conductive foam blocks: (**a**) the subject resting in a chair; (**b**) the subject walking; (**c**) the subject slowly jogging.

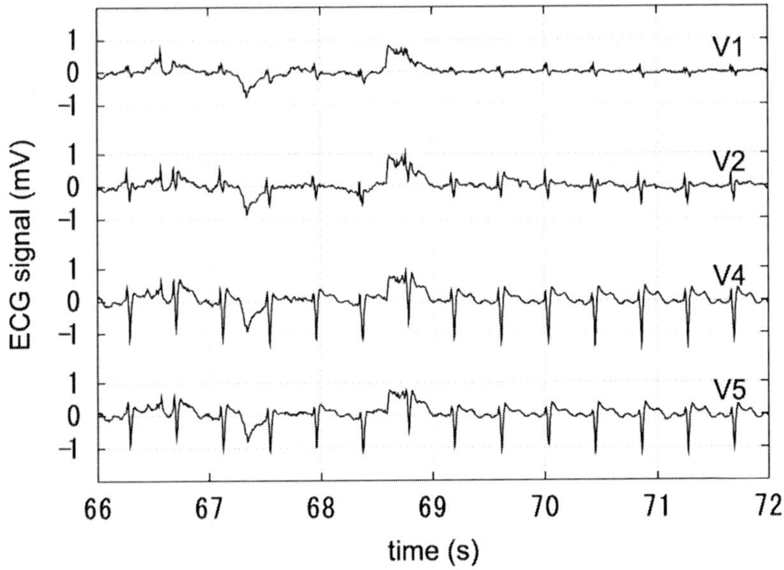

Figure 12. ECG signals measured by the smart shirt with the conductive foam blocks. The subject slowly jogged for 69 s and then rested in a standing position. The ECG signals became stable immediately when the subject stopped exercising.

We measured ECG signals using the smart shirt. The ECG signals from electrodes V1 and V2 were unstable when the conductive foam blocks were not attached to the smart shirt. In contrast, ECG signals from all electrodes could be measured if the conductive foam blocks were attached at each electrode. The ECG signals were also measured stably while the subject walked. However, the ECG signals from electrodes V1 and V2 contained large swing noise when the subject jogged slowly. In this case, the ECG signals from electrodes V4 and V5 also contained swing noise but we could detect peaks of the ECG signal.

We considered that signals from electrodes V1 and V2 were noisy because of the change in the contact conditions. Large vibrations due to exercise led to the large swing noise. Solutions to this problem might include the use of a petrosal conductive foam block that increases the contact pressure and/or the use of softer conductive foam that allows close contact. We will investigate these methods in future work.

The smart shirt should have washing durability and the electrodes should have long-term stability. These properties are being investigated by our group. Furthermore, for different subjects, we will investigate the comfort of the smart shirt and the conductive foam, and we will investigate the accuracy of the ECG measurement.

ACKNOWLEDGMENTS

A portable electrocardiograph in this study was provided by Bio Signal Co., Ltd., Osaka, Japan.

AUTHOR CONTRIBUTIONS

Yasunori Tada, Yusaku Amano, Tomonobu Sato, Shigeru Saito, and Masahiro Inoue designed the experiments; Yasunori Tada and Yusaku Amano performed the experiments; Yasunori Tada wrote the paper.

CONFLICTS OF INTEREST

The authors declare no conflict of interest.

REFERENCES

1. Paradiso, R.; Rossi, D.D. Advances in textile technologies for unobtrusive monitoring of vital parameters and movements. In Proceedings of the 28th IEEE EMBS Annual International Conference, New York City, NY, USA, 30 August–3 September 2006; pp. 392–395.

2. Karlsson, J.S.; Wiklund, U.; Berglin, L.; Östlund, N.; Karlsson, M.; Bäcklund, T.; Lindecranz, K.; Sandsjö, L. Wireless Monitoring of Heart Rate and Electromyographic Signals Using a Smart T-shirt. In Proceedings of the 5th International Workshop on Wearable Micro, and Nano Technologies for Personalised Health, pHealth, Valencia, Spain, 21–23 May 2008.

3. Vojtech, L.; Bortel, R.; Neruda, M.; Kozak, M. Wearable Textile Electrodes for ECG Measurement. Adv. Elect. Electr. Eng. **2013**, 11, 410–414.

4. Tada, Y.; Inoue, M.; Tokumaru, T. An evaluation of the characteristics of a stretchable ink wire suitable for the measurement of biological signals. J. Text. Inst. **2014**, 105, 692–700.

5. Tsukada, S.; Kasai, N.; Kawano, R.; Takagahara, K.; Fujii, K.; Sumitomo, K. Electrocardiogram Monitoring Simply by Wearing a Shirt—For Medical, Healthcare, Sports, and Entertainment. NTT Tech. Rev. **2014**, 12, 1–7.

6. Lage, J.; Catarino, A.; Carvalho, H.; Rocha, A. Smart Shirt with Embedded Vital Sign and Moisture Sensing. In Proceedings of The First International Conference on Smart Portable, Wearable, Implantable and Disability-oriented Devices, Brussels, Belgium, 21–26 June 2015; pp. 25–30.

7. Carvalho, H.; Catarino, A.P.; Rocha, A.; Postolache, O. Health monitoring using textile sensors and electrodes: An overview and integration of technologies. In Proceedings of the 2014 IEEE International Symposium on Medical Measurements and Applications (MeMeA), Lisbon, Portugal, 11–12 June 2014; pp. 1–6.

8. Morrison, T.; Silver, J.; Otis, B. A Single-chip Encrypted Wireless 12-Lead ECG Smart Shirt for Continuous Health Monitoring. In Proceedings of the 2014 Symposium on VLSI Circuits Digest of Technical Papers, Honolulu, HI, USA, 10–13 June 2014; pp. 1–2.

9. Reuters: ECG Shirt Warns Wearer of Heart Problems via Smartphone. Available online: http://www.personal-healthwatch.com/reuters-ECG-shirt.aspx (accessed on 25 August 2015).

10. Marozas, V.; Petrenas, A.; Daukantas, S.; Lukosevicius, A. A comparison of conductive textile-based and silver/silver chloride gel electrodes in exercise electrocardiogram recordings. J. Electrocardiol. **2011**, 44, 189–194.

11. Paul, G.; Torah, R.; Beedy, S.; Tudor, J. The development of screen printed conductive networks on textilesfor biopotential monitoring applications. Sens. Actuator A **2014**, 206, 35–41.

12. Seoane, F.; Mohino-Herranz, I.; Ferreira, J.; Alvarez, L.; Buendia, R.; Ayllón, D.; Llerena, C.; Gil-Pita, R. Wearable Biomedical Measurement Systems for Assessment of Mental Stress of Combatants in Real Time. Sensors **2014**, 14, 7120–7141.

13. Lobodzinski, S.S.; Laks, M.M. Comfortable textile-based electrocardiogram systems for very long-term monitoring. Cardiol. J. **2008**, 15, 477–480.

14. Sardini, E.; Serpelloni, M. Instrumented Wearable Belt for Wireless Health Monitoring. In Proceedings of the Eurosensors XXIV, Linz, Austria, 5–8 September 2010; pp. 580–583.

15. Sardini, E.; Serpelloni, M.; Ometto, M. Multi-parameters wireless shirt for physiological monitoring. In Proceedings of the 2011 IEEE International Workshop on Medical Measurements and Applications, Bari, Italy, 30–31 May 2011; pp. 316–321.

16. Trindade, I.G.; Martins, F.; Miguel, R.; Silva, M.S. Design and Integration of Wearable Devices in Textiles. Sens. Transducers **2014**, 183, 42–47.

17. Wagner, G.S.; Strauss, D.G. Marriott's Practical Electrocardiography, 12th ed.; Lippincott Williams & Wilkins: Philadelphia, PA, USA, 2013; pp. 28–30.

18. Inoue, M.; Itabashi, Y.; Tada, Y. Development of Bimodal Electrically Conductive Pastes with Ag Micro- and Nano-fillers for Printing Stretchable E-textile Systems. In Proceedings of the 20th European Microelectronics and Packaging Conference & Exhibition, Friedrichshafen, Germany, 14–16 September 2015.

19. Inoue, M.; Hayashi, Y.; Takizawa, H.; Suganuma, K. Formation mechanism of nanostructured Ag films from Ag_2O particles using a sonoprocess. Colloid Polym. Sci. **2010**, 288, 1061–1069.

CHAPTER 8

Wearable Electronics and Smart Textiles: A Critical Review

Matteo Stoppa and Alessandro Chiolerio *

Center for Space Human Robotics, Istituto Italiano di Tecnologia, Corso Trento 21, 10129 Torino, Italy

ABSTRACT

Electronic Textiles (e-textiles) are fabrics that feature electronics and interconnections woven into them, presenting physical flexibility and typical size that cannot be achieved with other existing electronic manufacturing techniques. Components and interconnections are intrinsic to the fabric and thus are less visible and not susceptible of becoming tangled or snagged by surrounding objects. E-textiles can also more easily adapt to fast changes in the computational and sensing requirements of any specific application, this one representing a useful feature for power management and context awareness. The vision behind wearable computing foresees future electronic systems to be an integral part of our everyday outfits. Such electronic devices have to meet special requirements concerning wearability. Wearable systems will be characterized by their ability to automatically recognize the activity and the behavioral status of their own user as well as of the situation around her/him, and to use this information to adjust the systems' configuration and functionality. This review focuses on recent advances in the field of Smart Textiles and pays particular attention to the materials and their manufacturing process. Each technique shows advantages and disadvantages and our aim is to highlight a possible trade-off between flexibility, ergonomics, low power consumption, integration and eventually autonomy.

KEYWORDS

smart textiles; electronic textiles; wearable systems

1. INTRODUCTION

The term "Smart Textiles" refers to a broad field of studies and products that extend the functionality and usefulness of common fabrics. Smart Textiles are defined as textile products such as fibers and filaments, yarns together with woven, knitted or non-woven structures, which can interact with the environment/user. The convergence of textiles and electronics (e-textiles) can be relevant for the development of smart materials that are capable of accomplishing a wide spectrum of functions, found in rigid and non-flexible electronic products nowadays. Smart Textiles will serve as a means of increasing social welfare and they might lead to important savings on welfare budget. They integrate a high level of intelligence and can be divided into three subgroups:

- Passive smart textiles: only able to sense the environment/user, based on sensors;
- Active smart textiles: reactive sensing to stimuli from the environment, integrating an actuator function and a sensing device;
- Very smart textiles: able to sense, react and adapt their behavior to the given circumstances.

Sensors provide a nervous system to detect signals, thus in a passive smart material, the existence of sensors is essential. The actuators act upon the detected signal either autonomously or from a central control unit [1]; together with the sensors, they are the essential element for active smart materials. Fabric-based sensing has been a large field of research in the biomedical and safety communities [2]. The fabric sensors can be used for electrocardiogram (ECG) [3], electromyography (EMG) [4], and electroencephalography (EEG) [5,6] sensing; fabrics incorporating thermocouples can be used for sensing temperature [7]; luminescent elements integrated in fabrics could be used for biophotonic sensing [8]; shape-sensitive fabrics can sense movement, and can be combined with EMG sensing to derive muscle fitness [9]. Carbon electrodes integrated into fabrics can be used to detect specific environmental or biomedical features such as oxygen, salinity, moisture, or contaminants [10,11].

Active functionality could include power generation or storage [12], human interface elements [13], radio frequency (RF) functionality, or assistive technology [14]. All electronic devices require power, and this is a significant design challenge for Smart Fabrics. Power generation can be achieved through piezoelectric [15] elements that harvest energy from motion or photovoltaic elements [16]. Human interfaces to active systems can be roughly grouped into two categories: input devices and annunciation or display devices. Input devices can include capacitive patches that function as pushbuttons [17], or shape-sensitive fabrics [18] that can record motion or flexing, pressure, and stretching or compression. Annunciation and display devices may include fabric speakers [17], electroluminescent yarns [19], or yarns that are processed to contain arrays of organic light emitting diodes (OLEDs) [20]. Fabrics can also include elements that provide bio-feedback [21] or simply vibrate. Fabric-based antennas are a relatively simple application of Smart Fabrics. Simple fabric antennas are merely conductive yarns of specific lengths that can be stitched or woven into non-conducting fabrics [22].

A study about intelligent textiles is at his first stage reduced to a study on smart materials. In a second phase, it is to be considered in which way these smart materials can be processed into a textile material. These smart materials are incorporated into the textile structure by different technologies (Figure 1). Among those we may list embroidering [23], sewing, non-woven textile, knitting [24], weaving [25], making a spinning [26], braiding [27], coating/laminating [28], printing [29] and chemical treatments [30] that provide specific features such as controlled hydrophobic behavior.

Innumerable combinations of these source materials result into a whole range of textiles but sometimes the commercial output is represented by garments that contain conventional cables, miniaturized electronic components and special connectors. As humans prefer to wear comfortable textiles rather than hard, rigid boxes, first efforts have been made to use the textiles themselves for electronic functions [31].

Smart Textiles present a challenge in several fields such as the medical, sport, and artistic communities, the military and aerospace. The early European Commission's 6th and 7th framework programs provided significant research and development funding for personal health monitoring through smart wearable systems and for projects targeting the integration of sensors, energy sources, processing, and communication inside the clothing. The list below shows the projects funded by the European Commission's 6th and 7th Framework programs that have focused on smart fabrics and interactive textiles (Table 1) [32,33].

Figure 1. Different kinds of textile/fabric manufacturing and treatment. (**a**) Embroidery; (**b**) sewing; (**c**) weaving; (**d**) non-woven; (**e**) knitting; (**f**) spinning; (**g**) breading; (**h**) coating/laminating; (**i**) printing and (**j**) chemical treatment.

Particular attention in this review is devoted to describing the materials and methodologies to develop smart textiles. Each scientific approach will be followed by a review of the related work carried out by companies, universities or research institutes.

2. FABRICATION TECHNIQUES

Over the past decade, many techniques and materials have been used in order to realize smart textiles. In the following section, together with metodologies, the relative projects are also presented.

2.1. Conductive Fibers

Initially, conductive threads were mainly used in technical areas: clean room garments, military apparel, medical application and electronics manufacturing [34]. Textile structures that exhibit conductivity or serve an electronic or computational function are called electro-textiles [35]. They can have a variety of functions, like antistatic applications [36], electromagnetic interference shielding (EMI) [37], electronic applications, infrared absorption or protective clothing in explosive areas [38].

Table 1. Smart Textiles project—EU FP6 and FP7.

Project Title	Description
WEALTHY Sept 2002–Feb 2005 http://www.wealthy-ist.com/	Pioneering research to develop and test comfortable smart fabrics for biological monitoring–ECG and respiration.
MyHeart Jan 2004–Oct 2007 http://www.hitech-projects.com/euprojects/myheart/	Development of Intelligent Biomedical Garments for monitoring, diagnosing and treatment of medical conditions
BIOTEX Oct 2005–Feb 2008 http://www.biotex-eu.com/	Sought to develop biochemical-sensing techniques that could integrate into textiles. Project aimed to develop textile sensing patches to target bodily fluid sensing.
PROETEX Feb 2006–Jan 2010 http://www.proetex.org/	Developed smart wearable garments for emergency disaster intervention personnel to improve safety, coordination, and efficiency.
STELLA Feb 2006–Jan 2010 http://www.stella-project.de/	Sought to develop flexible and stretchable textile substrates with electrical interconnects.
OFSETH Mar 2003–Jun 2009 http://www.ofseth.org/	Focused on how silica and polymer optical fibers can be used for sensing vital parameters while being compatible with a textile manufacturing process.
CONTEXT Jan 2006–Jun 2008 http://www.context-project.org/	Focused on development of contactless sensors in textiles for monitoring ECG and EMG.
WearIT@Work Jun 2004–Nov 2008 http://www.wearitatwork.com/	Aimed to prove the applicability of computer systems integrated to clothes, creating wearable interfaces for various industrial environments.
MICROFLEX May 2008–May 2012 http://microflex.ecs.soton.ac.uk/	Development of flexible materials in the form of high added value smart fabrics/textiles which are able to sense stimuli and react or adapt to them in a predetermined way.
DEPHOTEX Nov 2008–Oct 2011 http://www.dephotex.com/	Research and development of flexible photovoltaic textiles based on novel fibers
PLACE-it Feb 2010–Jun 2013 http://www.place-it-project.eu/	Development of a technology platform for lightweight, thin and conformable opto-electronic systems interconnect technology

The conventional process to produce metal fibers is wire drawing, a mechanical production process. This process is characterized by its various drawing steps, called coarse, medium, fine and carding train (Figure 2).

The drawing die, used to draw the fiber, consists of a steel mount with a core out of ceramics, carbide or diamond. The initial diameter of the metal wire varies depending on the material. For copper, for instance, it is usually is 8 mm, while for iron it is 5 mm. After drawing, the wire is annealed at temperatures ranging between 600 and 900 °C. Subsequently, they are quenched. The fine metal wire is then wrapped onto a revolving wire drawing cylinder [39].

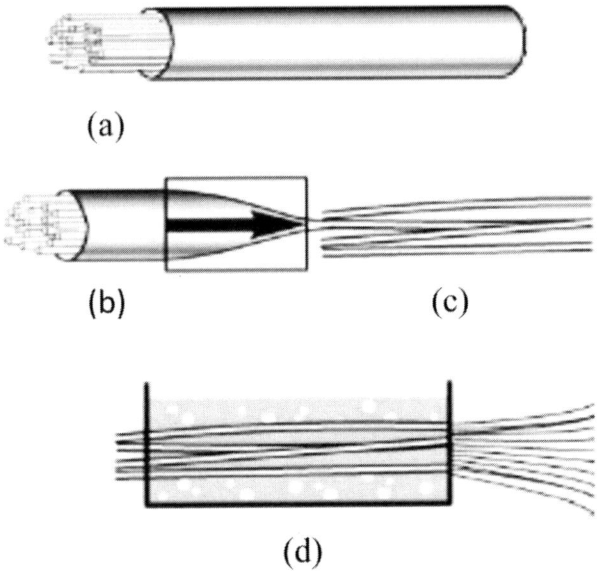

Figure 2. (**a**) Metal coated wire combined in iron tube; (**b**) Several diameter reductions of tube; (**c**) Bundling of tubes; (**d**) Leaching, realizing fibers.

Related Works

The Swiss company Elektrisola Feindraht AG (Escholzmatt, Switzerland) produces metal monofilaments that can be blended with all sorts of fibers or that can be directly used in weaving and knitting. Importantly, according to the material used, there are different electrical properties (see Table 2). The products range from copper (Cu) and silver-plated copper (Cu/Ag) filaments, brass (Ms) and silver-plated brass (Ms/Ag) filaments, aluminum (Al) filaments to copper-clad aluminum (CCA) filaments [40].

The company Swiss-Shield® (Flums, Switzerland) specializes in producing metal monofilaments which are incorporated into base yarns like cotton, polyester, polyamides and aramides. The metal monofilaments are made out of copper, brass, bronze, silver, gold, aluminum, for instance. The following Figure 3 shows a typical conductive yarn with base fibers and a metal monofilament twisted around them [41].

2.2. Treated Conductive Fibers

Instead of attaching electronics to textile substrates, the yarns of the textile can be functionalized with electronics. Electrically conductive fibers can also be produced by coating the fibers with metals, galvanic substances or metallic salts. Coatings can be applied to the surface of fibers, yarns, or even fabrics to create electrically conductive textiles. Common textile coating processes include electroless plating, evaporative deposition, sputtering, coating the textile with a conductive polymer [42].

Table 2. Electrical properties of metal monofilaments fibers.

Metal	Electrical Properties				
	Conductivity [S·m/mm²)	Resistivity [Ω·mm²/m]	Thermal Coefficient of Resistance [10^{-6} K^{-1}]		
			Min	Typ	Max
Cu	58.5	0.0171	3900	3930	4000
Cu/Ag	58.5	0.0171	3900	4100	4300
Ag 99%	62.5	0.0160	3800	3950	4100
Ms * 70	16.0	0.0625	1400	1500	1600
Ms/Ag	16.0	0.0625	1400	1500	1600
AgCu	57.5	0.0174	3800	3950	4100
Bronze	7.5	0.1333	600	650	700
Steel 304	1.4	0.7300		1020	
Steel 316L	1.3	0.7500		1020	

* German Milbe denomination, where "Ms" is accompanied by a number stating the composition in %Cu with respect to a Zn complement to 100%.

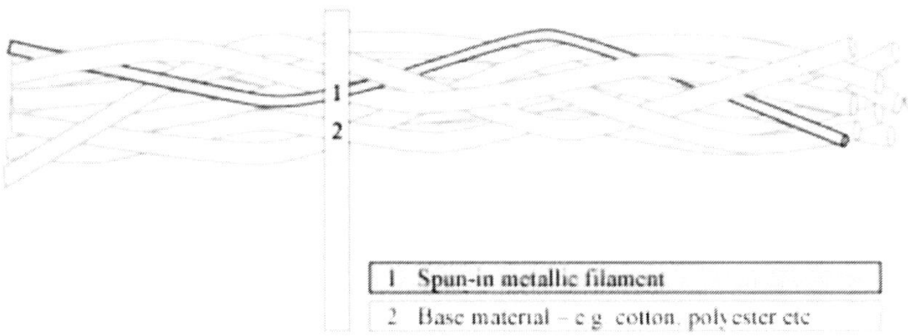

Figure 3. Schematic of conductive fiber twisted with the normal fibers.

In [43] a method to fabricate fibers with different material layers and material structuring is presented. The fabrication process is based on the conventional preform-based fiber-processing, easily yielding kilometers of functional fiber during the process.

Another relevant work is to use the crossing yarns in a textile to fabricate a transistor [44,45]. A schematic of a yarn-based transistor is shown in Figure

4. The resulting transistor shows an on-off current ratio of more than 1000, operated with a gate voltage of 1.5 V.

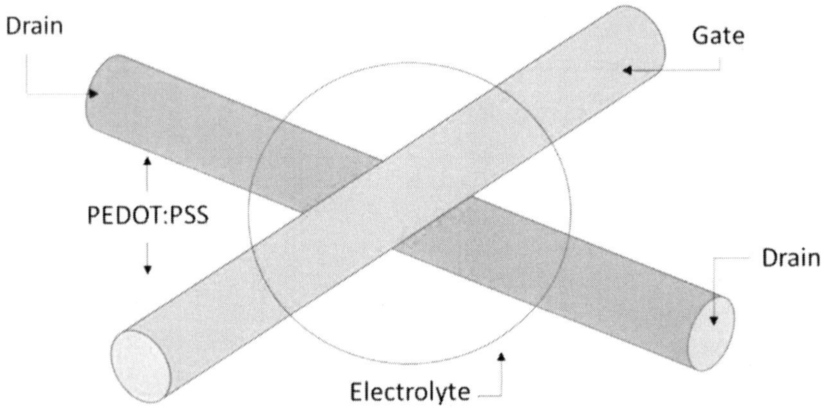

Figure 4. Yarn-based transistor.

Figure 3 represents two yarns coated with PEDOT:PSS, one serving as the gate contact for the transistor while the second serves as drain and source contact. At the crossing of the two yarns, an electrolyte is placed. A redox process at the interface between electrolyte and PEDOT:PSS turns the transistor on and off [44].

Related Works

The Textile Research Institute of Thuringia-Vogtland (TITV, Greiz, Germany) has succeeded in producing conductive threads by coating a conventional yarn with metal layers, called ELITEX®. They used Shieldex Nylon 66 threads that are coated with a thin silver layer as base material. With a specific conductivity of about 1.2×10^3 S·cm^{-1}, the threads have a specific resistance of about 8.34 Ω·mm^2/m. Hence, the resistivity is too high to conduct current [46].

2.3. Conductive Fabrics

There are different ways to produce electrically conductive fabrics. One method is to integrate conductive yarns in a textile structure, e.g., by weaving. However, the integration of conductive yarns in a structure is a complex and seldom a uniform process as it needs to be ensured that the electrically conductive fabric is comfortable to wear or soft in touch rather than hard and rigid. Conductivity can be established with different thread types (Figure 5):

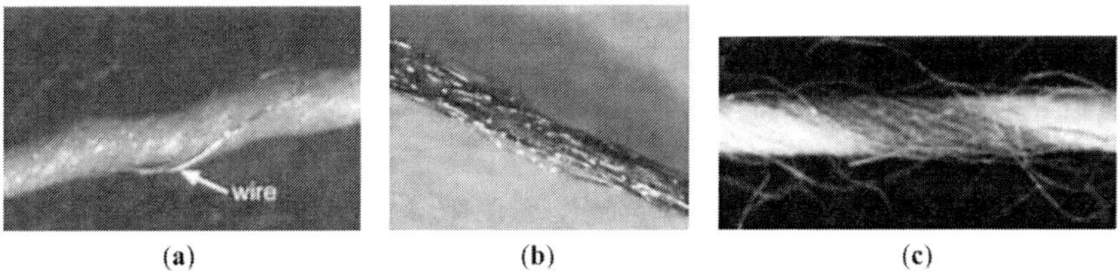

Figure 5. (**a**) Twisted metal wire: The metal wire is twisted around the polymer yarn; (**b**) Metal coating: The polymer yarn is physically/chemically coated with a thin metal layer; (**c**) Metal fibers: The conductive yarn consists of metal multifilaments [47].

However, woven fabric structures can provide a complex network that can be used as elaborated electrical circuits with numerous electrically conducting and non-conducting constituents, and be structured to have multiple layers and spaces to accommodate electronic devices.

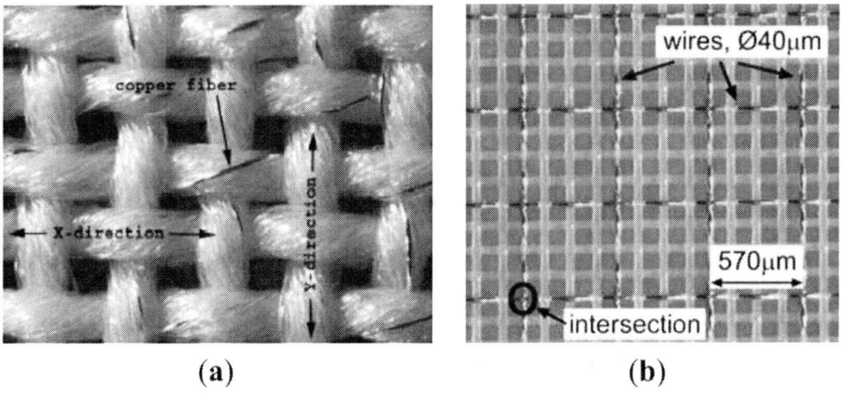

Figure 6. (**a**) Standard design of copper yarn twisted with polyester fibers; (**b**) PETEX.

Researchers at the Electronics Department and the Wearable Computing Laboratory at the ETH in Zürich produced a plain woven textile structure consisting of polyester yarns that are twisted with one copper thread. Initially, they started with a standard design (Figure 6a), then the researchers design a hybrid fabric called PETEX (Figure 6b) [47]. It consists of woven polyester monofilament yarn (PET) with diameter of 42 μm and copper alloy wires with diameter 50 ± 8 μm (AWG 461). Each copper wire itself is coated with a polyurethane varnish as electrical insulation. The copper wire grid in the textile features a spacing of 570 μm (mesh count in warp and in weft is 17.5 cm^{-1}).

With the PETEX the ETH researchers introduced a new approach to Smart Textiles and in particular a new manufacturing method. The aim was the possibility to realize a custom textile circuit (Figure 7). The wiring structure among circuit components is established by connecting the fabric embedded copper wires. Cuts must be placed at specific locations in the wiring in order to avoid short-circuits between copper wires. In particular, the procedure is as follows [47]:

- coating removal on copper wires at defined intersections with laser ablation;
- cutting the wires avoiding the signal leakage with laser;
- creating the interconnection with a drop of conductive adhesive;
- adding mechanical and electrical protection with an epoxy resin deposition.

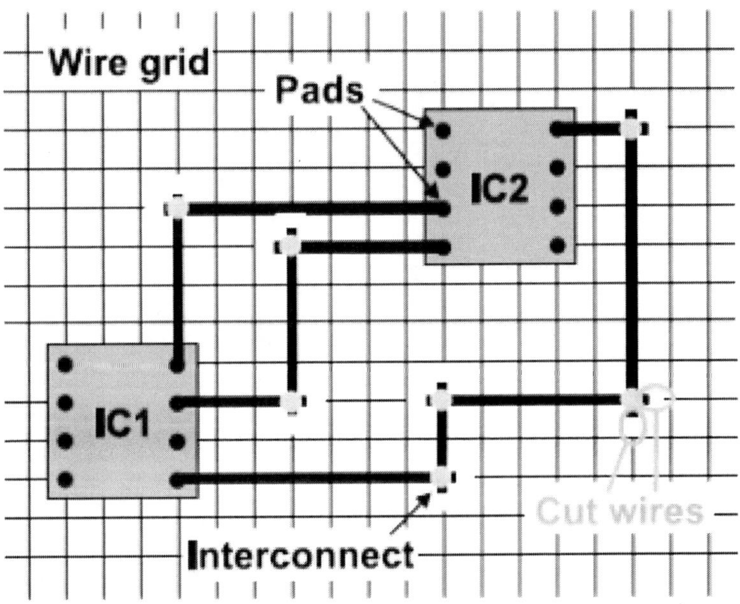

Figure 7. Approach to integrate circuits in a fabric with wire grid.

Related Works

The British company Baltex (Ilkeston, UK) uses the knitting technology to incorporate metal wires in textile structures. Their fabrics, which they commercialize under the name Feratec®, can be used mainly for two purposes, namely heatable textiles and electro-magnetic shielding materials [48].

The American company Thremshield LLC (Niagara Falls, NY, USA) produces metallized woven nylon fabrics in different shapes and profiles. The metals they use are silver, copper or a combination of copper and nickel [49].

The Danish company Chr. Dalsgaard Project Development (Aarhus, Denmark) works with the development of weaving electronics into fabrics, electronic conductors in clothing, operating panels in textiles (soft keyboards, displays, etc.) and micro-sensors. The conductive yarn they use is a copper thread, plated with a silver layer and coated with polyester [50].

Another possibility to achieve a conductive fabric is to attach a conductive structure to a ground structure by using the embroidery technique. In 2000, the Massachusetts Institute of Technology Media Laboratory researches were the first to propose a way of stitching patterns that can define circuit traces, component connection pads, or sensing surfaces designed with traditional CAD tools for circuit layout (Figure 8) [31].

(a) (b)

Figure 8. (**a**) Musical Jacket comprising a fabric keypad on one side, a MIDI synthesizer on the other side, speakers behind speaker grills in the pockets and fabric buses visible inside the jacket; (**b**) The fabric keypad with the circuit board placed behind it.

2.4. Conductive Inks

Interactive electronic textiles can also be produced by using conductive inks. First of all conductive inks must contain an appropriate highly conductive metal precursor such as Ag, Cu, and Au NPs and a carrier vehicle. Most of them are water based: water is the main ink component and to limit contaminants, it must be as pure as possible. These specialized inks can be printed onto various materials, among them textiles, to create electrically active patterns. Screen printing also makes integration with planar electronics simpler than with conductive yarn systems.

There are several technologies that can print conductive material on different substrate. Sheet-based inkjet and screen printing are best for low-volume, high-precision work.

Inkjets are flexible and versatile, and can be set up with relatively low effort [51]. Inkjets offer lower throughput of around 100 m^2/h and lower resolution (ca. 50 μm). It is well suited for low-viscosity, soluble materials like organic semiconductors. With high-viscosity materials, like organic dielectrics, and dispersed particles, like inorganic metal inks, difficulties due to nozzle clogging occur. Because ink is deposited via droplets, thickness and dispersion homogeneity is reduced. Simultaneously using many nozzles and pre-structuring the substrate allows improvements in productivity and resolution, respectively [52].

For inkjet printing, the inks should respect the following requirements [53]:

- high electrical conductivity;
- resistance to oxidation;
- dry out without clogging the nozzle during printing;
- good adhesion to the substrate;
- lower particle aggregation;
- suitable viscosity and surface tension.

Inks may also contain additives which are used to tune ink properties or to add specific properties thus increasing its performance [54].

After inkjet printing of a metal NP-based ink, in order to form a conductive printed pattern, particles must be sintered to create continuous connectivity between them and obtain electrical percolation. Sintering is the process of welding particles together at temperatures below the corresponding bulk metal melting point, involving surface diffusion phenomena rather than phase change between the solid and the liquid [55]. For instance, with inks based on gold NPs (1.5 nm diameter), the melting temperature was experimentally found to be as low as 380 °C, while for inks based on silver NPs (15 to 20 nm in diameter), a complete sintering can be obtained down to 180 °C [54,56].

Screen printing is appropriate for fabricating electrics and electronics due to its ability to produce patterned, thick layers from paste-like materials. The screen printing procedure, a stencil process, comprises the printing of a viscous paste through a patterned fabric screen and is usually followed by a drying process. The method can be applied to flat or cylindrical substrates. Depending on the substrate materials and the requirements for the printed structures, a high temperature densification can also be necessary (organic substrate $T < 150$ °C. Glass, ceramic and metal substrate $T > 500$ °C) [57].

This method can produce conducting lines from inorganic materials (e.g., for circuit boards and antennas), but also insulating and passivating layers. Generally, the throughput is about 50 m^2/h with a resolution lower than 100 μm. By optimizing of the process condition and material the resolution may decrease to 30 microns line/space on thin flexible substrates [58].

This versatile and comparatively simple method is used mainly for conductive and dielectric layers [59], but also organic semiconductors can be printed, e.g., for OPVCs and OFETs [60].

Recently the researchers of University of Southampton [61] have developed an innovative screen printed network of electrodes and associated conductive tracks on textiles for medical applications (Figure 9). A polyurethane paste is screen printed on to a woven textile to create a smooth, high surface energy interface layer and a silver paste is subsequently printed on top of this interface layer to provide a conductive track. Silver pastes have been printed on to non-woven textiles to create wearable health monitoring devices.

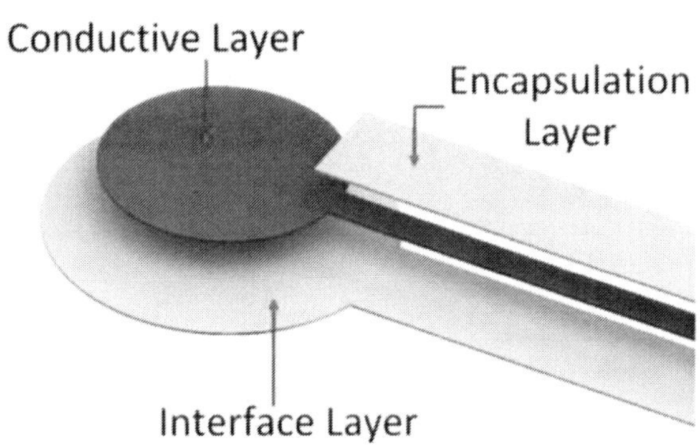

Figure 9. Screen printing fabrication for conductive tracks.

The researchers developed different bio-potential sensing systems with dry electrodes and conductive ink for signal traces in order to demonstrate that this technology could be used for biomedical application. In particular the application tests were: ECG, facial EMG (Figure 10) and forearm EMG.

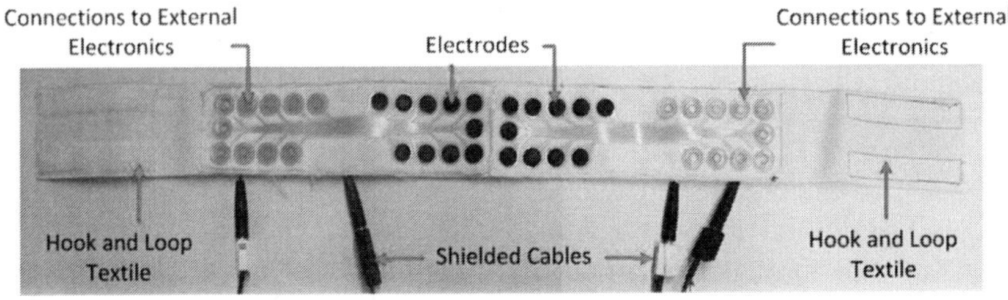

Figure 10. Textile headband for facial EMG.

Related Works

The National Textile Center of the North Carolina State University (Raleigh, NC, USA) is currently working on a project dealing with 'Printing Electric Circuits on Non-Woven Fabrics' used to produce a prototype for a physiological monitoring garment that measures ECG, heart-rate, respiration and temperature. In the scope of the project they work together with conductive ink manufacturers. For their experimental investigations and to succeed in producing samples of antennas printed on non-woven textile structures, they used Evolon® by Freudenberg KG (Weinheim, Germany), Tyvek® by DuPont™ (Wilmington, DE, USA) FiberWeb Resolution™ Print Media by BBA FiberWeb™ (Old Hickory, TN, USA) as well as conductive inks made by Precisia LLC (Clark, NJ, USA) and Creative Materials Inc. (Ayer, MA, USA) [62]. Table 3 shows the ink's sheet resistivity.

Table 3. Electrical properties of inks.

Inks	Sheet Resistivity (Ω/sq/mil)
CMI 112-15	0.010
DuPont 5025	0.010–0.012
DuPont 5096	<0.010

A group of researchers of Istituto Italiano di Tecnologia—Center for Space Human Robotics, Politecnico di Torino—Applied Science and Technology

Department, in collaboration with a spin-off company, Politronica Inkjet Printing S.r.l. (Torino, Italy), developed EMG sensor matrices by inkjet printing a silver nanoparticle-based ink on a polyimide flexible patch. Results indicate excellent behavior with respect to traditional bulk silver systems and a base conductivity of 35% with respect to bulk silver [63].

2.5. Conductive Materials as Sensors

Conductive textiles that change their electrical properties as a result of the environmental impact can be used as sensors. Typical examples are textiles that react to deformations such as pressure sensors, stretch sensors and breathing sensors. On the other hand, with smart textiles we have the further possibility to make bio-potential sensors.

2.5.1. Stretch Sensors

Stretch sensors are predominantly used for sensing and monitoring body parameters, as the textile is in contact with the skin over a large body area. This means that monitoring can take place at several locations on the body. For instance, these sensors can be used for determining: heart rate, respiration, movement and pressure blood [64]. A specific structure of textile sensors is that integrating fibers featuring piezo-resistive properties, enabling their use as strain or deformation sensor.

Figure 11. DataGlove™ VRLOGIC with flex sensors.

A first approach to integrate electronics into textile structures was certainly achieved by gloves wired to the computer that allows it to take input from user's hand gestures. Sensors in the glove detect (Figure 11) the wearer's hand movements. Four wires were used for each finger or tube to build up a circuit. The voltages coming out are varying depending on the finger position [65].

Related Works

A Flemish consortium of universities and companies, among them the textile department of Ghent University, developed a prototype suit called Intellitex. It is a biomedical suit meant for the long term monitoring of heart rate and respiration of children at the hospital. To measure the ECG, a three-electrode configuration is used. Two measurement electrodes are placed on a horizontal line on the thorax, a third one, acting as a reference (Right Drive Leg, RDL), is placed on the lower part of the abdomen [66].

Philips Research Laboratory (Redhill, UK), developed a stretch sensor integrated into a garment. The stretch sensor, which is produced out of conductive and elastic yarns knitted together, is based on the fact that the electrical resistance changes when stretching the sensing material. Thus, it can be used to control the volume of music or changing the track [67].

2.5.2. Pressure Sensors

Pressure sensors are commonly used either as switches and interfaces with electronic devices or also to monitor vital signs of the user.

Several technologies [68,69], have been developed to manufacture plane pressure sensors. The operating principle is that of changes in piezoelectric resonance frequency with the applied pressure or capacitance variations caused by an elastic foam overlaid with a matrix of conductive threads. For capacitive sensors, a change in parasitic capacitance and resistance can be compensated by the electronics, therefore the wiring has a marginal influence on the sensed signal [9].

The Wearable Computing Lab of ETH Zurich has developed a matrix with several capacitive pressure sensors for integration into a piece of clothing (Figure 12). With this method they are able to measure pressure on a human body and detect muscle activity on the upper arm. Applying this matrix on different body areas, it can provide more details for motion tracking or for the detection of physical state of the muscles [9].

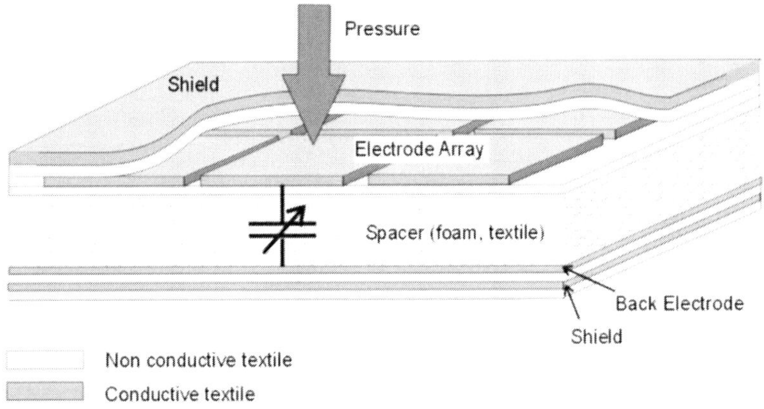

Figure 12. Scheme of sensor with an array of textile capacitors.

Related Works

The British company Eleksen Limited, formerly Electrotextiles (Tunstall, UK), commercializes a soft and flexible textile based sensory fabric under the tradename ElekTex® Smart Fabric Interfaces. It is a combination of conductive fibers and nylon. This combination results into a durable, reasonably priced, washable and even wearable 3D structure [70].

The American-based company Logitech Inc. (Le Lieu, Switzerland) manufactures a soft-touch KeyCaseTM keyboard that can wrap around a personal digital assistant (PDA) for storage and protection. The keyboard is lightweight and made out of textile [71].

The U.S. company Pressure Profile Systems, Inc. (Los Angeles, CA, USA) designs, develops and manufactures high performance multi-element pressure and tactile sensing systems, called Tactarray and ConTacts (Figure 13a) [72].

The team of the Design for Life Centre at Brunel University in Surrey (UK) has developed a fabric (the Sensory Fabric) that can be used by handicapped children to make themselves understood (Figure 13b). The Sensory Fabric consists of two layers of electrically conductive textile, divided by a layer of non-conductive mesh. When the textile is pressurized, one conductive layer comes into contact with the other, as a result of which an electric stream can flow [73].

The team of the Center for Micro-Bio Robotics of Istituto Italiano di Tecnologia (Pisa, Italy) produced a composite capacitive three-axial sensor based entirely on commercial conductive fabrics, demonstrating its high compliance and stability under manipulation [74].

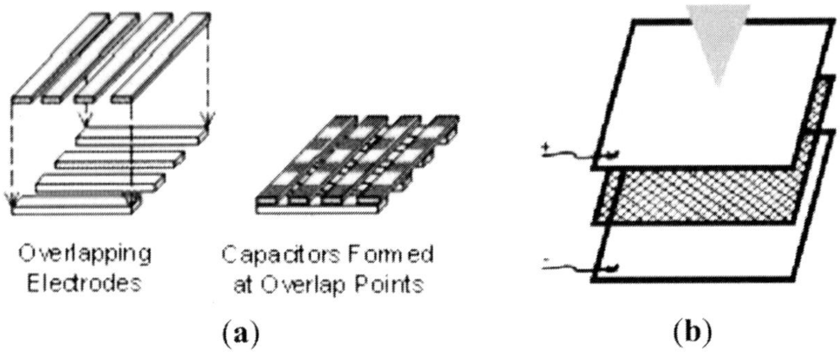

Figure 13. (**a**) Scheme of capacitor sensor developed by the U.S. Company Pressure Profile Systems, Inc. and (**b**) Design for Life Centre at Brunel University in Surrey.

2.5.3. Electrochemical Sensors

Recent insights into novel fabrication methodologies and electrochemical techniques have resulted in the demonstration of chemical sensors able to augment conventional physical measurements (i.e., heart rate, EEG, ECG, etc.) [75]. Recent insights have resulted in the development of a new generation of textile-based chemical sensors that are able to improve conventional physical sensors with more information. Flexible and textile-based screen printed electrochemical sensors may be candidates for non-invasive monitoring, but these devices cannot easily be attached to the body and in particular to the skin.

Related Works

The researchers of National Centre for Sensor Research, Dublin City University (Ireland), present examples of wearable chemical sensors that monitor the person and also their environment. In particular the chemical sensor was able to measure and analyze sweat in real-time on the body. They have developed a microchip version of the platform to measure changes in the pH of sweat (Figures 13a and 14a). The color change of the pH sensitive fabric was detected by placing a surface mount (SMT) LED and photodiode module on either side of the chip, aligned with the pH sensitive fabric. The final device (180 μm thick) is flexible and can adapt to the body [76].

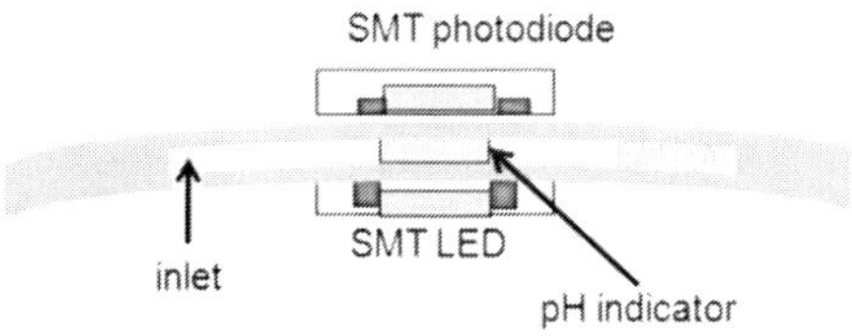

(a)

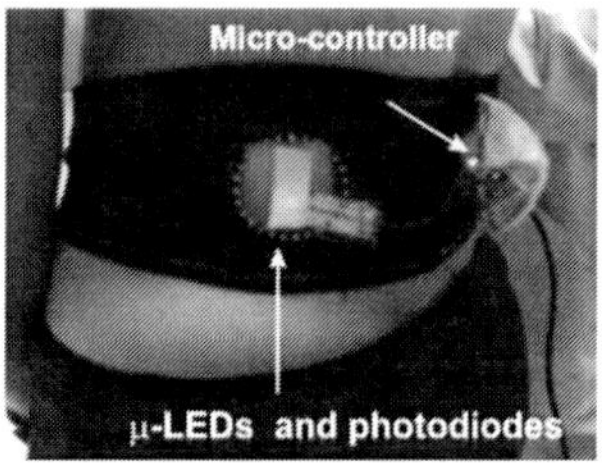

(b)

Figure 14. (**a**) Scheme of electrochemical sensor for pH analysis and (**b**) the system application.

2.5.4. Textile Energy Harvesting and Portable Power Supply System

Power supply technologies provide the electrical power for activating the components integrated in the electronic textile; this is still a critical issue in the field of wearable electronics. Although considerable progresses have been seen for wearable electronics, lithium rechargeable batteries, the power sources of the devices, have not kept pace with this progress due to their tenuous mechanical stability, causing them to remain as the limiting elements in the entire technology [77]. For these reasons, the aim is to develop wearable systems capable of accumulating energy dissipated by the body.

The supply of energy by the user's body during everyday actions through leg motions and body heat is also exploited by other research teams. For example, Infineon is currently trying, to recover energy by body movements to feed Mp3 players integrated in a jacket using piezoelectric materials [78]. In the UK the University of Bolton has developed a novel technology that integrates piezoelectric polymer substrate and photovoltaic coating system to create a film or fibre structure that is capable of harvesting energy from nature, including sun, rain, wind, wave and tide [79].

In a project funded by the Engineering and Physical Sciences Research Council (EPSRC), Researchers at the University of Southampton's School of Electronics and Computer Science (ECS) are developing an energy harvesting film in textiles using a rapid printing processes and active printed inks [80,81].

Georgia Tech researchers led by materials-science professor Zhong Lin Wang, have made a flexible fiber coated with zinc oxide nanowires that can convert mechanical energy into electricity. The researchers say the fibers should be able to harvest any kind of vibration or motion for electric current. Gold-plated zinc oxide nanowires, each about 3.5 micrometers tall, are grown on a flexible polymer fiber and these nanowires brush against untreated nanowires, which flex and generate current. Yarn spun from the fibers could lead to fabrics that convert body movements into electric current [82].

Another technology is represented by the solar clothing in which the solar energy is harvested through new generation of flexible solar cells [83,84]. Integration of flexible solar cells into clothing can provide power for portable electronic devices. Photovoltaic is the most advanced way of providing electricity far from any mains supply, although it suffers from the limits of ambient light intensity. Nevertheless the energy demand of portable devices is now low enough that clothing-integrated solar cells are able to power most mobile electronics [85].

Related Works

The ILLUM jacket (Figure 15) is based on technologies including printed electroluminescent ink and printed photovoltaic technology. The functional parts are placed outside the jacket and into several ergonomic panels, while at the front and the photovoltaic source at the shoulders and top of the back [86].

Thermotron of UNITIKA (Osaka, Japan) is a particular fabric able to converts sun light into thermal energy while storing heat without wasting it. Inside the Thermotron there are microparticles of zirconium carbide which allow the fabric to absorb and filter sunlight. The inner layer of the fabric withholds the heat generated and prevents it from becoming lost, thus providing a salutary effect on the human body [87].

An example of a battery capable of providing electrical power for interactive electronic textiles was recently developed by a German research team led by The Fraunhofer Institute for Reliability and Micro-integration (FhG-IZM, Berlin, Germany). This research team developed a small battery that can be printed on a substrate and fabricated at high production speeds in button-sized or coin-type format at a cost below one USD. The battery is fabricated by screen printing a thick layer of a silver-oxide based paste, then applying a thin sealing layer. The final result is a textile substrate with a printed 120 μm thick AgO-ZN battery. These batteries can be printed on a variety of substrates [88].

2.6. Planar Fashionable Circuit Board (P-FCB)

The P-FCB is one of the new technologies allowing implementation of a circuit board on a plain fabric patch for wearable electronics applications. It features a soft and flexible impression just as normal clothes.

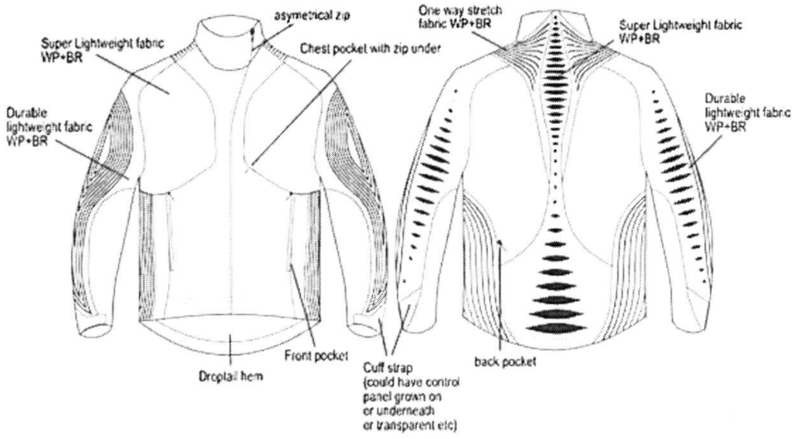

Figure 15. ILLUM jacket layout.

Figure 16. (**a**) Single-layered P-FCB system manufacturing process; (**b**) Implementation of multi-layer connection using eyelet.

The P-FCB substrate is fabricated using woven fabrics. The planar electrodes are deposited on the fabric patch directly by silk screening of conducting epoxy or by gold sputtering. First, the circuit board is silkscreen printed on the fabric patch. Then the IC is placed on the fabric and wire-bonded to the patterned electrodes. Finally, the IC is molded with non-conductive epoxy (Figure 16a) [89].

With these techniques, KAIST researchers also developed a multilayer circuit (Figure 16b). The previous research highlight how electrical and mechanical characteristics and valuable system design parameters are obtained, such as maximum power consumption, maximum current density, crosstalk between neighboring two lines, and durability [90].

2.7. Wearable Antenna

Thanks to the rapid progress on the fabrication of conductive textiles, a significant development of wearable antennas has begun, exploiting new flexible and conformable smart structures [91].

An antenna is essential, if the purpose is to develop a wearable and autonomous system. It allows one to transfer information from the sensors hosted inside the garment to a control unit or to monitor other electronic parameters.

A wearable antenna is thus the bond that integrates clothes into the communication system, making electronic devices less obtrusive. To achieve good results, wearable antennas have to be thin, lightweight, low maintenance, robust, inexpensive and easily integrated in radio frequency (RF) circuits. Planar structures, flexible conductive and dielectric materials are specific requirements for wearable antennas [92,93].

Several properties of the materials influence the behavior of the antenna. For instance, the bandwidth and the efficiency of a planar microstrip antenna are mainly determined by the permittivity and the thickness of the substrate [94,95]

In general, textiles present a very low dielectric constant that reduces the surface wave losses and increases the impedance bandwidth of the antenna. Therefore it is important to know how these characteristics influence the behavior of the antenna in order to minimize unwanted effects.

Another issue regards the movement of the body that can deform the spatial geometry of the antenna and affect its performance. When the textile fabric adapts to the surface topology it bends and deforms, causing changes to its electromagnetic properties and thus influencing the antenna performance [96]. Thus, a wearable FM antenna should be designed so as to be wider than the FM broadcast band (about 81–130 MHz) not to suffer from the detuning caused by the human body [97].

To summarize, the guidelines for a correct antenna project are shown below:

- choosing the correct positioning of the textile antenna [98,99];
- the textile antenna must be made with an accurate thickness stacking the different fabrics [99];
- the geometrical dimensions of the patch must remain stable [100];
- the connections between the layers must not affect the electrical properties and the connections with other parts of "e-garments" have to be stable and robust [101–103].

If some of these points are not followed, undesired effects may occur to the functioning of the device.

Related Works

Researchers at Katholieke Universiteit Leuven and Universiti Malaysia Perlis were the first to develop a fully textile waveguide antenna using a material inspired unit cell that is also used in composite right/left-handed transmission lines. The antenna is compact, robust and can be used for 2.45 and 5.4 GHz dual-band WLAN applications [104].

Patria (Halli, Finland) is a company with expertise in textile antenna design. It develops textile antennas composed by conventional or industrial fabrics, and typically conductive antenna parts are made out of modern conductive fibers [105].

Textiles RFID is a particular solution of antenna. In this sector TexTrace AG (Frick, Switzerland) provides the manufacturing line as well as the components for industrial in-house production of woven RFID labels. Integrate RFID and the label will provide added value from garment manufacturing through logistics to sales and after-sales management [106].

2.8. Stretchable Interconnection

Deformable electronic circuits, in particular in bio-medical application, are needed. E-textiles are the candidate technology for this scope but an important aspect regards the interconnections between components and devices. In most cases the approach used is to develop sinuous electroplated metallic wires in a stretchable substrate material [107].

Figure 17 presents a stretchable interconnection structure. This configuration, called "horseshoe-shape", is able to accommodate large deformation in response to a mechanical stress, preserving the electrical properties [108].

Indeed, with high-frequency signal the interconnection have to be stretchable and therefore the substrate and the conductors have to be stretchable as well. Usually, a polymeric material is chosen as encapsulating substrate, because it can be stretched and also because it may be bio-compatible [109].

Interconnections made with these techniques coupled with conductive yarns can go beyond actual issues regarding the lack of robustness in e-textiles circuit.

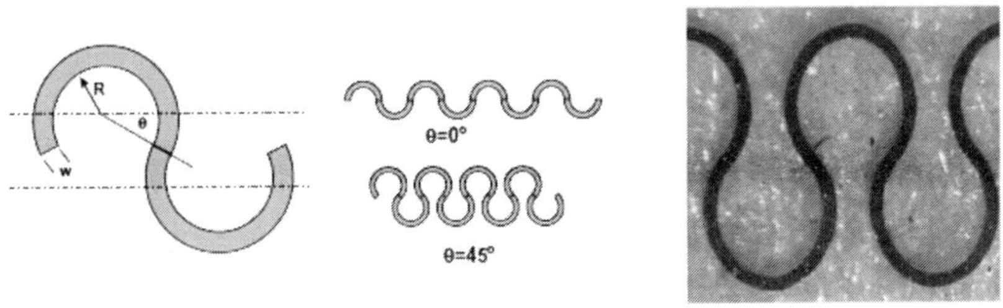

(a) (b)

Figure 17. (**a**) Notable measures of a horseshoe design: inner radius (R), joining angle (θ) and width of the metal track (w); (**b**) Horseshoe metal interconnects embedded into a Sylgard 186 PDMS matrix.

3. DISCUSSION

The methods presented in the previous section are quite different and each one has very specific feature such as conductivity, flexibility, biocompatibility, mechanical resistance and washability. However, only a few approaches are able to satisfy all these requirements at the same time. The section below has the aim to clarify some features about the materials and methods.

3.1. Conductive Wires

Especially for clothing, tactile properties such as stretch, recovery, drape, shear and handle are quite important. For this reason the fibers that are used should be fine and fabrics should have a low weight per unit area (not more than 300 g/m^2). These demands are inconsistent with the materials and geometries that are needed for a reasonable electrical conductivity, because the incorporation of elements such as metal wires within textiles increases stiffness and reduces elasticity [110].

Often in literature no distinction is made between metal wires and metal fibers. However, the company Sprint Metal (Hemer, Germany) defined metal fibers and wires according to their diameter. While a fine wire has a diameter between 30 μm to 1.4 mm, a metal fiber possesses a diameter of 2 to 40 μm [111].

The advantages of metal fibers are their strength, composition, biological inertness and ready availability in textile form at low costs. Due to its inertness it is not sensitive to washing or sweating.

However, they cannot provide uniform heating and their brittle characteristics can damage spinning machinery over time. Additionally, they are heavier than most textile fibers making homogeneous blends difficult to produce [112].

Using the 'conductive thread' approach, no additional step after manufacturing of the fabric is required to establish conductivity. The conductivity of these conductive threads lies in the range of 10–500 Ω/m [113].

For instance, Figure 4a shows the copper wire twisted around the yarn and their locations are not precisely defined within the fabric because of their helical path around the yarn [114]. The conductance of a wire in the fabric features about 17.8 Ω/m.

3.2. Treated Coatings

The advantage of coatings is that they are suitable for many fibers types and they produce good conductivity without significantly altering existing key substrate properties such as density, flexibility and handle. Nevertheless adhesion between the metal and the fibers as well as corrosion resistance can lead to problems.

It is possible to make a conductive fabric through a coating treatment even though the main applications for these textiles are protection against static electricity charge and electromagnetic interference (EMI). The surface resistivity (Rs) of normal textiles goes from more than 10^5 Ω/sq down to antistatically finished textiles and conductive textiles for EMI and heating applications, where Rs can be lower than 100 Ω/sq. It is usual to incorporate or impregnate organic antistatic agents [98].

At present the most important ways of producing conductive fibers are:

- incorporation of conductive fillers (carbon black, metal wires, graphite and metal powder or flakes of Al, Cu, Ni, Ag).
- blended fabrics using conductive metallic or polymeric fibers (such as polypyrrole, polyaniline) (Figures 16b and 18a) [115].

However, the surface resistance of polypyrrole-coated ($C_{pyrrole}$) polyester fabrics is another method to functionalize the textiles. Rs decreased marginally from 106 Ω/sq to 103 Ω/sq with increasing pyrrole concentration within 0.2 mg/mL until the concentration reached a value of about 0.4 mg/mL, above which the rate of decrease diminished. The decrease in surface resistance is due to the increase in the thickness of the conducting polymer layer [116].

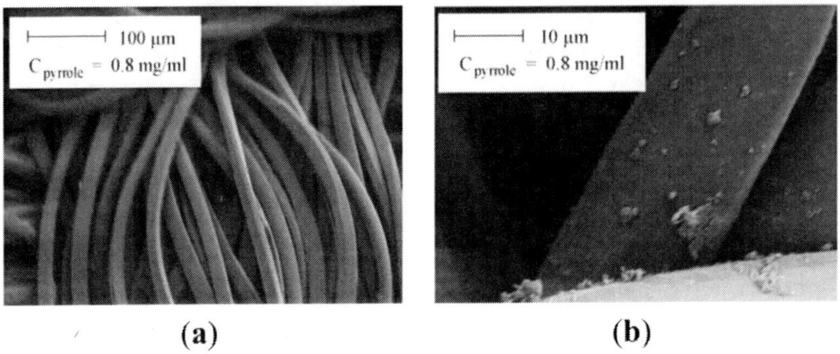

Figure 18. SEM image of the polypyrrole coated textile ($C_{pyrrole}$ = 0.8 mg/mL), (**a**) 100 μm scale and (**b**) 10 μm scale.

The main advantage of functionalized yarns is that the textile properties of the yarns are preserved, in particular they remain flexible.

3.3. Conductive Fabric PETEX

Normally, a fabric made with conductive fibers has a structure according to Figure 19. If the aim is to exploit the conductive fibers to create an electrical circuit, we must take into account the irregular shape of the fabric. It is not possible to obtain a geometric repeatability of the position of the contact points between the conductive fibers.

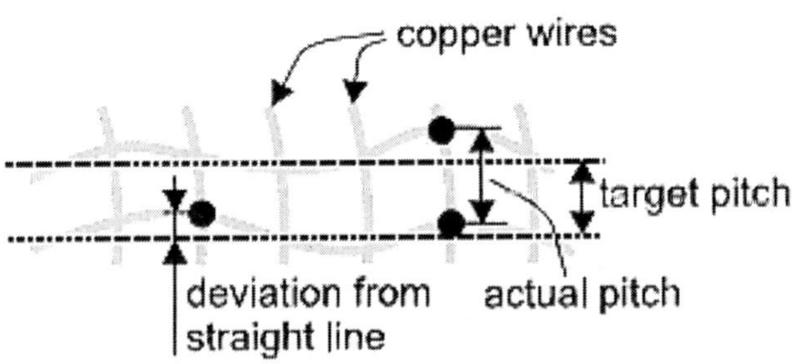

Figure 19. Misalignment of the contact points.

The solution adopted by the ETH researchers is to develop a synthetic fabric and it consists of woven polyester monofilament yarn (PET) with diameter

42 μm and copper alloy wires with diameter 50 ± 8 μm (AWG461). Each copper wire itself is coated with a polyurethane varnish as electrical insulation. The wires in warp direction possess a higher tensile strength compared to the wires in weft direction. These differences support a reliable manufacturing process and minimize breakage.

Figures 18b and 20a below, show the differences between a normal electronic textile structure and PETEX [47].

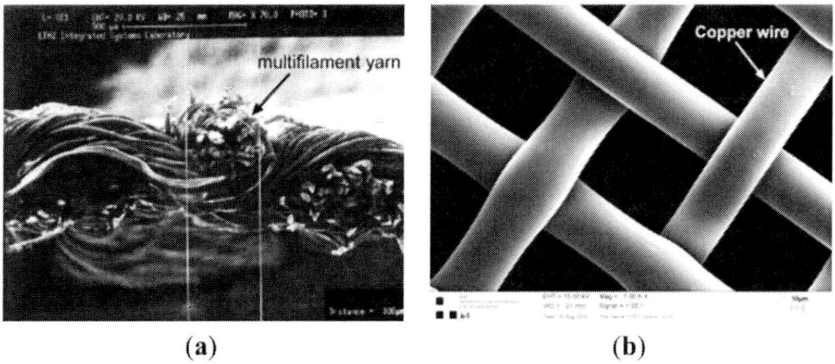

Figure 20. SEM image of fabric with copper fiber (**a**) and PETEX (**b**).

3.4. MIT CAD Embroidery

This technique allows precisely specifying the circuit layout and stitching pattern in a computer-aided design (CAD) environment, from which any number of articles can be sewn under machine control. This process also allows control and integration of yarns with different electrical properties, for instance, different resistances. Embroidery offers advantages over knitting or weaving. Conductive thread and yarn embroidery can be accomplished on single or multiple layers of fabric or can be applied on various types of textile and apparel products in one step [39].

3.5. Conductive Inks

Direct printing of conductive tracks on a fabric is a versatile technique but sometimes shows its limits. Most conductive inks and pastes are based on silver filler and suffer from brittleness. During printing, one of the limits of the silver paste is the thickness. For instance, several passes are needed to achieve a layer thick 40 μm and the electrical resistance depends on the number of passes applied (Figure 21).

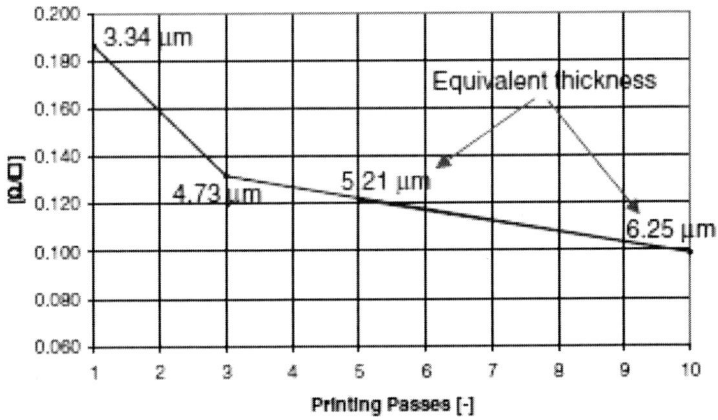

Figure 21. Sheet resistance vs. printing passes.

Another problem is the mechanical resistance. If you create a condition as shown in Figure 22a, the risk is to have a break in the conductivity. In particular the electrical conductivity changes both for the number of passes and for the number of creasing interactions (Figure 22b) [47].

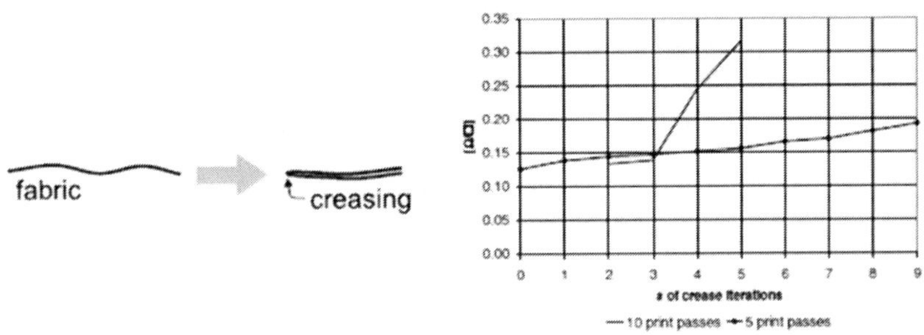

(a) (b)

Figure 22. (**a**) Creasing of the fabric with printed transmission lines; (**b**) DC resistance changes vs. number of creasing iterations.

The researchers of North Carolina Institute, developed a new method to control the durability of the printed circuits, which includes coating of the printed lines. The viscosity of the ink dictated the performance of the printed media during washing trials.

The results showed that higher viscosity inks stay on the surface, resulting in better short-term results. (Figure 23) The ink used was produced by Creative Materials and showed rather a smooth and lower impedance profile than other commercial conductive inks. This ink is based on silver particles and the lines printed on Evolon® show 42 times higher resistivity than standard PCBs (0.50 mΩ/m). However, Evolon® like structures with the highest viscosity ink represent the best solution in terms of durability and conductivity also after washing. This substrate-ink combination protected by a TPU shows the best compromise to become a low-cost electronic textile solution for wearable computing interconnects [117].

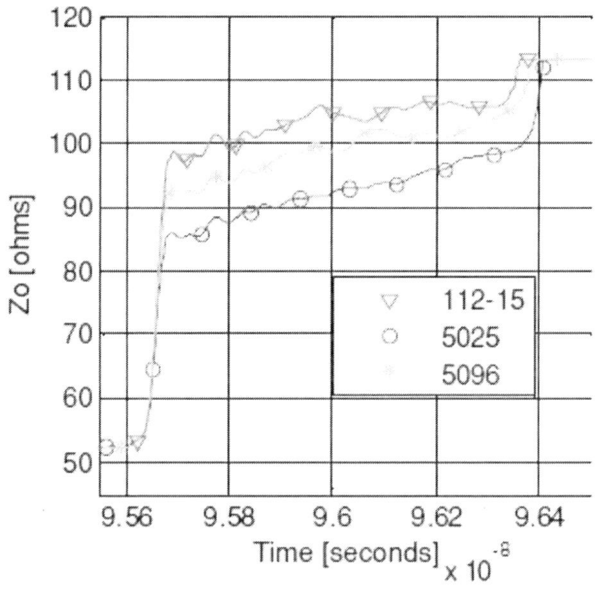

Figure 23. Evolon impedence profile before five washes. The three lines show the comparison between different conductive inks (Dupont 5025, Dupont 5096, Creative Materials 112-15).

3.6. Stretchable Sensors

Usually, textile materials composed of fibers form complex networks of conducting parts that make multiple contacts. During deformation a number of mechanisms take place:

- the number of contact points changes;
- fibers are extended;
- fiber cross-section is decreased.

These factors listed above may compromise the stretch sensors qualities [118]. Polymers are interesting when developing strain sensors with large strain [1]. The stretch sensor developed by Mattmann et al. [119] consists of a mixture of 50 wt-% thermoplastic elastomer (TPE) and 50 wt-% carbon black particles and is fiber-shaped with a diameter of 0.315 mm. This sensor configuration was characterized using a strain tester and measuring the resistance (extension-retraction cycles): it showed a linear resistance response to strain, a small hysteresis, no ageing effects and a small dependence on the strain velocity. The attachment of the sensor to the textile is realized using a silicone film.

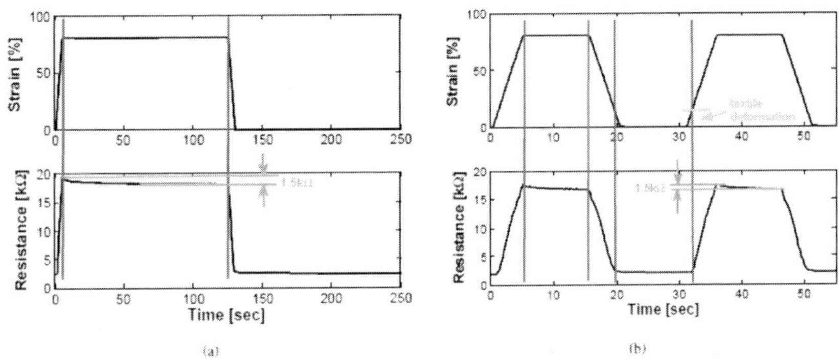

Figure 24. (**a**) Typical response of sensor to a given strain (sensor length 2 cm) waiting time 2 min; (**b**) waiting time 10 s.

The graphs (Figure 24) below show the behavior of the sensor and the typical resistance vs. time plot. The sensor was cycled between 0% and 80% strain at a speed of 200 mm/min and waiting times at minimal and maximal strain of 2 min. In the lower plot the measured resistance is shown which varying between 2 kΩ and 19 kΩ. Instead, Figure 25 shows a linear rise in resistance when applying strain and only a small hysteresis.

Commercially available strain gauges reach a higher than 80% linearity but at a reduced working range of less than 1%. The sensor presented above has a high sensitivity of 1.25 kΩ/mm with a stable sensor properties while in continuous usage. Furthermore the sensor is washable.

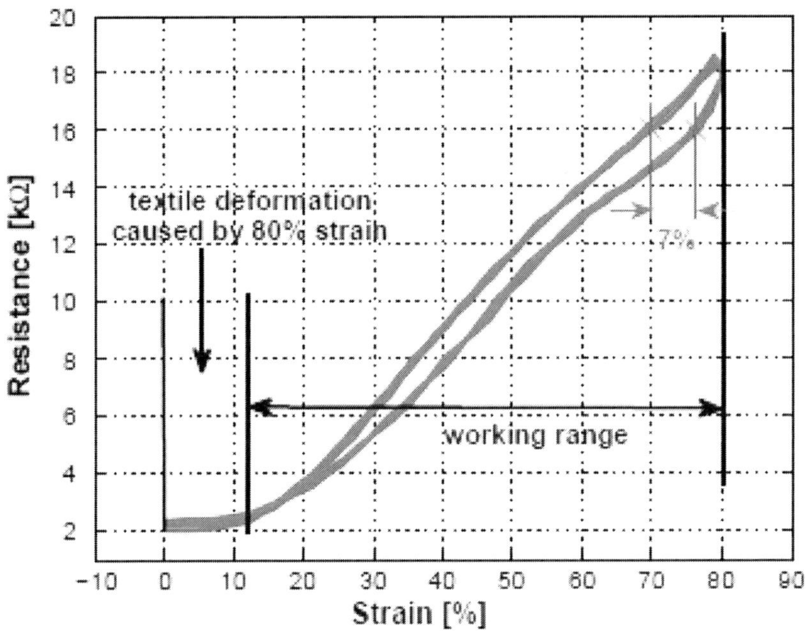

Figure 25. Typical response of sensor to a given strain.

A similar approach is that used by Shyr et al., in which they developed three different geometries of a strain-resistance sensors using Lycra fiber and carbon coated polymide fiber: flat, tubular and belt webbings. It was found that tensile hysteresis and contact resistance significantly influence the tensile elasticity and the resistance sensitivity of the webbings, respectively. Their results showed that all the webbings had a good linear elasticity relationship between tensile load and strain when stretched with the aim to reduce the tensile hysteresis, decreasing the friction among the conductive yarns [118]. The stability of resistance sensitivity and the elasticity are two relevant properties for e-textiles when used as a strain-resistance sensor.

3.7. Pressure Sensors

The pressure sensor developed by Meyer et al. [9] has a spatial resolution of 2 × 2 cm and an average error below 4% within the measurement range 0 to 10 N/cm^2. It is a pressure sensor based on capacitive behavior. The measured capacitance of a single electrode of the prototype with the textile spacer varies from 3.5 pF uncompressed to 5.8 pF at a pressure of 5 N/cm^2 for the textile spacer with 6 mm thickness. Figure 26 shows the capacitance vs. pressure with a hysteresis up to 30%. The high value of hysteresis is due to the dielectric material interposed between the two conductive faces capacitor.

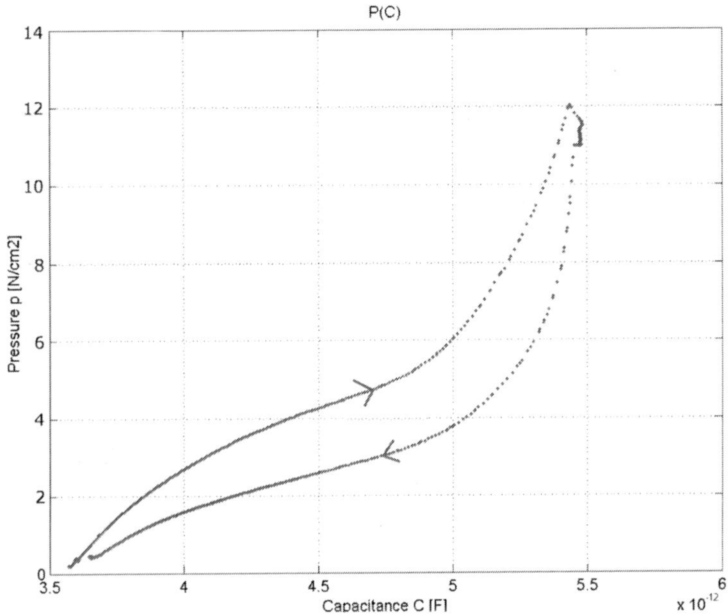

Figure 26. Hysteresis of a singular pressure electrode.

An ideal pressure sensor must ensure repeatable measures over time with a low hysteresis value. Therefore, the textile pressure sensors are applicable in areas where high accuracy is not required.

3.8. Electrochemical Sensors

One of the common disadvantages of chemical/biochemical sensors is the need to calibrate the response from time to time due to either small changes in the activity of the sensor itself and/or the contamination of the sensor surface during operation. This has seriously limited the application of these sensors in continuous monitoring scenarios for long time. The major obstacles still remain, preventing the realization of practical, robust wearable sensors suitable for routine use by the general population.

Among these limitations, we have the lack of appropriate methods to seamlessly integrate electronics and wireless transmitters into the sensor package. The graph below (Figure 27) shows the performance of the wearable sweat sensor during an exercise trial. Reference pH measurements are taken along with heart rate and breathing rate data. The analysis were taken by placing the electrode on a reference fabric patch, as performed in previous studies [120].

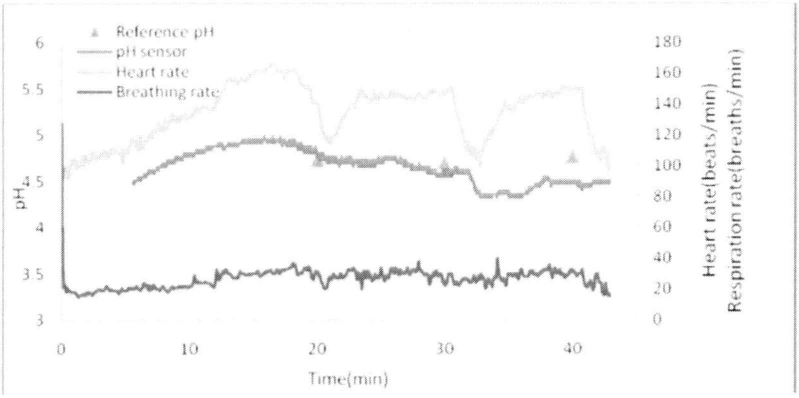

Figure 27. Wearable sweat sensors performance.

3.9. Power Supply

In the following Table 4, an overview of different possible energy sources and the amount of energy that they can generate is given [120,121].

Table 4. Comparison of different energy source and their energy capability.

Energy Source	Available Amount of Energy	Remarks
Primary batteries (Li)	400 Wh/kg, 200 Wh/L	
Secondary batteries (Li-Ion)	75 Wh/kg, 200 Wh/L	Lifetime: 2000 cycles
Si solar cells	20 W/m^2	Light source needed
Recovery of body heat	0.01 W/m^2	
Power harvesting from breathing	0.4 W	Uncomfortable for wearer
Power harvesting from walking	0.25 W	Continuous walking necessary
Microcombustion	10–50 W/cm^3	Fuel needed

From these data, we can assume that the energy harvesting coming from our body is very small although it is always present. If the wearable technologies will take advantage of these energy sources, they must feature a low electrical consumption and a high efficiency harvesting system.

3.10. Planar Fashionable Circuit Board (P-FCB)

The measured sheet resistance of the conducting film on the P-FCB is 44 mΩ/m, as determined by the four-point probe method. Although no noticeable degradation has been observed, even after 50 laundry cycles, more durability examinations will be required with an automatic testing system.

The resistance measurement results of the proposed via and conductive adhesive and bandwidth of a P-FCB transmission line 15 cm long and 1mm wide are shown in Figures 26b and 28a. Its bandwidth is 80 MHz, and this is sufficient to deal with bio-signal processing. The analysis were performed with a vector network analyzer [122]. To get an average value, as many as 100 connections were formed and measured. The resistances of the proposed via and conductive adhesive measured 0.24 W and 0.34 W, respectively [123].

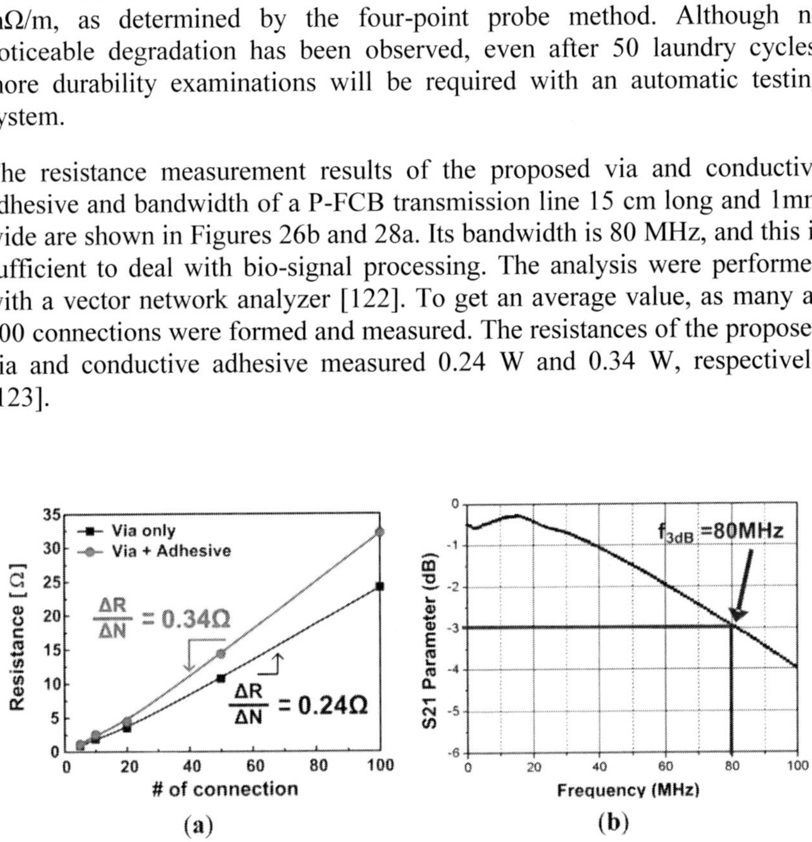

Figure 28. (**a**) Measurement result of the proposed via and conductive adhesive resistance; (**b**) Bandwidth of a P-FCB transmission line (15 cm long, 1 mm wide).

Mechanical strength of the P-FCB was tested through washing tests; after 50+ cycles, the electrical characteristics change was negligible [123]. As a result, the monitoring shirt can be easily made at low cost, and the wearability and flexibility are improved.

3.11. Wearable Antenna

Antennas for wearable applications need to be comfortable to wear and traditional technology with rigid substrates is not well suited to this.

In this sector the dielectric properties of the material that make up the antenna are more relevant. A wrong choice of material would compromise the performance of the device.

The constitutive parameter of dielectrics is the permittivity, ε, a complex value parameter and it usually expressed by Equation (1) below:

$$\varepsilon_r: \quad \varepsilon = \varepsilon_0 \cdot \varepsilon_r = \varepsilon_0 \cdot (\varepsilon_r' - \mathrm{j} \cdot \varepsilon_r'') \tag{1}$$

where ε_0 is the permittivity of vacuum, 8.854×10^{-12} F/m [124]. The dielectric properties are a function of different physical parameters: temperature, frequency, surface roughness, moisture content, purity and homogeneity of the material.

Table 5 below shows the dielectric constant $\boldsymbol{\varepsilon}'_r$ and the ratio of the imaginary part to the real part, also known as $\tan \delta = \boldsymbol{\varepsilon}''_r/\boldsymbol{\varepsilon}'r$ [94].

Table 5. Dielectric Properties of normal fabrics.

Nonconductive Fabric	ε'_r	tan δ
Condura ®	1.90	0.0098
Cotton	1.60	0.0400
Polyester	1.90	0.0045
Quartzel ®	1.95	0.0004
Cordura/Lycra®	1.50	0.0093

The low dielectric constant (between 1 and 2) reduces the surface wave losses that are tied to guided wave propagation within the substrate. Therefore, lowering the dielectric constant increases spatial waves and hence increases the impedance bandwidth of the antenna [125].

In addition, the moisture can affect the dielectric behavior of textiles. In particular, when water is absorbed by the fibres, an increase of the dielectric constant and loss tangent occurs. A textile cover or superficial treatment may then provide a sufficient protection from the humidity and varying climatic conditions [100].

Researchers of Tampere University of Technology, in collaboration with Patria, developed a robust textile antenna able to operate in different environmental conditions. They demonstrated that different functional textile layers can improve and optimize the antenna performance. In this occasion they succeeded to protect the antenna against moisture and others environmental factors [126].

The thickness of the dielectric material changes the bandwidth of the planar antenna. It is necessary to find a compromise between efficiency and bandwidth, well as to influences the geometric sizing of the antenna [125].

For the antenna design, the relevant parameter is the conductivity of the fabric (σ), which unit is Siemens per meter (S/m). It is expressed by this Equation (2):

$$\sigma = 1/(\rho_s \cdot t) \quad (2)$$

where ρ_S is the surface resistivity and t is the thickness of the fabric. An issue regarding the conducibility of conductive fabrics for antenna is the homogeneity of σ value; In fact, there are several discontinuities inside the fabric and these affect the current flow [127].

Another aspect that influences the conductivity is the structure of the fabric. The current flow depends on the type of textile structure used and to minimize the conductive losses it is better to align the conductive paths with the current direction [98].

Finally, a correct choice of materials combined with textile structures can improve the performance of the textile antenna equaling the features guaranteed by traditional antennas.

3.12. Stretchable Interconnections

Stretchability of the electrical conductors is a requirement that can be satisfied with different designs. The design optimization is done using FEM modeling (Figure 29) and other tests. Gonzalez et al. evaluated the "horseshoe-shape" to be the optimal solution for this purpose [108]. In this way the damage in the conductor is significantly decreased with a larger deformability. They demonstrated that in case of multiple and narrow metallisations the stretchability of the circuit reaches 100% of the deformability.

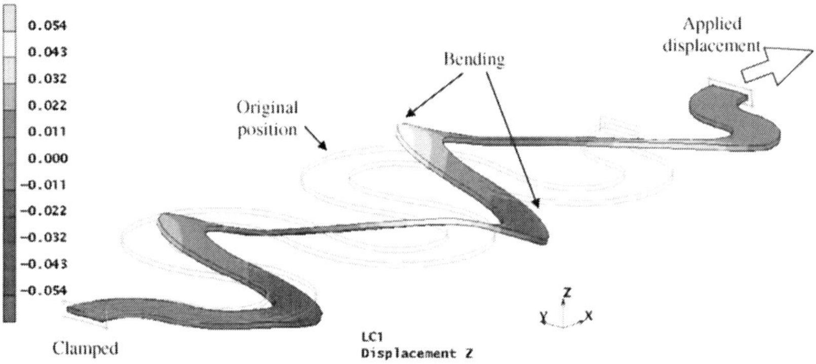

Figure 29. FEM simulation of the "horseshoe-shape".

Hu et al. used a different approach to obtain a stretchable conductive textile. With a simple "dipping and drying" process using single-walled carbon nanotube (SWNT) ink, they produced highly conductive textiles with conductivities of 125 $S \cdot cm^{-1}$ and sheet resistances less than 1 Ω/sq. In this way, such conductive textiles preserve the same stretchability of the normal fabric. Indeed, the porous structure of the textiles facilitates accessibility from any electrolyte and such porous and stretchable conductors find a wide range of applications in e-textiles field [128].

4. CONCLUSION/OUTLOOK

Textiles represent an attractive class of substrates for realizing wearable biosensors. Electronic textiles, or smart textiles, describe the convergence of electronics and textiles into fabrics which are able to sense, compute, communicate and actuate. As many different electronic systems can be connected to any clothing, a wearable system becomes more versatile, and the user can change its look depending on environmental changes and individual preference.

The vision of wearable computing describes future electronic systems as an integral part of our everyday clothing serving as intelligent personal assistants. Therefore, such wearable sensors must maintain their sensing capabilities under the demands of normal wear, which can impose severe mechanical deformation of the underlying garment/substrate.

One promising approach to reduce the rigidity of electronic textiles and enhance its wearability is to replace PCBs by flexible electronics.

In this review we wanted to explain how it is possible to develop a Smart Textile. Some methods show advantages with respect to others, but in our opinion and in according to the consulting company Smart Garment People (Lancashire, UK), while some manufacturers are very experienced with electronics and others with textiles, very few do both well.

Current advances in textile technologies, new materials, nanotechnology and miniaturized electronics are making wearable systems more feasible but the final key factor for user acceptance of wearable devices is the fit comfort. We are convinced that this goal can only be achieved by addressing mechanical resistance, and durability of the materials in what is recognized to be a harsh environment for electronics: the human body and society.

Finally, we consider relevant that the development of smart textiles requires a multidisciplinary approach in which knowledge of circuit design, smart materials, micro-electronics and chemistry are fundamentally integrated with a deep understanding of textile fabrication.

AUTHOR CONTRIBUTIONS

Authors contributed equally to the manuscript in its final form.

CONFLICTS OF INTEREST

The authors declare no conflict of interest.

REFERENCES

1. Langereis, G.R.; Bouwstra, S.; Chen, W. Sensors, Actuators and Computing Architecture Systems for Smart Textiles. In Smart Textiles for Protection; Chapman, R., Ed.; Woodhead Publishing: Cambridge, UK, 2012; Volume 1, pp. 190–213.

2. Custodio, V.; Herrera, F.J.; López, G.; Moreno, J.I. A review on architectures and communications technologies for wearable health-monitoring systems. Sensors 2012, 12, 13907–13946.

3. Coosemans, J.; Hermans, B.; Puers, R. Integrating wireless ECG monitoring in textiles. Sens. Actuators A Phys. 2006, 130–131, 48–53.

4. Linz, T.; Gourmelon, L.; Langereis, G. Contactless EMG sensors embroidered onto textile. Proceedings of the 4th International Workshop on Wearable and Implantable Body Sensor Networks, Aachen, Germany, 26–28 March 2007; Volume 13, pp. 29–34.

5. Löfhede, J.; Seoane, F. Thordstein, Soft textile electrodes for EEG monitoring. Proceedings of 2010 the 10th IEEE International Conference on Information Technology and Applications in Biomedicine (ITAB), Corfu, Greece, 2–5 November 2010; pp. 1–4.

6. Löfhede, J.; Seoane, F.; Thordstein, M. Textile electrodes for EEG recording—A pilot study. Sensors 2012, 12, 16907–16919.

7. Sibinski, M.; Jakubowska, M.; Sloma, M. Flexible temperature sensors on fibers. Sensors 2010, 10, 7934–7946.

8. Omenetto, F.; Kaplan, D.; Amsden, J.; Dal Negro, L. Silk Based Biophotonic Sensors. Patent US 2013/0330710, 2013.

9. Meyer, J.; Lukowicz, P.; Tröster, G. Textile Pressure Sensor for Muscle Activity and Motion Detection. Proceeding of the 10th IEEE International Symposium on Wearable Computers, Montreux, Switzerland, 11–14 October 2006.

10. Coyle, S.; Lau, K.-T.; Moyna, N.; O'Gorman, D.; Diamond, D.; Di Francesco, F.; Costanzo, D.; Salvo, P.; Trivella, M.G.; De Rossi, D.E.; et al. BIOTEX—Biosensing textiles for personalised healthcare management. IEEE Trans. Inf. Technol. Biomed. 2010, 14, 364–370.

11. Zadeh, E. Flexible biochemical sensor array for laboratory-on-chip applications. Proceeding of the International Workshop on Computer Architecture for Machine Perception and Sensing, Montreal, QC, Canada, 18–20 September 2006; pp. 65–66.

12. Vatansever, D.; Siores, E.; Hadimani, R.; Shah, T. Smart Woven Fabrics in Renewable Energy Generation. In Advances in Modern Woven Fabrics Technology; Vassiliadis, S., Ed.; InTech: Rijeka, Croatia, 2011; pp. 23–38.

13. Baurley, S. Interactive and experiential design in smart textile products and applications. Pers. Ubiquitous Comput. 2004, 8, 274–281.

14. Black, S. Trends in Smart Medical Textiles. In Smart Textiles for Medicine and Healthcare: Materials, Systems and Applications; Van Langenhove, L., Ed.; University of Ghent: Ghent, Belgium, 2007; Volume 1, pp. 10–22.

15. Edmison, J.; Jones, M.; Nakad, Z.; Martin, T. Using piezoelectric materials for wearable electronic textiles. Proceedings of the 6th International Symposium on Wearable Computers (ISWC), Seattle, WA, USA, 7–10 October 2002; pp. 41–48.

16. Bedeloglu, A.; Demir, A.; Bozkurt, Y.; Sariciftci, N.S. A Photovoltaic Fiber Design for Smart Textiles. Text. Res. J. 2009, 80, 1065–1074.

17. How to Get What You Want? Available online: http://www.kobakant.at/DIY/?p=48 (accessed on 11 January 2014).

18. Pacelli, M.; Loriga, G.; Taccini, N.; Paradiso, R. Sensing Fabrics for Monitoring Physiological and Biomechanical Variables: E-textile solutions. Proceeding of the IEEE/EMBS International Summer School on Medical Devices and Biosensors, St. Catharine's College, Cambridge, UK, 19–22 August 2007; pp. 1–4.

19. Dias, T. Development and Analysis of Novel Electroluminescent Yarns and Fabrics for Localised Automotive Interior Illumination: El Yarns and Fabrics. Text. Res. J. 2012, 82, 1164–1176.

20. Janietz, S.; Gruber, B.; Schattauer, S.; Schulze, K. Integration of OLEDs in Textiles. Adv. Sci. Technol. 2012, 80, 14–21.

21. Munro, B.J.; Steele, J.R.; Campbell, T.E.; Wallace, G.G. Wearable Textile Biofeedback Systems: Are They too Intelligent for the Wearer? In Wearable eHealth Systems for Personalised Health Management: State of the Art and Future Challenges; Lymberis, A., De Rossi, D., Eds.; IOS Press—STM Publishing House: Amsterdam, The Netherlands, 2005; Volume 108, pp. 271–277.

22. Salonen, P.; Hurme, L. A novel fabric WLAN antenna for wearable applications. Proceeding of IEEE International Symposium on Antennas and Propagation Society, Columbus, OH, USA, 22–27 June 2003; Volume 2, pp. 700–703.

23. Popular Embroidery Techniques Used to Decorate Fabrics. Available online: http://nanetteparker.hubpages.com/hub/Popular-Embroidery-Techniques-Used-to-Decorate-Fabrics (accessed on 4 February 2014).

24. Creative Sewing. Available online: http://www.creativesewing.co.nz/ (accessed on 4 February 2014).

25. Loominous. Available online: http://www.loominous.co.uk/ studio.html (accessed on 4 February 2014).

26. Cornell University—Fabrics of Our Livelihoods. Available online: http://smallfarms.cornell.edu/2011/07/04/fabrics-of-our-livelihoods/ (accessed on 4 February 2014).

27. CMI. Available online: https://www.colonialmills. com/PublicStore/catalog/Braiding-Process,156.aspx (accessed on 4 February 2014).

28. Textile Innovation Knowledge Platform. Available online: http://www.tikp.co.uk/knowledge/technology/coating-and-laminating/laminating (accessed on 4 February 2014).

29. Custom Fabric Printing. Available online: http:// sophiasdecor.blogspot.it/2012/09/inside-spoonflower-custom-fabric.html (accessed on 4 February 2014).

30. Durable water repellent. Available online: http://en.wikipedia.org/wiki/Durable_water_repellent (accessed on 4 February 2014).

31. Post, E.R.; Orth, M.; Russo, P.R.; Gershenfeld, N. E-broidery: Design and fabrication of textile-based computing. IBM Syst. J. 2000, 39, 840–860.

32. Systenx Project. Available online: http://www.systex.org/projects (accessed on 10 September 2013).

33. Proetex Project. Available online: http://www.proetex.org/related_projects.htm (accessed on 10 September 2013).

34. Resistat Fiber Collection. Available online: http://www.resistat.com/ (accessed on 12 January 2014).

35. Redström, M.; Redström, J.; Mazé, R. IT + Textiles, 1st ed.; The Interactive Institute: Boràs, Sweden, 2005; pp. 59–93.

36. Sophitex Ltd. Available online: http://www.sophitex.com (accessed on 31 January 2014).

37. LessEMF. Available online: http://www.lessemf.com/fabric.html (accessed on 15 July 2013).

38. McFarland, E.G.; Carr, W.W.; Sarma, D.S.; Dorrity, J.L. Effects of Moisture and Fiber Type on Infrared Absorption of Fabrics. Text. Res. J. 1999, 69, 607–615.

39. Mac, T.; Houis, S.; Gries, T. Metal Fibers. Proceeding of the International Conference on Shape Memory and Superelastic Technologies, Baden-Baden, Germany, 3–7 October 2004; Volume 47.

40. Elektrisola Feindraht AG. Available online: www.textile-wire.com (accessed on 11 January 2014).

41. Spoerry & Co AG, Swiss Shield®. Available online: www.swiss-shield.ch (accessed on 12 January 2014).

42. Gimpel, S.; Moehring, U.; Mueller, H.; Neudeck, A.; Scheibner, W. The galvanic and electrochemical modification of textiles. Band-und Flechtind. 2003, 40, 115–120.

43. Abouraddy, A.F.; Bayindir, M.; Benoit, G.; Hart, S.D.; Kuriki, K.; Orf, N.; Shapira, O.; Sorin, F.; Temelkuran, B.; Fink, Y. Towards multimaterial multifunctional fibres that see, hear, sense and communicate. Nat. Mater. 2007, 6, 336–347.

44. Hamedi, M.; Forchheimer, R.; Inganäs, O. Towards woven logic from organic electronic fibres. Nature Mater. 2007, 6, 357–362.

45. Müller, C.; Hamedi, M.; Karlsson, R.; Jansson, R.; Marcilla, R.; Hedhammar, M.; Inganäs, O. Woven electrochemical transistors on silk fibers. Adv. Mater. 2011, 6, 898–901.

46. Scheibner, W.; Feustel, M.; Moehring, U.; Hofmann, J.; Linz, T. Textile electrical connecting cables. Band-und Flechtind. 2003, 40, 76–78.

47. Locher, I. Technologies for System-on-Textile Integration. Ph.D. Thesis, Swiss Federal Institute of Technology (ETH), Zurich, Switzerland, 2006.

48. Baltex. Available online: http://www.baltex.co.uk (accessed on 11 November 2013).

49. Thermshield. Available online: http://www. thermshield. com/ThermshieldPages/MetallizedFabric.html (accessed on 12 January 2014).

50. Ohmatex, Smart Textile Technology. Available online: http://www.ohmatex.dk (accessed on 3 June 2013).

51. Parashkov, R.E.; Becker, E. Large area electronics using printing methods. Proc. IEEE 2005, 93, 1321–1329.

52. Blayo, A.; Pineaux, B. Printing processes and their potential for RFID printing. Proceeding of Joint sOc-UESAI Conference, Grenoble, France, 7–10 October 2005; pp. 27–30.

53. Stassi, S.; Cauda, V.; Canavese, G.; Manfredi, D.; Roppolo, I.; Martino, P.; Chiolerio, A. Nanosized Gold and Silver Spherical, Spiky,

and Multi-Branched Particles. In Handbook of Nanomaterials Properties; Bhushan, B., Ed.; Springer-Verlag: Berlin/Heidelberg, Germany, 2014; pp. 199–202.

54. Tiberto, P.; Barrera, G.; Celegato, F.; Coisson, M.; Chiolerio, A.; Martino, P.; Pandolfi, P.; Allia, P. Magnetic properties of jet-printer inks containing dispersed magnetite nanoparticles. Eur. Phys. J. B. 2013, 83, 173–179.

55. Lee, H.-H.; Chou, K.-S.; Huang, K.-C. Inkjet printing of nanosized silver colloids. Nanotechnology 2005, 16, 2436–2441.

56. Camarchia, V.; Chiolerio, A.; Cotto, M.; Fang, J.; Ghione, G.; Pandolfi, P.; Pirola, M.; Pirola, M.; Quaglia, R.; Ramella, C. Demonstration of inkjet-printed silver nanoparticle microstrip lines on alumina for RF power modules. Org. Electron. 2014, 15, 91–98.

57. Printing, S.; Sauer, M.; Meilchen, S.; Kalleder, A.; Mennig, M.; Schmidt, H. Screen printing. In Sol-Gel Technologies for Glass Producers and Users; Michel, A., Aegerter, A., Mennig, M., Eds.; Springer: Norwell, MA, USA, 2004; Volume 1.

58. Numakura, D. Advanced Screen Printing—Practical Approaches for Printable & Flexible Electronics; DKN Research: Haverhill, MA, USA, 2009.

59. Das, R.N.; Lin, H.T.; Lauffer, J.M.; Markovich, V.R. Printable electronics: Towards materials development and device fabrication. Circuit World 2011, 37, 38–45.

60. Ionescu, C.; Bonfert, D.; Codreanu, N.D.; Svasta, P. Investigations on current capabilities of PEDOT:PSS conductors. Proceeding of the 35th International Spring Seminar on Electronics Technology, Bad Aussee, Austria, 9–13 May 2012; pp. 59–64.

61. Gordon, P.; Torah, R; Beeby, S.; Tudor, J. The development of screen printed conductive networks on textiles for biopotential monitoring applications. Sens. Actuators A Phys. 2006, 206, 35–41.

62. Karaguzel, B.; Merritt, C.R.; Kang, T.; Wilson, J.M.; Nagle, H.T.; Grant, E.; Pourdeyhimi, B. Flexible, durable printed electrical circuits. J. Text. Inst. 2009, 100, 1–9.

63. Scalisi, R.G.; Paleari, M.; Favetto, A.; Stoppa, M.; Ariano, P.; Pandolfi, P.; Chiolerio, A. A new paradigm in surface electromyography: Inkjet printed flexible electrodes. Submitted.

64. Pacelli, M.; Loriga, G.; Taccini, N.; Paradiso, R. Sensing Fabrics for Monitoring Physiological and Biomechanical Variables: E-textile

solutions. Proceedings of the 3rd IEEE-EMBS International Summer School and Symposium on Medical Devices and Biosensors MIT, Boston, MA, USA, 4–6 September 2006.

65. Huang, C.-T.; Shen, C.-L.; Tang, C.-F.; Chang, S.-H. A wearable yarn-based piezo-resistive sensor. Sens. Actuators A Phys. 2008, 141, 396–403.

66. Hertleer, C.; Grabowska, M.; Van Langenhove, L.; Catrysse, M.; Hermans, B.; Puers, R.; Kalmar, A.; Van Egmond, H.; Matthys, D. Toward a Smart Suit. Proceeding of Wearable Electronic and Smart Textiles, Leeds, UK, 11 June 2004.

67. Gough, P. Electronics and Clothes: Watt to Wear? Proceeding of Wearable Electronic and Smart Textiles, Leeds, UK, 11 June 2004.

68. Ashruf, C.M.A. Thin flexible pressure sensors. Sens. Rev. 2002, 22, 322–327.

69. Rothmaier, M.; Luong, M.P.; Clemens, F. Textile Pressure Sensor Made of Flexible Plastic Optical Fibers. Sensors 2008, 8, 4318–4329.

70. Eleksen Ltd. Available online: http://www.eleksen.com (accessed on 12 January 2014).

71. Logitech Inc. Available online: http://www.logitech.com (accessed on 12 January 2014).

72. Pressure Profile Systems, Inc. Available online: http://www.pressureprofile.com/presStrip.php (accessed on 11 January 2014).

73. Swallow, S.; Peta Thompson, A. Available online: http:// www.intelligenttextiles.com (accessed on 1 January 2014).

74. Viry, L.; Levi, A.; Totaro, M.; Mondini, A.; Mattoli, V.; Mazzolai, B.; Beccai, L. Flexible three-axial force sensor for soft and highly sensitive artificial touch. Adv. Mater. 2014, 2659–2664.

75. Windmiller, J.R.; Wang, J. Wearable Electrochemical Sensors and Biosensors: A Review. Electroanalysis 2012, 1, 29–46.

76. Benito-Lopez, F.; Coyle, S.; Byrne, R.; Smeaton, A.; O'Connor, N.; Diamond, D. Pump Less Wearable Microfluidic Device for Real Time pH Sweat Monitoring. Proceeding of the Eurosensors XXIII Conference, Lucerne, Switzerland, 6–9 September 2009.

77. Nishide, H.; Oyaizu, K. Materials science. Toward flexible batteries. Science 2008, 319, 737–738.

78. Infineon Technologies AG. Available online: http://www.infineon.com (accessed on 2 February 2014).

79. Swallow, L.M.; Luo, J.K.; Siores, E.; Patel, I.; Dodds, D. A piezoelectric fibre composite based energy harvesting device for potential wearable applications. Smart Mater. Struct. 2008, 17.

80. Wei, Y.; Torah, R.; Yang, K. Screen printing of a capacitive cantilever-based motion sensor on fabric using a novel sacrificial layer process for smart fabric applications. Meas. Sci. Technol. 2013, 24.

81. Wei, Y.; Torah, R.; Yang, K.; Beeby, S.; Tudor, J. Screen printed capacitive free-standing cantilever beams used as a motion detector for wearable sensors. Procedia Eng. 2012, 47, 165–169.

82. Bai, S.; Zhang, L.; Xu, Q.; Zheng, Y.; Qin, Y.; Wang, Z. Two dimensional woven nanogenerator. Nano Energy 2013, 2, 1–5.

83. Lee, Y.-H.; Kim, J.-S.; Noh, J.; Lee, I.; Kim, H.J.; Choi, S.; Seo, J.; Jeon, S.; Kim, T.-S.; Lee, J.-Y.; et al. Wearable Textile Battery Rechargeable by Solar Energy. NanoLetters 2013, 13, 5753–5761.

84. Chiechi, R.C.; Havenith, R.W.A.; Hummelen, J.C.; Koster, L.J.A.; Loi, M.A. Modern plastic solar cells: Materials, mechanisms and modeling. Mater. Today 2013, 16, 281–289.

85. Schubert, M.B.; Werner, J.H. Flexible solar cells for clothing. Mater. Today 2006, 9, 42–50.

86. Consultancy Goose Design, "PDD Illum Project". Available online: http://www.gooseuk.com/ (accessed on 11 November 2013).

87. Unitika LTD. Available online: http://www.unitika.co.jp/e (accessed on 2 February 2014).

88. Meoli, D. Interactive Electronic Textiles: Technologies, Applications, Opportunities, and Market Potentia. MS.c. Thesis, North Carolina State University, Rayleigh, NC, USA, 2002.

89. Kim, H.; Kim, Y.; Yoo, H.J. A 1.12 mW continuous healthcare monitor chip integrated on a planar-fashionable circuit board. Proceeding of the IEEE International Solid-State Circuits Conference (ISSCC) Digest of Technical Papers, San Francisco, CA, USA, 3–7 February 2008; pp. 150–151.

90. Seulki, L.; Binhee, K.; Yoo, H.J. Planar Fashionable Circuit Board Technology and Its Applications. J. Semicond. Technol. Sci. 2009, 9, 174–180.

91. Giddens, H.; Paul, D.L.; Hilton, G.S.; McGeehan, J.P. Influence of body proximity on the efficiency of a wearable textile patch antenna. Proceeding of the 6th European Conference Antennas & Propagation (EuCAP), Prague, Czech, 26–30 March 2012; pp. 1353–1357.

92. Zhang, L.; Wang, Z.; Psychoudakis, D.; Volakis, J.L. Flexible Textile Antennas for Body-Worn Communication. Proceedings of IEEE International Workshop on Antenna Technology, Tucson, ZA, USA, 5–7 March 2012; pp. 205–208.

93. Grupta, B.; Sankaralingam, S.; Dhar, S. Development of Wearable and Implantable Antennas in the Last Decade: A Review. Proceedings of Mediterranean Microwave Symposium (MMS), Guzelyurt, Turkey, 25–27 August 2010; pp. 251–267.

94. Liu, N.; Lu, Y.; Qiu, S.; Li, P. Electromagnetic Properties of Electro-Textile for Wearable Antennas Applications. Front. Electr. Electron. Eng. China 2011, 6, 553–566.

95. Brebels, S.; Ryckaert, J.; Boris, C.; Donnay, S.; De Raedt, W.; Beyne, E.; Mertens, R.P. SOP Integration and Codesign of Antennas. IEEE Trans. Adv. Packag. 2004, 27, 341–351.

96. Moretti, A. Estudo do Brim Santista Visando Aplicações em Antenas têxteis. MS.c. Thesis, Universidade Estadual de Campinas, Campinas, Brazil, 2011.

97. Roh, J.-S.; Chi, Y.-S.; Lee, J.-H.; Tak, Y.; Nam, S.; Kang, T.J. Embroidered Wearable Multiresonant Folded Dipole Antenna for FM Reception. IEEE Antennas Wirel. Propag. Lett. 2010, 9, 803–806.

98. Ouyang, Y.; Chappell, W.J. High Frequency Properties of Electrotextiles for Wearable Antenna Applications. IEEE Trans. Antenna Propag. 2008, 56, 381–389.

99. Kaija, T.; Lilja, J.; Salonen, P. Exposing Textile Antennas for Hash Environment. Proceedings of 2010 Military Communications Conference, San Jose, CA, USA, 31 October–3 November 2010; pp. 737–742.

100. Hertleer, C.; Laere, A.V.; Rogier, H.; Langenhove, L.V. Influence of Relative Humidity on Textile Antenna Performance. Text. Res. J. 2009, 80, 177–183.

101. Matthews, J.C.G.; Pettitt, G. Development of Flexible, Wearable Antennas. Proceedings of EuCAP 2009: 3rd European Conference on Antennas and Propagation, Berlin, Germany, 23–27 March 2009; pp. 273–277.

102. Tronquo, A.; Rogier, H.; Hertleer, C.; Langenhove, L.V. Applying Textile Materials for the Design of Antennas for Wireless Body Area Networks. Proceedings of EuCap 2006: First European Conference on Antennas and Propagation, Nice, France, 6–10 November 2006.

103. Hertleer, C.; Rogier, H.; Member, S.; Vallozzi, L.; Langenhove, L.V. A Textile Antenna for Off-Body Communication Integrated into Protective Clothing for Firefighters. IEEE Trans. Adv. Packag. 2009, 57, 919–925.

104. Yan, S.; Soh, P.J. Wearable dual-band composite right/left-handed waveguide textile antenna for WLAN applications. Electron. Lett. 2014, 50, 424–426.

105. Patria. Available online: http://www.patria.fi/ (accessed on 7 June 2014).

106. TEXTRACE. Available online: http://www.textrace.com/ (accessed on 7 June 2014).

107. Brosteaux, D.; Gonzalez, M.; Vanfleteren, J. Design and Fabrication of Elastic Interconnections for Stretchable Electronic Circuits. IEEE Electron Device Lett. 2007, 28, 552–554.

108. Gonzalez, M.; Axisa, F.; Vanden Bulcke, M.; Brosteaux, D.; Vandevelde, B.; Vanfleteren, J. Design of metal interconnects for stretchable electronic circuits. Microelectron. Reliab. 2008, 48, 825–832.

109. Huyghe, B.; Rogier, H. Design and manufacturing of stretchable high-frequency interconnects. IEEE Trans. Adv. Packag. 2008, 31, 802–808.

110. STATEX. Available online: http://www.statex.de/index.php/en/ (accessed on 4 July 2013).

111. UGITECH S.A. Available online: "http://www.sprintmetal.com (accessed on 10 September 2013).

112. Bekaert Fibre Technologies. Available online: http://www. bekaert. com (accessed on 25 September 2013).

113. Locher, I.; Kirstein, T.; Tröster, G. Routing methods adapted to e-textiles. Proceeding of the 37th International Symposium on Microelectronics (IMAPS), Long Beach, CA, USA, 16–18 November 2004.

114. Cottet, D.; Grzyb, J.; Kirstein, T.; Tröster, G. Electrical characterization of textile transmission lines. IEEE Trans. Adv. Packag. 2003, 26, 182–190.

115. Bhat, N.V.; Seshadri, D.T.; Radhakrishnan, S. Preparation, characterization, and performance of conductive fabrics: Cotton + PANi. Text. Res. J. 2004, 74, 155–166.

116. Tong, L.; Lijing, W.; Xungai, W.; Akif, K. Polymerising pyrrole on polyester textiles and controlling the conductivity through coating thickness. Thin Solid Films 2005, 479, 777–782.

117. Cochrane, C.; Koncar, V.; Lewandowski, M.; Dufour, C. Design and development of a flexible strain sensor for textile structures based on a conductive polymer composite. Sensors 2007, 7, 473–492.

118. Shyr, T.W.; Shie, J.W.; Jhuang, Y.E. The effect of tensile hysteresis and contact resistance on the performance of strain-resistant elastic-conductive webbing. Sensors 2001, 11, 1693–1705.

119. Mattmann, C.; Clemens, F.; Tröster, G. Sensor for Measuring Strain in Textile Wearable. Sensors 2008, 8, 3719–3732.

120. Coyle, S.; Lau, K.; Moyna, N.; Diamond, D.; Di Francesco, F.; Constanzo, D.; Salvo, P.; Trivella, M.; De Rossi, D.; Taccini, N.; et al. BIOTEX—Biosensing textiles for personalised healthcare management. IEEE Trans. Inf. Technol. BioMed. 2010, 14, 374–370.

121. Gorlatova, M.; Sarik, J.; Cong, M. Movers and Shakers: Kinetic Energy Harvesting for the Internet of Things. In Technical Report; Electrical Engineering Department, Columbia University: New York, NY, USA, 2013.

122. Seulki, L.; Binhee, K.; Taehwan, R.; Sunjoo, H.; Hoi-Jun, Y. Arm-band type textile-MP3 player with multi-layer Planar Fashionable Circuit Board (P-FCB) techniques. Proceeding of the International Symposium on Wearable Computers (ISWC), Seoul, South Korea, 10–13 October 2010.

123. Kim, Y.; Kim, H.; Yoo, J. Electrical characterization of printed circuits on the fabric. IEEE Trans. Adv. Packag. 2010, 33, 196–205.

124. Baker-Jarvis, J.; Janezic, M.D.; DeGroot, D.C. High-Frequency Dielectric Measurements. IEEE Trans. Instrum. Meas. 2010, 13, 24–31.

125. Salvado, R.; Loss, C.; Gonçalves, R.; Pinho, P. Textile materials for the design of wearable antennas: A survey. Sensors 2012, 12, 15841–15857.

126. Lilja, J.; Salonen, P.; Kaija, T.; De Maagt, P. Design and Manufacturing of Robust Textile Antennas for Harsh Environments Antennas and Propagation. IEEE Trans. Antenna Propag. 2012, 60, 4130–4140.

127. Salonen, P.; Rahmat-Samii, Y.; Hurme, H.; Kivikoski, M. Effect of Conductive Material on Wearable Antenna Performance: A Case Study of WLAN Antennas. Proceedings of IEEE Antennas and Propagation Society International Symposium, Monterey, CA, USA, 20–25 June 2004; pp. 455–458.

128. Hu, L.; Pasta, M.; La Mantia, F.; Cui, L.; Jeong, S.; Deshazer, H.D.; Choi, J.W.; Han, S.M.; Cui, Y. Stretchable, porous, and conductive energy textiles. Nano Lett. 2010, 10, 708–714.

CHAPTER 9

Medical Textiles as Vascular Implants and Their Success to Mimic Natural Arteries

Charanpreet Singh [1], Cynthia S. Wong [1,†] and Xungai Wang [1,2,*]

[1] Australian Future Fibres Research and Innovation Centre, Institute for Frontier Materials, Deakin University, Geelong, VIC 3216, Australia

[2] School of Textile Science and Engineering, Wuhan Textile University, Wuhan 430073, China

ABSTRACT

Vascular implants belong to a specialised class of medical textiles. The basic purpose of a vascular implant (graft and stent) is to act as an artificial conduit or substitute for a diseased artery. However, the long-term healing function depends on its ability to mimic the mechanical and biological behaviour of the artery. This requires a thorough understanding of the structure and function of an artery, which can then be translated into a synthetic structure based on the capabilities of the manufacturing method utilised. Common textile manufacturing techniques, such as weaving, knitting, braiding, and electrospinning, are frequently used to design vascular implants for research and commercial purposes for the past decades. However, the ability to match attributes of a vascular substitute to those of a native artery still remains a challenge. The synthetic implants have been found to cause disturbance in biological, biomechanical, and hemodynamic parameters at the implant site, which has been widely attributed to their structural design. In this work, we reviewed the design aspect of textile vascular implants and compared them to the structure of a natural artery as a basis for assessing the level of success as an implant. The outcome of this work is expected to encourage future design strategies for developing improved long lasting vascular implants.

KEYWORDS

vascular stent; graft; artery; weaving; knitting; electrospinning; braiding; compliance; non-linearity; anisotropy

1. INTRODUCTION

The use of textiles for medical applications can be traced back to the early ages in wound care applications such as sutures and wound dressings. The suitability of textiles, whether in fibre or fabric form, lies in their structural flexibility, whereby some exhibit properties similar to human tissues, which are also composed of fibrous components. Another advantage of textile-based substrates is their design flexibility (from fibre to fabric stage), which can be modified to emulate the mechanical behaviour (elasticity, strength, stiffness, fluid permeability) of native biological tissue. These properties give textiles an edge over other materials (metals and plastics) in the area of soft tissue repair, for example, cardiovascular implants, which are used to replace/repair diseased arteries.

Cardiovascular implant market (grafts and stents) is growing at a fast pace due to increasing number of patients with vascular diseases and limited biological replacement options available. Thus, synthetic implants offer off-the-shelf solution in a range of design specifications. During the last two decades, a significant amount of research and industry effort has been put into developing vascular implants intended for various anatomical locations. Among the soft tissue implants, design optimisation of vascular implants is considered as one of the most complex tasks and has therefore been a continuing challenge for biomedical device engineers. The reason for this is the inability to match biomechanical behaviour of a synthetic implant to that of an artery. This is due to the heterogeneous structure of an artery, which imparts unique mechanical features (non-linearity, anisotropy, viscoelasticity, compliance) in the vascular wall. An artery shows very low stress response at low pressures and exhibits a steep increase in elastic modulus as the pressure is increased, a property known as non-linearity or incremental elastic modulus (Figure 1). Simultaneously, geometrical arrangement of structural components in the vessel wall imparts anisotropy characteristic to its mechanical behaviour. On the other hand, a synthetic arterial substitute (woven Dacron® graft) is a homogenous structure, which exhibits low elasticity and relatively higher linear stress response at similar levels of strain (Figure 1) [1]. The viscoelasticity property is another important component that determines the hemodynamic behaviour of an arterial vessel and is an equally difficult feature to mimic in a synthetic implant. When an artery is subjected to cyclic inflation-deflation stresses, it does not react instantaneously to these stresses [2,3]. Instead, the artery, due to it being viscoelastic, produces a delayed response known as hysteresis when subjected to changes in pressure and volume (Figure 1). In addition to the biomechanical properties mentioned above, compliance, defined as radial extensibility of an artery in the physiological pressure range (80–120 mmHg) plays a decisive role in vascular mechanics. In a synthetic implant, this property is directly related to structural construction and material property, and hence highly variable. An unmatched compliance between an artery and implant is a common cause of long-term complications which result in ultimate failure of the surgical procedure due to altered pressure and flow dynamics [4,5]. Therefore, the design feature of medical textiles is critical when they are being considered for vascular implant applications.

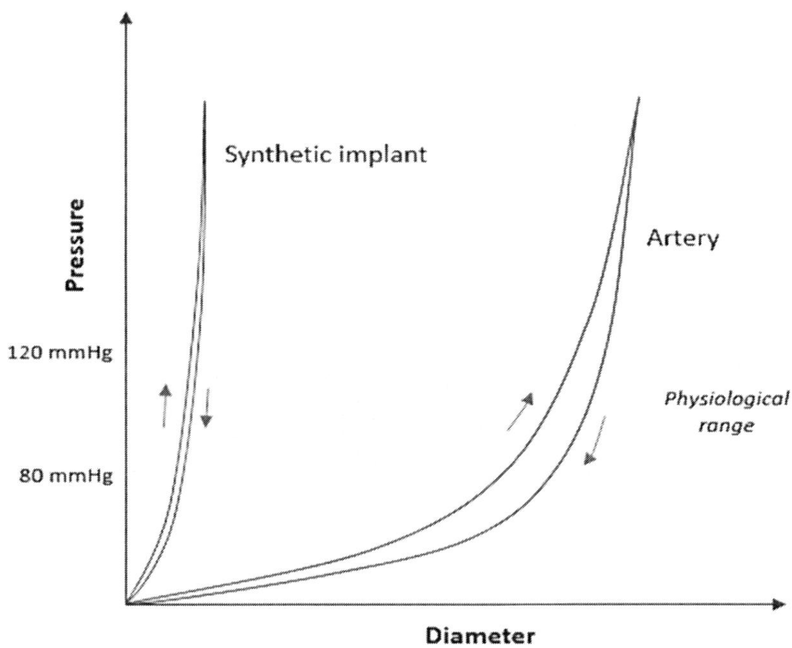

Figure 1. Comparison of pressure-diameter curves between an artery and a synthetic implant.

Synthetic vascular implants are currently manufactured using standard textile manufacturing techniques such as weaving, knitting, braiding and electrospinning. A number of reviews have reported on the mechanical property comparisons of different types of vascular grafts and their clinical performance [6,7,8]. In 1986, Pourdeyhimi and Wagner presented an extensive review focusing on structures of the synthetic grafts to explain the reported clinical observations of these grafts [9,10]. However, recent studies on grafts and stents rarely consider the effect manufacturing techniques have on the structure. This paper will discuss the structure of vascular implants produced by different textile manufacturing methods and analyse them with respect to the arterial wall structure. The analysis is done while following the development timeline of these manufacturing techniques in the vascular implant industry. This approach will provide an understanding of the degree of success, which has been achieved by textile-based implants in mimicking the mechanics of native arteries since their first clinical introduction. Some innovative design concepts, which have attempted to reduce the mechanical property mismatch between the implant and the host artery, are also discussed. These works are thought to encourage the design of longer lasting vascular implants in future.

2. ARTERY: STRUCTURE AND MECHANICAL BEHAVIOUR

The primary step towards designing a vascular implant with improved mechanical response is to understand the arterial structure, its components and mechanical role of each component. An arterial wall is composed of three major layers namely, tunica intima, tunica media, and tunica adventitia [11]. The intima or the endothelial layer comprises of a single layer of endothelial cells. The tunica media is the thickest layer and composed of circumferentially arranged elastic fibres, smooth muscle cells, and collagen fibres. The outermost adventitia layer comprises of large diameter collagen fibres oriented longitudinally as wavy bundles. The mechanical behaviour of an artery is based mainly on the thickness of media layer and its main structural components (elastin and collagen), which differ significantly in their elastic modulus (elastin = 0.6–1 MPa, collagen = 1 GPa) [2,3]. The function of both these components was investigated by Roach and Burton by selectively dissolving them and comparing the mechanical behaviour of the altered artery [1]. The study demonstrated that the initial slope in a typical stress-strain curve of an artery was contributed entirely by elastin fibres, while the final slope was due to collagen fibre stress response (Figure 2). The mid region (upturning region) involves successive transition or shifting of load from elastin to collagen and corresponds to the normal in vivo operating range (80–120 mmHg) of an artery. The non-linearity or incremental elastic modulus property is due to the wavy and random configuration of elastin and collagen fibres when unpressurised. With increase in pressure, elastin and collagen fibres start to straighten progressively. Elastin fibres become nearly straight at lower end of physiological pressure (80 mmHg). A further increase in pressure results in stretching of elastin and successive straightening as well as stretching of collagen fibres until top end of the physiological range (120 mmHg) is reached. An increase in pressure beyond this region (upturning region) results in fully stretched collagen and elastin fibres where stress response of collagen fibres dominates the arterial behaviour. This non-linear stress-strain behaviour is considered to be the key to elastic stability in arteries, which protects them from developing pathological conditions of aneurysms and 'blowout' at high pressures. The mechanical behaviour of arteries is also controlled by relative proportion of collagen and elastin fibres. High elastin content in ascending aorta (41% of dry weight) compared to descending aorta (30% of dry weight) is the reason behind a decreasing compliance trend observed while moving downstream in an aortic vessel [12,13].

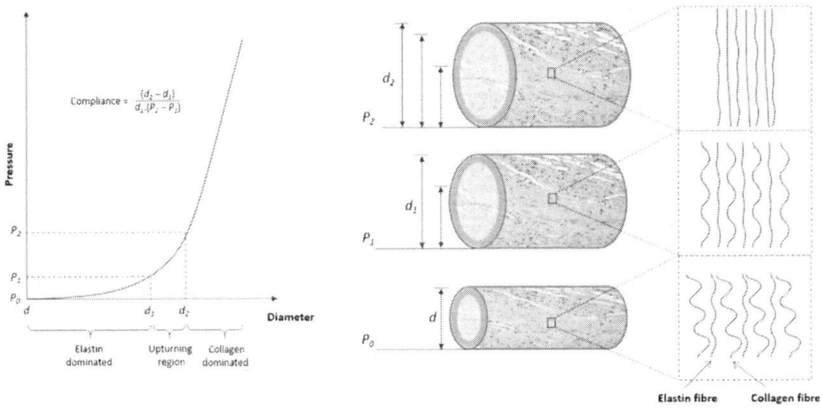

Figure 2. Role of fibrous components (elastin and collagen) in shaping the pressure-diameter relation of an artery.

The mechanical performance of an artery in the upturning region is expressed in terms of compliance value (percentage increase in diameter for a given increase in pressure) (Figure 2). A high compliance value indicates optimum level of structural performance for maintaining pulsatile blood flow. The property of an artery to support pulsatile flow is also known as the windkessel function, and is critically important in arteries in proximity of the heart such as the aorta [14]. During systole, the heart pumps nearly 60–100 mL of blood into the aorta while there is no supply during diastole. The windkessel function assists in temporarily storing a portion (approx. 50%) of systolic blood volume in aorta which can be later used during diastole to maintain consistent blood flow throughout the arterial network (Figure 3). During this process, the cross-sectional area of aorta (ascending aorta) can increase by 11% [15]. The basis of windkessel function lies in the compliance property, as a non-compliant (stiff) aortic vessel cannot expand sufficiently to store blood.

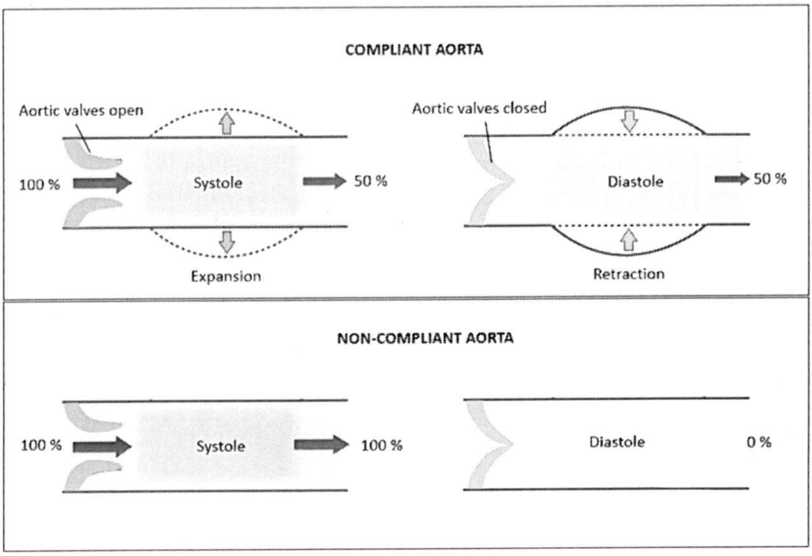

Figure 3. The role of compliance in the windkessel function of aorta.

Anisotropy is another important characteristic of arterial wall attributed to circumferential orientation of collagen fibres and their unequal orthogonality in unstressed state. Arterial tissue is 40% stiffer in circumferential compared to axial direction and approximately 100% stiffer when inflated to physiological pressures [16]. Viscoelasticity or stress relaxation is a decisive component of arterial elasticity property. When an artery is subjected to cyclic inflation-deflation stresses, viscoelasticity causes hysteresis in pressure-volume curve. The area enclosed by hysteresis loop represents the energy lost in each cycle, which indicates that a major component of strain energy is recovered elastically each time arterial wall is distended [2]. The dissipation of strain energy due to viscoelasticity assists in attenuating forward pressure pulses, which propagate along arteries as waves of circumferential distention. Additionally, viscoelastic property has been suggested to improve the fatigue life of arteries by reducing dynamic stresses and strains in the wall [17].

3. TEXTILE STRUCTURES AS VASCULAR IMPLANTS

The incorporation of textiles as vascular implants started in 1952 with the pioneering work of Voorhees and colleagues who replaced diseased aortic vessels of dogs with woven Vinyon-N (a polyvinylchloride) tubes [18,19]. Within few years of Voorhees's work, a number of studies reported clinical trials with different types of materials (Nylon, Teflon®, Dacron®, Orlon®), and constructions (woven, knitted, braided) in various diameters (6–20 mm) [20,21]. The role of manufacturing technique should be considered when deciding on parameters such as the handling and in vivo behaviour of the grafts. The definition of an ideal vascular implant requires it to be (1) biocompatible, (2) non-thrombogenic, (3) compliant, (4) fatigue resistant, (5) flexible yet robust, (6) readily available, and (7) easy to manufacture. Among these, attributes 3–7 are influenced by the manufacturing method used and structural design of the implant, while attributes 1–2 are contributed by material selection. There are a number of techniques used to manufacture medical textiles and those that are used for vascular implants are mainly weaving, knitting, braiding and electrospinning. Each of these techniques presents specific characteristics and benefits which led to varying suitability for different types of vascular implants.

3.1. Weaving

A woven graft is manufactured by interlacing two sets of yarn (warp and weft) oriented at 90° to each other. These grafts are currently available in different types of weave designs namely, plain, twill, and satin (Figure 4). The main characteristics which were considered important while introducing these grafts in clinical practice were surface smoothness, handling ease, non-reactivity, water permeability, bursting strength, suture retention strength, and biological healing response [22,23,24,25,26]. On the other hand, biomechanical optimisation of graft design was largely overlooked in the last six decades since the conception of this great innovation. This is also indicated by the fact that the basic construction (a homogenous single layer woven fabric) of a woven graft has remained the same since its introduction in the 1950s.

A commercial woven graft can be considered similar to a windcheater jacket from structural and material perspective, and bears no relationship to the composite design principle of an artery. The dissimilarity is apparent when the fibre itself is analysed, whereby the elastic behaviour of polyester fibres corresponds to collagen fibres while no graft component performs the role of elastin fibre, i.e., low stress elastic response (Table 1). Furthermore, at the structural level, the straight yarn interlacement in woven grafts is unlike the helically arranged wavy configuration of collagen fibres, which induces non-linearity property in an artery. The orthogonal arrangement of both yarn components results in low axial elongation and poor radial compliance properties in these grafts [27]. The desired axial elasticity is often achieved through graft crimping, which also helps maintain tubular shape of graft during bending. However, optimum compliance is still an unachieved goal even with numerous proofs of its clinical importance [28,29,30].

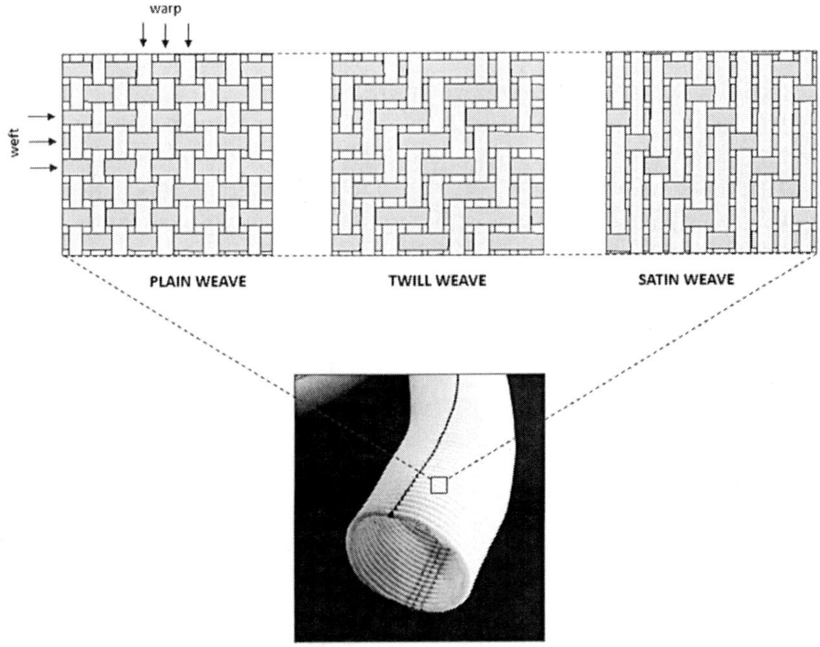

Figure 4. Structural design patterns of a woven Dacron® graft.

Table 1. Elastic modulus of components of an artery [31] and a woven Dacron® graft. (* Stainless steel strut in a woven stent-graft).

Structure	Component	Elastic Modulus (MPa)
Artery	Elastin	0.6–1
	Collagen	1000
	Smooth Muscle	0.1
Woven graft	Dacron® Polyester	800–900
	Stainless steel*	190×10^3–210×10^3

The consequences of poor compliance can be fatal as it can change the transmission characteristics of pulse waves [4,32,33]. Moreover, a compliant host artery expands more than the stiff graft, which creates an abnormal stretch on the suture line. The continuous stretch subsequently produces structural fatigue of the arterial wall at the anastomosis [5]. A significant amount of research has been done on optimising construction and handling properties of woven grafts [34,35,36,37,38,39] but radial compliance/elasticity is still an unresolved issue [27,40] and has a direct influence on long term patency of these grafts (Table 2).

Table 2. Comparison of compliance property of Dacron® grafts with natural blood vessels. Adapted from [33].

Structure	Compliance (mmHg × 10^{-2})
Artery	7.4
Vein	2.7
Dacron® Polyester (Woven)	1.9
Dacron® Polyester (Knitted)	2.3

There are very few studies that report initiatives to match graft mechanics to the host artery. These include studies published soon after the work of Voorhees, which attempted to tackle graft plasticity by using elastic filaments (Lycra® spandex) [41], and crimped synthetic filaments (Helanca® nylon and Dacron® polyester) [42,43]. Their concept was based on mimicking the role of wavy collagen and elastin fibres by using crimped and elastic yarns, respectively. Later, the combined use of both these components in plain weave constructions was successfully reported [44,45]. This approach improved the compliance nearly 17 times compared to a commercial woven graft (0.0324 ± 0.0083 kPa^{-1} vs. 0.00186 ± 0.0005 kPa^{-1}) and the new compliance matched that of a human common carotid artery (0.0238 ± 0.0132 kPa^{-1}) [44]. However, from the design perspective, woven graft was still in its primitive single layer fabric stage.

In 2011, an innovative woven graft design concept was reported by Chen et al. based on the coaxial graft concept of Sonoda et al. [46,47]. In this work, a bilayer woven graft prototype was developed in which the inner layer was constructed from low modulus yarns (poly-trimethylene terephthalate) while the outer layer made from high modulus yarns (polyester) was stitched in a crimped form to the inner layer (Figure 5). This type of structure can mimic the layered structure of blood vessels in a way that the strain at lower pressures is absorbed by the inner layer while a further increase in pressure will cause the outer layer to uncrimp and "join-in" with the inner layer to increase the overall elastic modulus of the graft. The authors reported a similar observation where elastic modulus of the new graft remained low, up to 80 mmHg and increased rapidly afterwards [46]. However, a drawback of this design is reduced bending flexibility or kink resistance due to the crimped outer layer, which may limit its clinical applicability.

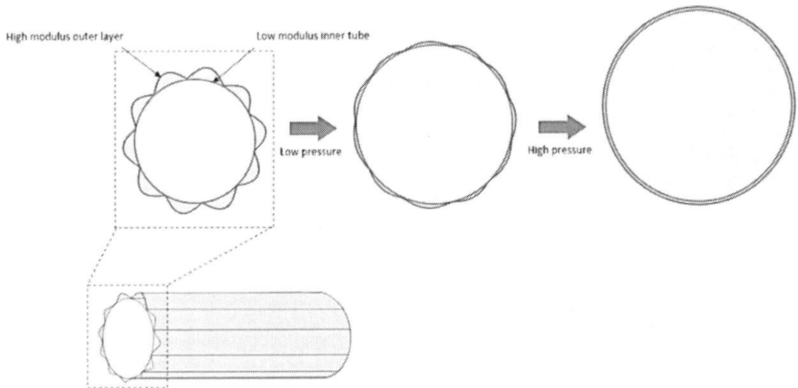

Figure 5. An explanation of thebilayer woven graft design concept proposed by Chen et al. [46].

While in vitro studies can provide proof of concept confirmation for new designs, in vivo studies, whether on animals or human beings, provide more convincing evidence of their performance in actual clinical environment. One of the early long-term animal studies established the advantage of bi-component graft design (spandex and Dacron®) in improving graft mechanical properties [45]. The graft was sufficiently compliant (6.9%, 80/100 mmHg; 9.5%, 80/200 mmHg) even after one year of implantation and did not show any plastic dilatation at explantation after three years. Unfortunately, no published in vivo trial data is available for rest of the new designs discussed above [44,46,47].

3.2. Knitting

Knitted grafts have a looped filament construction in which a continuous interconnecting chain of yarn loops spirals around the graft circumference. Knitted structures are softer, more flexible, compliant, and have better handling characteristics than woven structures. The most common types of knits, which are used for graft design, are the weft knit and warp knit constructions (Figure 6). Warp knitted structures have less stretch than weft knits, and therefore are inherently more dimensionally stable. The knitted grafts were first introduced in clinical practice in 1955 with the intent to remove the seam problem in a woven graft [48]. Later, in 1958, the renowned heart surgeon Dr. Michael DeBakey firmly established the clinical usefulness of knitted Dacron® grafts [49,50]. The trend of using Dacron® as a standard graft material started thereafter owing to the better long-term biostability of Dacron® compared to other available materials (Nylon and Orlon®). Improved stretch widthwise (radial compliance) compared to woven grafts was a promising observation in knitted grafts. However, high porosity and long-term dilatation were also reported [26,41]. The reason being absence of structural heterogeneity in knitted grafts as in arteries which protects them from undergoing fatigue dilation. Since then, several studies have been published reporting the use of different coating techniques, constituent materials, and knitted patterns with a motive to improve handling, fatigue and biological performance of these grafts [51,52,53,54,55,56,57,58,59]. Based on the inherent structural flexibility of knitted structures, many investigations have reported their use as elastic tubular substrates [60,61,62,63,64]. This property also formed the reason behind the use of monofilament knitted mesh structures as vascular stents [65,66,67,68,69]. However, commercially available knitted implants do not vary much in their structure, which is mainly a plain knit single layer homogenous structure bearing resemblance to an upholstery fabric rather than to an artery. In addition, Dacron® has been the only material for commercial knitted grafts since its use was first proposed by DeBakey [49]. Although Dacron® grafts are currently used in the clinics, there is a significant mismatch between their mechanical behaviour in comparison to natural arteries (Table 2) [40].

Similar to woven grafts, knitted grafts also underwent design trials to improve their elastic behaviour. The first report on use of spandex filament knitted graft as a dog abdominal aorta (diameter 8–10 mm) was presented by Wagner et al. [41]. However, the solo use of spandex fibre as graft material was observed to cause long-term dilatation defect in the graft and is attributed to homogenous single layer structure of the graft. Some of the latest studies tried to use a composite polyester/spandex filament yarn to improve graft elasticity [62,64]. This type of material composition allows load sharing among both components and can prevent dilatation issues if structural design pattern is also modified. However, these studies only report basic improvements in mechanical properties of a plain weft knit structure and lack the ability to be considered as a significant design improvement to mimic arterial mechanics in a knitted graft.

In a recent study, an innovative knitted stent-graft design was reported which closely mimics the natural artery mechanical behaviour [70]. The design is based on the concept of longitudinal structural segmentation or metamerism in which the knitted tube is divided into multiple low and high modulus segments arranged in alternating sequence (Figure 7). The low modulus (knitted polyurethane) sections tend to remain contracted (reduced diameter) when unpressurised while high modulus (knitted polyester) maintain the as-knit configuration. Therefore, at low internal pressure, the expansion of low modulus segments controls the stress response of the knitted tube until their circumference equals that of high modulus segments. At high pressures, the combined response of both the segments increases the stress response sharply, exhibiting an incremental elastic modulus property similar to natural arteries. The low modulus segments act as intermittent "buffer zones" which assist in radial expansion as well as provide a kink-free configuration to the knitted tube. The compliance of this new design (volumetric: 0.056 ± 0.006 mL/mmHg; radial: 9.8×10^{-4} $mmHg^{-1}$) is nearly 7 and 15 times better compared to a conventional knitted stent (radial: 1.45×10^{-4} $mmHg^{-1}$) and a commercial woven Dacron® graft (volumetric: 0.0038 ± 0.002 mL/mmHg), respectively, and falls well within the physiological range of aortic vessel [70,71]. However, the in vivo performance of this design is still unavailable to demonstrate its clinical performance.

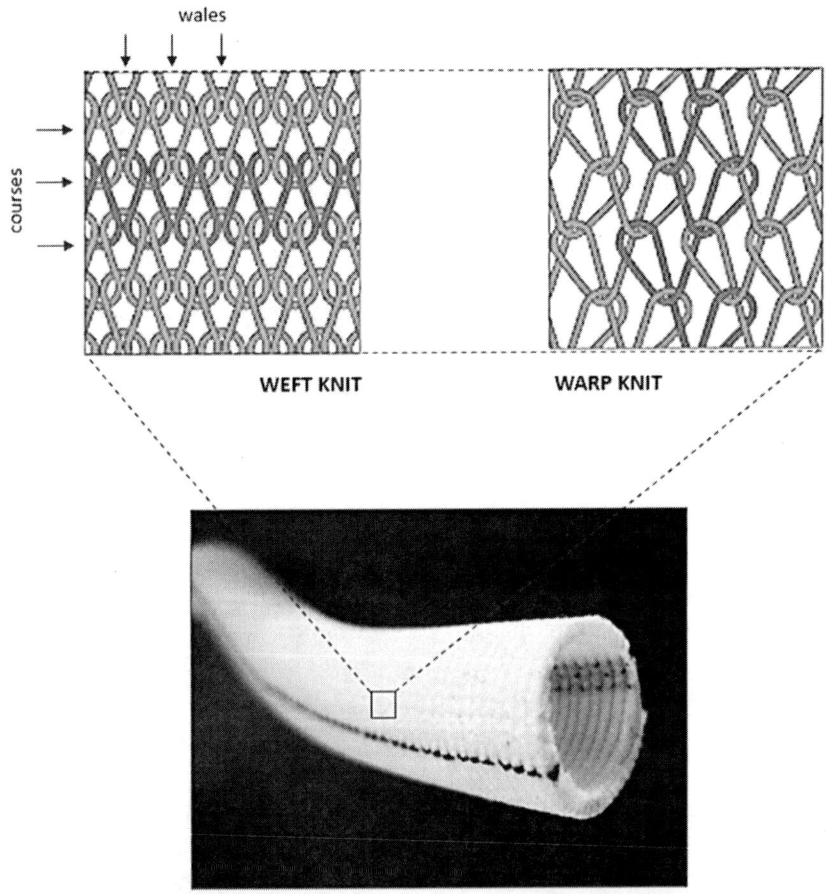

Figure 6. Structural design patterns of a knitted Dacron® graft.

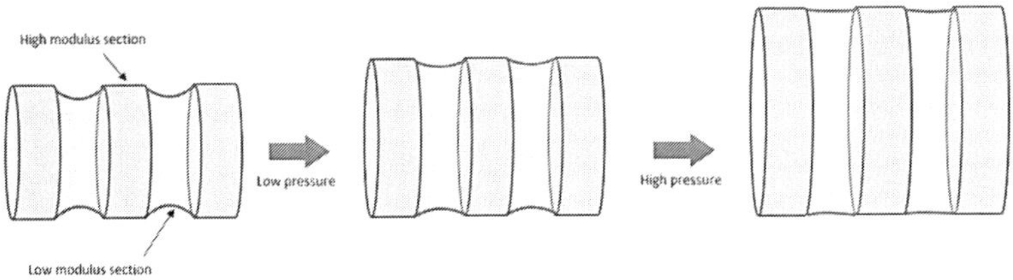

Figure 7. The segmented design concept as proposed by Singh and Wang to improve the compliance property of a knitted vascular implant [70,71].

3.3. Braiding

The application of braided textiles in traditional application areas (apparel, upholstery) is very limited compared to knitted and woven structures and is mainly found in technical textiles (industrial, sports, automotive). Braiding technique involves the use of three or more component yarns, which are intertwined at an angle to each other. In other words, braided structures are similar to traditional woven structures but with an angle bias. The angled mesh structure allows easy radial expansion of braided tubes. This property formed the basis of using braiding in the early years of textile graft innovation [72,73,74,75].

The advent of endovascular technology in the 1990s led to the search for user-friendly stent designs [76]. The ability of tubular braided structures to compress easily and recoil back made them a promising candidate for self-expanding vascular stents [77]. Low bending rigidity of braided mesh tube was also an additional advantage for stent application. However, radial expansion in braided tubes occurred at the expense of axial shortening and elastic recovery was limited to open mesh braids only. Also, braided stents were not very successful in endograft application as it required suturing/bonding of a tubular graft to stent wires and, hence, limited the movement of constituent wires at their cross-over points which ultimately led to stent failure (wire breakage) under cyclic, in vivo conditions [78]. This limitation shifted braided stents towards stent applications which do not require graft covering, suitable for peripheral atherosclerotic arteries, which is also their current main application area (Table 3) [79].

The commercially available braided stents are made from various metallic alloys (Table 3), which make them highly non-compliant and mechanically similar to a braided plumbing hose (Figure 8). Since low porosity is an important requirement in flow diverter stents, the number of wires is generally kept high (up to 96) to increase mesh density [80,81,82,83]. Similarly, compression of atherosclerotic region requires high radial compression strength, which is generally achieved by increasing mesh density. However, an inverse effect of high mesh density is increased stent stiffness (radial and longitudinal), which has a negative impact on the host artery hemodynamics [84,85]. The arterial length covered by the stent becomes non-compliant and results in pressure attenuation across this region and localised increase in pulse-wave velocity. A stiff metallic stent also causes straightening of the host artery and may induce kinks [86]. The development timeline of braided vascular stents to date mainly includes trials with various polymeric filaments [65,87,88,89,90,91,92,93,94,95,96] and stent parameter optimisation studies [89,96,97,98,99,100], while improvement in conventional braided design (a single component homogenous construction) to match arterial mechanics was rarely investigated.

Table 3. Categorisation of commercial braided stents according to their application area, material and design feature.

Trade Name	Manufacturer	Application	Material	Design Feature
PIPELINE®	ev3 Inc.	Flow diverter stent	Cobalt-Chromium + Platinum	Single layer braided tube
p64®	Phenox GmbH		Nitinol	
LEO® PLUS	Balt Extrusion		Nitinol	
SILK®			Nitinol + Platinum	
LVIS Device®	MicroVention Inc.		Nitinol	
WALLSTENT™	Boston Scientific Co.	Carotid stent	Elgiloy®	
ROADSAVER®	Terumo		Nitinol	
SUPERA®	Abbott Vascular	Peripheral stent	Nitinol	
Agili-D®	Altura Medical	Abdominal endograft	Metal alloy	

The research contribution towards improving braided stent mechanics is very limited but still worth mentioning. The concept of using a multilayer braid design was introduced to reduce porosity of the stent while maintaining the bending flexibility, which otherwise becomes too low in a high mesh density single layer stent to be able to conform well to the arterial wall. The use of multiple layer braided stent has been trialled in many clinical studies and is commercially available as CARDIATIS® multilayer stent [101,102,103]. Another new design concept was reported recently which is a single layer braided structure but with two different zones/sections along its length [104,105]. The central zone (70% of stent length) is a tightly braided high mesh density section while the margins have low mesh density. This allows reduced blood flow to the diseased portion of the blood vessel while maintaining flexibility in rest of the "non-functional" stent length. The commercial version of this design is available as E-volution® stent (Jotec GmbH, Germany). Marchand et al. tried to improve the radial compliance of a braided heart valve stent by optimising shape-setting parameters of the stent material (nitinol alloy) [106]. The modified heat treatment process was reported to improve radial expandability of the stent when tested in an in vitro test experiment.

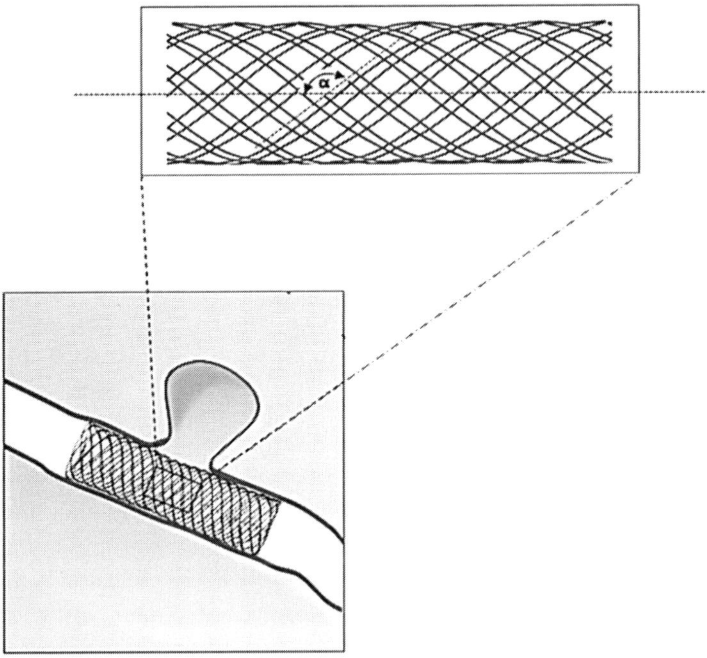

Figure 8. Structural geometry of a braided metallic stent (α = braid helix angle).

3.4. Electrospinning

Since their introduction, Dacron® grafts (woven and knitted) have dominated the vascular graft market in large diameter (15–30 mm) blood vessel replacement. Dacron® grafts have also been used as replacements for small diameter (<6 mm) arteries (coronary, below-the-knee, tibial, and peroneal). However, there were a number of limitations identified such as owing to mismatched compliance, graft thrombosis, and anastomotic intimal hyperplasia (Table 4) [6,23,107,108,109,110]. One of the most commonly used strategies to address some of these limitations is by coating the synthetic grafts with protein such as collagen. As shown in Table 5, the general trend observed is that grafts coated with collagen, fibronectin or heparin demonstrated better cell attachment in vitro and improved patency in vivo [111,112,113,114].

Table 4. Compliance and cumulative patency of different arterial grafts implanted in a small diameter (femoropopliteal artery) position. Adapted from [115]; PTFE, Polytetrafluoroethylene.

Graft Type	Compliance (% mmHg × 10^{-2})	Structure	Patency % (30/180 days)	Patency % (1/2 years)
Human femoral artery	5.9	–	–	–
Saphenous vein	4.4	Natural tissue	94/93	88/84
Umbilical vein	3.7	Natural tissue	97/93	83/80
PTFE	1.6	Extruded (non-textile)	85/81	60/42
Dacron®	1.9	Woven	88/72	65/42

The graft structure (porosity, fibre diameters, pore connectivity, surface area, and compliance) is important as it influences cell growth behaviour. This led to the development of a new textile fibre spinning technique known as electrospinning, which is capable of producing fibres to the scale of native collagen and elastin fibres. In an electrospinning process, a strong electric field is generated between a polymer solution (delivered through a syringe needle) and a metallic collector. When the voltage reaches a critical value, the charge overcomes the surface tension of the polymer solution drop at the needle tip and polymer jet is generated. While travelling towards the metallic collector the drawing force exerted by electric field and simultaneous evaporation of the solvent results in reduction in the diameter of the jet. The collected dry fibres form a nonwoven mesh of nanometre to micrometre diameter fibres (Figure 9). The process can be adjusted to control fibre diameter to some extent by varying the charge density and polymer solution concentration. The feature of producing very fine fibres increases the surface area significantly in electrospun meshes, which is a beneficial property for improving cell growth [116].

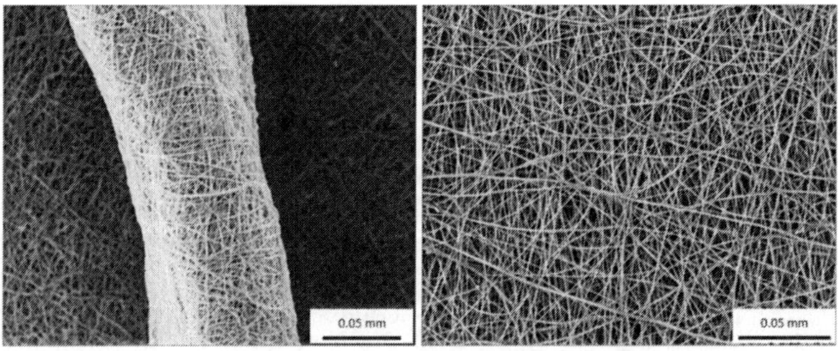

(a) (b)

Figure 9. The wall thickness (**a**) and surface (**b**) view of an electrospun mesh.

The structural construction of an electrospun fibre scaffold can closely mimic the natural extracellular matrix structure [117]. Electrospinning also enables the use of a wide range of natural and synthetic polymers [118], thus, increasing the possibility of matching both the biological and mechanical properties similar to an artery (Table 5) [119]. The electrospinning of natural proteins like collagen and elastin have demonstrated the usefulness of this technique to bring synthetic vascular grafts near to their biological counterparts [120,121]. From a structural design perspective also, electrospinning provides a range of possibilities in mimicking the multilayered construction of arterial wall [122,123,124,125]. The use of different polymeric components in individual layers provides more flexibility in controlling the graft mechanical response from each structural layer.

4. DISCUSSION

The woven graft development timeline suggests that this technology has shown biomechanical benefits of using new structural components but there are very limited attempts which focus on improvising conventional weave designs to suit the arterial site. Knitted structures appear to be a suitable candidate for vascular implant application owing to their inherent structural and design flexibility. Also, anisotropic elasticity property (axial > circumferential) in knitted structures is similar to that of the native artery, which is an advantage over woven structures. Currently, some of the latest developments in woven and knitted grafts use biological coatings to improved tissue growth and design modifications to match the implant site anatomy, while the lack of innovation from material and biomechanical aspect is evident in new commercial products (Table 6). The future products may require a combined input from advanced textile designing and biomechanics together in order to realise the full clinical potential of textile grafts. The helical arrangement of constituent filaments/wires give braided structures a design advantage over woven and knitted structures to mimic the helical geometry of collagen and elastin fibres in an artery. However, the focus on this aspect of braided stent improvement is mostly unrealised. Since, braided stents are deployed in their fully expanded state (helix angle approaching 90°), the low stress radial expansion property owing to helical geometry is completely lost (Figure 8). The use of elastic filaments and design of bio-component structures can be a future prospect for developing compliant braided stent devices. Expanded polytetrafluoroethylene (ePTFE) is an inert fluorocarbon polymer (stiffness: 0.5 GPa, tensile strength: 14–18 MPa), which is developed by heating, stretching, and extruding process resulting in a porous polymeric structure. Non-textile grafts made from ePTFE are currently used in the clinics as medium diameter grafts (7–9 mm) for peripheral vascular diseases. However, ePTFE grafts are not viable as conduits for small diameter (<6 mm) vessel replacement due to a high rate of occlusion. Although ePTFE is a chemically inert material, the structural stiffness and low radial compliance of these grafts contributes to their poor long-term patency (Table 4). There have been numerous investigations into endothelial cell seeding and surface functionalization of ePFTE grafts to improve their clinical performance as small diameter grafts. Current efforts have yet to produce a small diameter synthetic graft that is comparable to an autologous graft. Electrospinning technique has proven to be a promising option for small diameter grafts in many in vitro and animal studies. In comparison to established textile manufacturing methods (weaving, knitting, braiding), this technology is equipped with much higher levels of design flexibility in terms of material variety (natural, synthetic), structural heterogeneity (multi-layer, multi-component), and therapeutic ability (drug delivery). However, development of an off-the-shelf electrospun graft product capable of achieving rapid cell coverage with minimised risk of thrombosis, intimal hyperplasia, and mechanical failure has yet to be achieved.

Radial compliance is unarguably the most vital mechanical feature requirement from a synthetic vascular implant. In conjunction with compliance, non-linearity and anisotropy are two critical mechanical characteristics, which a vascular graft must also possess in order to successfully mimic natural artery mechanics. Non-linearity prevents the late development of pathological conditions related to graft failure at high pressures [2,3]. Anisotropic mechanical property prevents excessive stimulation of anastomotic region, while being compliant in radial direction to prevent any flow disturbances [125]. In an artery, both these properties are achieved owing to the multilayered heterogeneous structure of arterial wall, which a single-layer single-component isotropic textile structure is unable to mimic.

Table 5. Biological responses in various types of synthetic vascular grafts created from different manufacturing techniques.

Technique	Material	Scaffold Dimensions	Biological Response	Mechanical Testing	Ref.
Knitted	PET	4 mm diameter	ECs: Better cell attachment was observed on precoated grafts in the following order: collagen (3.5×10^5 cells/cm^2) > fibrin (2.8×10^5 cells/cm^2) > fibronectin (2.4×10^5 cells/cm^2) > laminin = untreated (1.3×10^5 cells/cm^2)	–	[126]
Knitted	PET	4 mm diameter cut into fusiform patches of 5 cm length × 8 mm width	*In vivo*: Implanted in sheep for 4 weeks and intimal hyperplasia assessed. Fluoropolymer coated PET sealed with gelatin produced the least amount of hyperplasia followed by PTFE, carbonlined PTFE and gelatin sealed PET	–	[127]

Table 5. *Cont.*

Technique	Material	Scaffold Dimensions	Biological Response	Mechanical Testing	Ref.
Knitted	PET	–	*In vivo*: Randomised clinical trial of 209 patients (femoropopliteal bypass). Patency at 3 years of collagen-coated heparin bonded PET (54%) was better than PTFE (44%). No difference in patency at 5 years between PET and PTFE	–	[128]
Woven and Electrospinning	PET and PU	Flat	*In vitro* (HUVECs): All 3 materials showed better cell attachment when coated with Collagen Type I/III as compared to their uncoated counterparts. Cell coverage on coated PTFE (34.6%) > PET (19.6%) > PU (17.5%); When exposed to shear stress of 1 Pa for 1 h, cell retention was highest on coated PTFE. No difference was observed across the 3 materials when shear stress increased to 2 Pa	–	[129]
Electrospinning	PCL and PU	NS	*In vitro* (HUVECs): Good cell attachment (~ 61%). Cells produced proteins such as PECAM and vWF, indicating normal phenotype and function	–	[130]
Electrospinning	Silk fibroin	1.5 mm diameter	*In vivo*: Implanted in rats for 7 days to 6 months. No graft degradation or foreign body reaction observed. 95% of grafts remained unblocked. After 1 month implantation, ECs were observed on the luminal surface of the grafts. Cell coverage continue to increase at 6 months	–	[131]
Electrospinning	Silk	5 mm diameter	*In vitro*: ECs and SMCs attached and proliferated on the scaffold. ECs were observed on the surface of the scaffold while SMCs migrated into the scaffold	Tensile strength: 2.42 MPa; Elastic modulus: 2.45 MPa; Mean burst pressure: 811 mmHg	[132]
Electrospinning	PCL	2 mm diameter	*In vivo*: Implanted in rats for 24 weeks. No narrowing of the grafts (stenosis) in the PCL group. ECs coverage confluent at 12 weeks in the PCL group *vs.* ePTFE group (incomplete at 24 weeks).	–	[133]
Electrospinning	PCL	NS	*In vitro*: A confluent layer of oriented SMCs in the direction of aligned fibres after 7 days culture	–	[134]

Table 5. *Cont.*

Technique	Material	Scaffold Dimensions	Biological Response	Mechanical Testing	Ref.
Electrospinning	Collagen Type 1, elastin and PLGA	Tubular (4.75 mm diameter, 12 cm long)	*In vitro*: Mean of 72% for ECs and 82% for SMCs viability over 7-day culture period; *In vivo* (mice): No systemic or neurological toxicity, normal blood count, transient local inflammation at implantation site	Burst pressure: 1425 mm Hg; Compliance: 12%–14% for scaffold *vs.* 9% for native vessel; Maximal axial and circumferential strain: 40% strain	[119]
Electrospinning	Collagen and PCL	Tubular (4.75 mm diameter, 12 cm long)	ECs: Cytoskeletal organisation and focal adhesion via actin and vinculin staining respectively were better developed when cultured on smaller sized fibres; SMCs: Infiltration of cells into scaffolds with fibres > 1 µm in diameter during a 4-week culture	Tensile strength: Increasing fibre diameter (0.27 µm to 4.45 µm) decreased tensile strength from 3.15 MPa to 0.75 MPa; Elongation at break: Increased with increasing fibre diameter (90% for 0.27 µm fibres to 734% for 4.45 µm fibres); Maximum load: Decreased from 25.75 N (0.27 µm fibres) to 8.63 N (4.45 µm fibres)	[135]
Knitting	Elastin solubilised proteins and PET	Flat	HUVECs: No cytotoxicity, 43% cell attachment for elastin solubilised protein-PET *vs.* 94% for PET	–	[136]
Electrospinning	Elastin and PDO	6 mm diameter	*In vitro*: Fibroblasts cultured for 7 days on PDO:Elastin graft showed migration into the fibrous materials *vs.* no migration in PDO graft	PDO:Elastin ratio of 50:50 produced compliance that mimics native femoral artery	[137]
Electrospinning	PCL, PDO and Silk	NS	*In vitro*: Risk of clotting using human monocytes – PCL < Silk < PDO. The risk in PCL is comparable to ePTFE (currently used grafts in clinics)	–	[138]
Electrospinning	Chitosan and PVA	NS	*In vitro*: Good fibroblast growth was observed with continual proliferation up to 21 days	–	[139]

EC = endothelial cell; ePTFE = expanded polytetrafluoroethylene; HUVEC = human umbilical vein endothelial cell; NS = not specified; PCL = polycaprolactone; PDO = polydioxanone; PET = polyethylene terephthalate; PLGA = poly(lactic-co-glycolic acid); PTFE = polytetrafluoroethylene; PU = polyurethane; PVA = polyvinyl alcohol; SMC = smooth muscle cell.

Table 6. A list of selected commercially available Dacron® grafts and their design features.

Structure	Material	Trade Name	Special Design Feature	Intended Improvement	Manufacturer	Application Area
Woven	Dacron® & PTFE	FUSION BIOLINE®	Two layer. Heparin coated	Improves patency and healing response	Macquet	Peripheral
	Dacron®	HEMASHIELD® PLATINUM	Multiple branched. Double velour, Collagen coated	Enhances healing response		Aortic
		VASCUTEK® GELWEAVE™ Pre-curved	Pre-curved design	Matches aortic arch anatomy	Terumo	
		VASCUTEK® SIENA™	Extended trunk and collar design	Suits hybrid surgery procedures		
		VASCUTEK® GELWEAVE™ Plexus	Multiple branched. Gelatin impregnated	Suits complete aortic arch surgery		
		VASCUTEK® GELWEAVE™ Valsalva	Extended skirt design	Matches aortic root anatomy		
		VASCUTEK® GELWEAVE™ Ante-Flo	Extra branch	Reduces surgery time. Lowers risk of neurological deficits		
		E-VITA™ OPEN PLUS	Extended stented trunk	Suits hybrid surgery procedures	Jotec	
Knitted		INTEGRAD® SILVER	Silver impregnated	Reduces graft infection	Macquet	Aortic, Peripheral
		HEMAGARD® Ultrathin	Collagen coated, Wall thickness = 0.35 mm	Improves healing response, Reduces dilatation. Increases suture strength		Aortic
		HEMASHIELD® GOLD	Collagen coated, Double velour surface	Improves healing response, Reduces dilatation. Increases suture strength		
		VASCUTEK® GELSEAL™	Gelatin impregnated	Improves healing	Terumo	Aortic, Peripheral
		VASCUTEK® GELSOFT™ ERS	Gelatin impregnated. Radially supported	Improves healing and handling		
		FLOWNIT BIOSEAL®	Texturised yarn, Collagen impregnated	Low dilatation. Enhances tissue incorporation	Jotec	Aortic

5. CONCLUSIONS

This paper reviewed the design aspect of medical textiles (woven, knitted, braided, electrospun) intended for vascular implant applications. The three dimensional structure of an arterial wall and its unique mechanical properties (anisotropy, non-linearity, compliance, viscoelasticity) have been widely researched and reported. However, these features remain widely unconsidered while designing synthetic vascular implants. Therefore, the vascular implants used currently in clinics serve the function of a rigid non-distensible conduit but lack the ability to revive lost biomechanical function of the diseased artery. Currently available textile vascular implants are not significantly different from those introduced six decades ago. Their structural geometry is analogous to traditional textile structures rather than to an arterial vessel. This difference is highlighted when late clinical complications arise from behavioural mismatch at the artery-implant anastomosis. These observations ultimately raise the importance of understanding the structure and biomechanics of an artery before adapting a textile structure from its conventional application area to a biological environment consisting of complex structure and functions such as the blood vessel. A synthetic vascular implant with structural characteristics that closely resemble a native artery will present with less complications, which, in turn, translates to a longer-lasting implant.

AUTHOR CONTRIBUTIONS

C.S., C.S.W. and X.W. planned the manuscript. C.S. and C.S.W. wrote the manuscript. X.W. commented on and modified the manuscript.

CONFLICTS OF INTEREST

The authors declare no conflict of interest.

REFERENCES

1. Roach, M.R., Burton, A.C. The reason for the shape of the distensibility curves of arteries. Can. J. Biochem. Physiol. 1957, 35, 681–690.

2. Shadwick, R.E. Elasticity in Arteries. Am. Sci. 1998, 86, 535–541.

3. Shadwick, R.E. Mechanical design in arteries. J. Exp. Biol. 1999, 202, 3305–3313.

4. Abbott, W.M.; Megerman, J.; Hasson, J.E.; L'Italien, G.; Warnock, D.F. Effect of compliance mismatch on vascular graft patency. J. Vasc. Surg. 1987, 5, 376–382.

5. Weston, M.W.; Rhee, K.; Tarbell, J.M. Compliance and diameter mismatch affect the wall shear rate distribution near an end-to-end anastomosis. J. Biomech. 1996, 29, 187–198.

6. Chlupác, J.; Filová, E.; Bačáková, L. Blood vessel replacement: 50 years of development and tissue engineering paradigms in vascular surgery. Physiol. Res. 2009, 58, 119–140.

7. Kannan, R.Y.; Salacinski, H.J.; Butler, P.E.; Hamilton, G.; Seifalian, A.M. Current status of prosthetic bypass grafts: A review. J. Biomed. Mater. Res. B. 2005, 74, 570–581.

8. Salacinski, H.J.; Goldner, S.; Giudiceandrea, A.; Hamilton, G.; Seifalian, A.M.; Edwards, A.; Carson, R.J. The mechanical behavior of vascular grafts: A review. J. Biomater. Appl. 2001, 15, 241–278.

9. Pourdeyhimi, B.; Text, C. A review of structural and material properties of vascular grafts. J. Biomater. Appl. 1987, 2, 163–204.

10. Pourdeyhimi, B.; Wagner, D. On the correlation between the failure of vascular grafts and their structural and material properties: A critical analysis. J. Biomed. Mater. Res. 1986, 20, 375–409.

11. Silver, F.; Snowhill, P.; Foran, D. Mechanical behavior of vessel wall: A comparative study of aorta, vena cava, and carotid artery. Ann. Biomed. Eng. 2003, 31, 793–803.

12. Fischer, G.M.; Llaurado, J.G. Collagen and elastin content in canine arteries selected from functionally different vascular beds. Circ. Res. 1966, 19, 394–399.

13. Kassab, G.S. Biomechanics of the cardiovascular system: The aorta as an illustratory example. J. R. Soc. Interface 2006, 3, 719–740.

14. Belz, G.G. Elastic properties and Windkessel function of the human aorta. Cardiovasc. Drugs Ther. 1995, 9, 73–83.

15. Greenfield, J.C.; Patel, D.J. Relation between pressure and diameter in the ascending aorta of man. Circ. Res. 1962, 10, 778–781.

16. Lillie, M.A.; Shadwick, R.E.; Gosline, J.M. Mechanical anisotropy of inflated elastic tissue from the pig aorta. J. Biomech. 2010, 43, 2070–2078.

17. Zhang, W.; Liu, Y.; Kassab, G.S. Viscoelasticity reduces the dynamic stresses and strains in the vessel wall: Implications for vessel fatigue. Am. J. Physiol. 2007, 293, H2355–H2360.

18. Blakemore, A.H.; Voorhees, A.B. The use of tubes constructed from vinyon "N" cloth in bridging arterial defects—Experimental and clinical. Ann. Surg. 1954, 140, 324–333.

19. Voorhees, A.B.J.; Jaretzki, A.I.; Blakemore, A.H. The use of tubes constructed from Vinyon "N" cloth in bridging arterial defects: A preliminary report. Ann. Surg. 1952, 135, 332–336.

20. Harris, E.J.; Shumacker, H.B., Jr.; Siderys, H.; Moore, T.C.; Grice, P.F. Pliable plastic aortic grafts: Experimental comparison of a number of materials. AMA Arch. Surg. 1955, 71, 449–459.

21. Hufnagel, C.A.; Rabil, P. Replacement of arterial segments, utilizing flexible orlon prostheses. AMA Arch. Surg. 1955, 70, 105–110.

22. Deterling, R.A., Jr.; Bhonslay, S.B. Use of vessel grafts and plastic prostheses for relief of superior vena caval obstruction. Surgery. 1955, 38, 1008–1026.

23. Harrison, J.H. Synthetic materials as vascular prostheses. I. A comparative study in small vessels of nylon, dacron, orlon, ivalon sponge and teflon. Am. J. Surg. 1958, 95, 3–15.

24. Julian, O.C.; Deterling, R.A., Jr.; Su, H.H.; Dye, W.S.; Belio, M.L. Dacron tube and bifurcation arterial prostheses produced to specification. Experimental and clinical use. Surgery 1957, 41, 50–61.

25. Knox, G.; Begg, C.F. Evaluation of the need for porosity in synthetic arterial prostheses. A case report and evaluation of dacron aortic grafts in dogs. Surgery 1957, 42, 922–927.

26. Harrison, J.H. Synthetic materials as vascular prostheses: II. A comparative study of nylon, dacron, orlon, ivalon sponge and teflon in large blood vessels with tensile strength studies. Am. J. Surg. 1958, 95, 16–24.

27. Tremblay, D.; Zigras, T.; Cartier, R.; Leduc, L.; Butany, J.; Mongrain, R.; Leask, R.L. A comparison of mechanical properties of materials used in aortic arch reconstruction. Ann. Thorac. Surg. 2009, 88, 1484–1491.

28. Lantelme, P.; Dzudie, A.; Milon, H.; Bricca, G.; Legedz, L.; Chevalier, J.M.; Feugier, P. Effect of abdominal aortic grafts on aortic stiffness and central hemodynamics. J. Hypertens. 2009, 27, 1268–1276.

29. Kim, S.Y.; Hinkamp, T.J.; Jacobs, W.R.; Lichtenberg, R.C.; Posniak, H.; Pifarré, R. Effect of an inelastic aortic synthetic vascular graft on exercise hemodynamics. Ann. Thorac. Surg. 1995, 59, 981–989.

30. Obrien, T.; Morris, L.; McGloughlin, T. Evidence suggests rigid aortic grafts increase systolic blood pressure: Results of a preliminary study. Med. Eng. Phys. 2008, 30, 109–115.

31. Fung, Y.C. Biomechanics: Mechanical Properties of Living Tissues; Springer: New York, NY, USA, 2013.

32. Tai, N.R.; Salacinski, H.J.; Edwards, A.; Hamilton, G.; Seifalian, A.M. Compliance properties of conduits used in vascular reconstruction. Br. J. Surg. 2000, 87, 1516–1524.

33. Kidson, I.G. The effect of wall mechanical properties on patency of arterial grafts. Ann. R. Coll. Surg. Engl. 1983, 65, 24–29.

34. Anderson, K.; Seyam, A. Developing seamless shaped woven medical products. J. Med. Eng. Technol. 2004, 28, 110–116.

35. Kotsuka, Y.; Ezure, M.; Tanaka, K.; Takamoto, S.; Kuwahara, K.; Ikeda, H. Ultrathin-wall vascular grafts for endovascular surgery. J. Artif. Organs 2002, 5, 108–112.

36. De Paulis, R.; Scaffa, R.; Maselli, D.; Salica, A.; Bellisario, A.; Weltert, L. A third generation of ascending aorta dacron graft: Preliminary experience. Ann. Thorac. Surg. 2008, 85, 305–309.

37. Abdessalem, S.B.; Durand, B.; Akesbi, S.; Chakfe, N. Blood flow in a polyester textile vascular prosthesis: Experimental and numerical study. Text. Res. J. 2001, 71, 178–183.

38. Hajjaji, R.; Abdessalem, S.B.; Ganghoffer, J.F. The influence of textile vascular prosthesis crimping on graft longitudinal elasticity and flexibility. J. Mech. Behav. Biomed. Mat. 2012, 16, 73–80.

39. Moghe, A.K.; Gupta, B.S. Small-diameter blood vessels by weaving: Prototyping and modelling. J. Text. Inst. 2008, 99, 467–477.

40. Hasegawa, M.; Azuma, T. Mechanical properties of synthetic arterial grafts. J. Biomech. 1979, 12, 509–517.

41. Wagner, M.; Reul, G.; Teresi, J.; Kayser, K.L. Experimental observations on a new and inherently elastic material for sutures and vascular prostheses: Lycra. Am. J. Sur. 1966, 111, 838–841.

42. Szilagyi, D.; France, L.C.; Smith, R.F.; Whitcomb, J.G. The clinical use of an elastic dacron prosthesis. AMA Arch. Surg. 1958, 77, 538–551.

43. Szilagyi, D.; Whitcomb, J.G.; Shonnard, C.P. Replacement of long and narrow arterial segments: II. Experimental studies with an elastic

("Helanca") seamless woven nylon prosthesis. AMA Arch. Surg. 1957, 74, 944–953.

44. Gupta, B.S.; Kasyanov, V.A. Biomechanics of human common carotid artery and design of novel hybrid textile compliant vascular grafts. J. Biomed. Mater. Res. 1997, 34, 341–349.

45. Rosenberg, N.; Simpson, A.N.; Brown, R.E. A circumferentially elastic arterial prosthesis: Three-year studies of a dacron-spandex graft in the dog. J. Surg. Res. 1983, 34, 7–16.

46. Chen, Y.; Ding, X.; Li, Y.; King, M.W.; Gao, J.; Zhao, X. A bilayer prototype woven vascular prosthesis with improved radial compliance. J. Text. Inst. 2011, 103, 106–111.

47. Sonoda, H.; Takamizawa, K.; Nakayama, Y.; Yasui, H.; Matsuda, T. Small-diameter compliant arterial graft prosthesis: Design concept of coaxial double tubular graft and its fabrication. J. Biomed. Mater. Res. 2001, 55, 266–276.

48. Sanger, P.W.; Taylor, F.H.; McCall, R.E.; Duchesne, R.; Lepage, G. Seamless synthetic arterial grafts; preliminary report on experimental and clinical experiences. J. Am. Med. Assoc. 1956, 160, 1403–1404.

49. De Bakey, M.; Cooley, D.A. Successful resection of aneurysm of distal aortic arch and replacement by graft. J. Am. Med. Assoc. 1954, 155, 1398–1403.

50. De Bakey, M.E.; Cooley, D.A.; Crawford, E.; Morris, G.C., Jr. Clinical application of a new flexible knitted dacron arterial substitute. AMA Arch. Surg. 1958, 77, 713–724.

51. Chakfe, N.; Dieval, F.; Wang, L.; Thaveau, F.; Rinckenbach, S.; Edah-Tally, S.; Mathieu, D.; Le Magnen, J.F.; Riepe, G.; Kretz, J.G.; Durand, B. In vitro approach to the dilative behavior of knitted vascular prosthetic grafts. Ann. Vasc. Surg. 2008, 22, 402–411.

52. Chakfé, N.; Riepe, G.; Dieval, F.; Le Magnen, J.F.; Wang, L.; Urban, E.; Beaufigeau, M.; Durand, B.; Imig, H.; Kretz, J.G. Longitudinal ruptures of polyester knitted vascular prostheses. J. Vasc. Surg. 2001, 33, 1015–1021.

53. Dieval, F.; Chakfé, N.; Wang, L.; Riepe, G.; Thaveau, F.; Heintz, C.; Mathieu, D.; Le Magnen, J.F.; Kretz, J.G.; Durand, B. Mechanisms of rupture of knitted polyester vascular prostheses: An in vitro analysis of virgin prostheses. Eur. J. Vasc. Endovasc. Surg. 2003, 26, 429–436.

54. Ichikawa, H.; Ishikawa, S.; Ohtaki, A.; Otani, Y.; Sakata, K.; Takahashi, T.; Aizaki, M.; Sato, Y.; Morishita, Y. Clinical experience

with a knitted Dacron graft impregnated with gelatin (Gelseal Triaxial®). Artif. Organs 1996, 5, 59–63.

55. Kudo, F.A.; Nishibe, T.; Miyazaki, K.; Flores, J.; Yasuda, K. Albumin-coated knitted Dacron aortic prostheses. Study of prosperative inflammatory reactions. Int. Angiol. 2002, 21, 214–217.

56. Marais, Y.; Guidoin, R.; Deng, X.; King, M.W.; Martin, L.; Roy, R. The Dialine II graft: A new collagen-impregnated warp-knitted polyester arterial prosthesis. Ann. Vasc. Surg. 1997, 11, 133–140.

57. Mary, C.; Marois, Y.; King, M.W.; Hong, T.; Laroche, G.; Douville, Y.; Martin, L.; Guidoin, R. In vitro and in vivo studies of a polyester arterial prosthesis with a warp-knitted sharkskin structure. J. Biomed. Mater. Res. 1997, 35, 459–472.

58. Meister, R.H.; Schweiger, H.; Lang, W. Knitted double-velour Dacron prostheses in aortobifemoral position—Long-term performance of different coating materials. Vasa J. Vasc. Disease. 1998, 27, 236–239.

59. Phaneuf, M.D.; Dempsey, D.J.; Bide, M.J.; Quist, W.C.; LoGerfo, F.W. Coating of Dacron vascular grafts with an ionic polyurethane: a novel sealant with protein binding properties. Biomaterials 2001, 22, 463–469.

60. Sirry, M.S.; Zilla, P.; Franz, T. A computational study of structural designs for a small-diameter composite vascular graft promoting tissue regeneration. Cardiovasc. Eng. Technol. 2010, 1, 269–281.

61. Wang, F.J.; Mohammed, A.; Li, C.J.; Wang, L. Promising poly(ε-caprolactone) composite reinforced with weft-knitted polyester for small-diameter vascular graft application. Adv. Mater. Sci. Eng. 2014, 2014, 1–9.

62. Xu, W.; Zhou, F.; Ouyang, C.; Ye, W.; Yao, M.; Xu, B. Mechanical properties of small-diameter polyurethane vascular grafts reinforced by weft-knitted tubular fabric. J. Biomed. Mater. Res. 2010, 92, 1–8.

63. Yagi, T.; Sato, M.; Nakazawa, Y.; Tanaka, K.; Sata, M.; Itoh, K.; Takagi, Y.; Asakura, T. Preparation of double-raschel knitted silk vascular grafts and evaluation of short-term function in a rat abdominal aorta. J. Artif. Organs 2011, 14, 89–99.

64. Yang, H.; Zhu, G.; Zhang, Z.; Wang, Z.; Fang, J.; Xu, W. Influence of weft-knitted tubular fabric on radial mechanical property of coaxial three-layer small-diameter vascular graft. J. Biomed. Mater. Res. B 2012, 100, 342–349.

65. Freitas, A.F.D.P.; de Araujo, M.D.; Zu, W.W.; Fangueiro, R.M.E. Development of weft-knitted and braided polypropylene stents for arterial implant. J. Text. Inst. 2010, 101, 1027–1034.

66. Li, G.; Li, Y.; Lan, P.; He, X.; Hu, H. Polydioxanone weft-knitted intestinal stents: fabrication and mechanics optimization. Text. Res. J. 2013, 83, 2129–2141.

67. Moodley, L.; Franz, T.; Human, P.; Wolf, M.F.; Bezuidenhout, D.; Scherman, J.; Zilla, P. Protective constriction of coronary vein grafts with knitted nitinol. Eur. J. Cardiothorac. Surg. 2013, 44, 64–71.

68. Weinandy, S.; Rongen, L.; Schreiber, F.; Cornelissen, C.; Flanagan, T.C.; Mahnken, A.; Gries, T.; Schmitz-Rode, T.; Jockenhoevel, S. The BioStent: Novel concept for a viable stent structure. Tissue Eng. A 2012, 18, 1818–1826.

69. Zilla, P.; Moodley, L.; Wolf, M.F.; Bezuidenhout, D.; Sirry, M.S.; Rafiee, N.; Lichtenberg, W.; Black, M.; Franz, T. Knitted nitinol represents a new generation of constrictive external vein graft meshes. J. Vasc. Surg. 2011, 54, 1439–1450.

70. Singh, C.; Wang, X. A biomimetic approach for designing stent-graft structures: Caterpillar cuticle as design model. J. Mech. Behav. Biomed. Mater. 2014, 30, 16–29.

71. Singh, C.; Wang, X. A new design concept for knitted external vein-graft support mesh. J. Mech. Behav. Biomed. Mater. 2015, 48, 125–133.

72. Self, M.M.; Cooley, D.A.; Debakey, M.E.; Creech, O., Jr. The use of braided nylon tubes for aortic replacement. Ann. Surg. 1955, 142, 836–843.

73. Edwards, W.S.; Tapp, J.S. Braided textiles tubes as arterial grafts. Surgery 1995, 38, 61–67.

74. Dale, W.; Niguidula, F.N. Study of elasticized dacron as arterial prosthesis: Experimental comparison with other plastics, homologous arteries, and autogenous veins. AMA Arch. Surg. 1959, 78, 246–259.

75. Whittlesey, R.H.; Sekerak, N.M.; Blunt, J.W., Jr.; Hudack, S.S. Observations of elastic synthetic fabric sleeves as vascular prostheses. AMA Arch. Surg. 1956, 73, 440–454.

76. Parodi, J.C.; Palmaz, J.C.; Barone, H.D. Transfemoral intraluminal graft implantation for abdominal aortic aneurysms. Ann. Vasc. Surg. 1991, 5, 491–499.

77. Wilson, G.; Klement, P.; Kato, Y.P.; Martin, J.B.; Khan, I.J.; Alcime, R.; Dereume, J.-P.; Macgregor, D.C.; Pinchuk, L. A self-expanding bifurcated endovascular graft for abdominal Aortic aneurysm repair an initial study in a canine model. ASAIO J. 1996, 42, M386–M393.

78. Georg, Y.; Settembre, N.; Marchand, C.; Lejay, A.; Thaveau, F.; Durand, B.; Chakfe, N. Poor long-term stability of the Corvita abdominal stentgraft. Eur. J. Vasc. Endovasc. Surg. 2014, 47, 160–163.

79. Raillat, C.; Rousseau, H.; Joffre, F.; Roux, D. Treatment of iliac artery stenoses with the Wallstent endoprosthesis. Am. J. Roentgenol. 1990, 154, 613–616.

80. Kallmes, D.F.; Ding, Y.H.; Dai, D.; Kadirvel, R.; Lewis, D.A.; Cloft, H.J. A new endoluminal, flow-disrupting device for treatment of saccular aneurysms. Stroke 2007, 38, 2346–2352.

81. Kallmes, D.F.; Ding, Y.H.; Dai, D.; Kadirvel, R.; Lewis, D.A.; Cloft, H.J. A Second-Generation, Endoluminal, Flow-Disrupting Device for Treatment of Saccular Aneurysms. Am. J. Neuroradiol. 2009, 30, 1153–1158.

82. Lieber, B.; Stancampiano, A.; Wakhloo, A. Alteration of hemodynamics in aneurysm models by stenting: Influence of stent porosity. Ann. Biomed. Eng. 1997, 25, 460–469.

83. Mallik, A.S.; Nuss, K.; Kronen, P.W.; Klein, K.; Karol, A.; von Rechenberg, B.; Rüfenacht, D.A.; Wanke, I.; Kulcsár, Z. A new-generation, low-permeability flow diverting device for treatment of saccular aneurysms. Eur. Radiol. 2014, 24, 12–18.

84. Alderson, H.; Zamir, M. Effects of stent stiffness on local haemodynamics with particular reference to wave reflections. J. Biomech. 2004, 37, 339–348.

85. Nii, K.; Tsutsumi, M.; Aikawa, H.; Hamaguchi, S.; Etou, H.; Sakamoto, K.; Kazekawa, K. Incidence of hemodynamic depression after carotid artery stenting using different self-expandable stent types. Neurol. Med. Chir. 2011, 51, 556–560.

86. Tanaka, N.; Martin, J.B.; Tokunaga, K.; Abe, T.; Uchiyama, Y.; Hayabuchi, N.; Berkefeld, J.; Rüfenacht, D.A. Conformity of carotid stents with vascular anatomy: Evaluation in carotid Models. Am. J. Neuroradiol. 2004, 25, 604–607.

87. Hyun Kim, J.; Jin Kang, T.; Yu, W.-R. Simulation of mechanical behavior of temperature-responsive braided stents made of shape memory polyurethanes. J. Biomech. 2010, 43, 632–643.

88. Irsale, S.; Adanur, S. Design and characterization of polymeric stents. J. Ind. Text. 2006, 35, 189–200.

89. Lin, J.H.; Wen, S.P.; Lou, C.W.; Ueng, K.C. Compression behaviors of stainless steel braided coronary stents and polyvinyl alcohol/stainless steel braided coronary stents. Adv. Mater. Res. 2014, 910, 274–278.

90. Mikkonen, J.; Uurto, I.; Isotalo, T.; Kotsar, A.; Tammela, T.L.J.; Talja, M.; Salenius, J.P.; Törmälä, P.; Kellomäki, M. Drug-eluting bioabsorbable stents—An in vitro study. Acta Biomater. 2009, 5, 2894–2900.

91. Nuutinen, J.P.; Välimaa, T.; Clerc, C.; Törmälä, P. Mechanical properties and in vitro degradation of bioresorbable knitted stents. J. Biomater. Sci. Polym. Ed. 2002, 13, 1313–1323.

92. Rebelo, R.; Vila, N.; Fangueiro, R.; Carvalho, S.; Henriques, M. Development of braided fiber-based stents. Stud. Health Technol. Inform. 2014, 207, 135–144.

93. Ueng, K.-C.; Wen, S.-P.; Lou, C.-W.; Lin, J.-H. Braiding structure stability and section treatment evaluations of braided coronary stents made of stainless steel and bio-absorbable polyvinyl alcohol via a braiding technique. Fib. Polym. 2015, 16, 675–684.

94. van der Giessen, W.J.; Slager, C.J.; Van Beusekom, H.M.M.; Van Ingen Schenau, D.S.; Huijts, R.A.; Schuurbiers, J.C.; de Klein, W.J.; Serruys, P.W.; Verdouw, P.D. Development of a Polymer endovascular prosthesis and its implantation in porcine arteries. J. Interv. Cardiol. 1992, 5, 175–186.

95. Wang, X.; Li, Q.; Hu, X.; Ma, L.; You, C.; Zheng, Y.; Sun, H.; Han, C.; Gao, C. Fabrication and characterization of poly(L-lactide-co-glycolide) knitted mesh-reinforced collagen-chitosan hybrid scaffolds for dermal tissue engineering. J. Mech. Behav. Biomed. Mater. 2012, 8, 204–215.

96. Yuksekkaya, M.E.; Adanur, S. Analysis of polymeric braided tubular structures intended for medical applications. Text. Res. J. 2009, 79, 99–109.

97. De Beule, M.; Van Cauter, S.; Mortier, P.; Van Loo, D.; Van Impe, R.; Verdonck, P.; Verhegghe, B. Virtual optimization of self-expandable braided wire stents. Med. Eng. Phys. 2009, 31, 448–453.

98. Kim, J.H.; Kang, T.J.; Yu, W.R. Mechanical modeling of self-expandable stent fabricated using braiding technology. J. Biomech. 2008, 41, 3202–3212.

99. Makoyeva, A.; Bing, F.; Darsaut, T.E.; Salazkin, I.; Raymond, J. The varying porosity of braided self-expanding stents and flow diverters: An experimental study. Am. J. Neuroradiol. 2013, 34, 596–602.

100. Sui, W.; Jiang, J.; Chen, N. Effects of braiding parameters on radial compressive property of intraluminal stents. Appl. Mech. Mater. 2013, 268, 213–216.

101. Balderi, A.; Antonietti, A.; Pedrazzini, F.; Ferro, L.; Leotta, L.; Peano, E.; Grosso, M. Treatment of a hepatic artery aneurysm by endovascular exclusion using the multilayer cardiatis stent. Cardiovasc. Intervent. Radiol. 2010, 33, 1282–1286.

102. Henry, M.; Polydorou, A.; Frid, N.; Gruffaz, P.; Cavet, A.; Henry, I.; Hugel, M.; Rüfenacht, D.A.; Augsburger, L.; De Beule, M.; et al. Treatment of renal artery aneurysm with the multilayer stent. J. Endovasc. Ther. 2008, 15, 231–236.

103. Müller-Eschner, M.; Kortes, N.; Rehnitz, C.; Sumkauskaite, M.; Rengier, F.; Böckler, D.; Kauczor, H.-U.; Radeleff, B. Endovascular repair of a complex splenic artery aneurysm using a multilayer stent. Cardiovasc. Intervent. Radiol. 2015, 38, 494–497.

104. Ahlhelm, F.; Kaufmann, R.; Ahlhelm, D.; Ong, M.; Roth, C.; Reith, W. Carotid artery stenting using a novel self-expanding braided nickel–titanium stent: Feasibility and safety porcine trial. Cardiovasc. Intervent. Radiol. 2009, 32, 1019–1027.

105. Ahlhelm, F.; Roth, C.; Kaufmann, R.; Schulte-Altedorneburg, G.; Romeike, B.M.; Reith, W. Treatment of wide-necked intracranial aneurysms with a novel self-expanding two-zonal endovascular stent device. Neuroradiology 2007, 49, 1023–1028.

106. Marchand, C.; Heim, F.; Durand, B.; Chafke, N. Nitinol stent for percutaneous heart valve implantation: Material shape setting. Mater. Manuf. Process. 2011, 26, 181–187.

107. Massell, T.B.; Heringman, E.; Greenstone, S.M. Woven Dacron and woven Teflon prostheses: Use for small artery replacement. Arch. Surg. 1962, 84, 73–79.

108. Cate, W.R., Jr. Bypass grafts in small arteries: An experimental comparison of freeze-dried arterial homografts, autogenous vein and crimped dacron. Am. Surg. 1959, 25, 523–527.

109. Desai, M.; Seifalian, A.M.; Hamilton, G. Role of prosthetic conduits in coronary artery bypass grafting. Eur. J. Cardiothorac. Surg. 2011, 40, 394–398.

110. Seifalian, A.M.; Tiwari, A.; Hamilton, G.; Salacinski, H.J. Improving the clinical patency of prosthetic vascular and coronary bypass grafts: The role of seeding and tissue engineering. Artif. Organs. 2002, 26, 307–320.

111. Burkel, W.E.; Vinter, D.W.; Ford, J.W.; Kahn, R.H.; Graham, L.M.; Stanley, J.C. Sequential studies of healing in endothelial seeded vascular prostheses: Histologic and ultrastructure characteristics of graft incorporation. J. Surg. Res. 1981, 30, 305–324.

112. Fujita, Y.; Wu, M.H.D.; Ishida, A.; Shi, Q.; Walker, M.; Hammond, W.P.; Sauvage, L.R. Accelerated healing of Dacron grafts seeded by preclotting with autologous bone marrow blood. Ann. Vasc. Surg. 1999, 13, 402–412.

113. Pasquinelli, G.; Freyrie, A.; Preda, P.; Curti, T.; D'Addato, M.; Laschi, R. Healing of prosthetic arterial grafts. Scanning Microsc. 1990, 4, 351–362.

114. Tolan, M.; Wells, F.; Kendall, S.; Large, S.; Wallwork, J. Clinical experience with a collagen impregnated woven Dacron graft. J. Cardiovasc. Surg. 1995, 36, 323–327.

115. Walden, R.; L'Italien, G.J.; Megerman, J.; Abbott, W.M. Matched elastic properties and successful arterial grafting. Arch. Surg. 1980, 115, 1166–1169.

116. Li, W.-J.; Laurencin, C.T.; Caterson, E.J.; Tuan, R.S.; Ko, F.K. Electrospun nanofibrous structure: A novel scaffold for tissue engineering. J. Biomed. Mater. Res. 2002, 60, 613–621.

117. Xu, C.Y.; Inai, R.; Kotaki, M.; Ramakrishna, S. Aligned biodegradable nanofibrous structure: A potential scaffold for blood vessel engineering. Biomaterials. 2004, 25, 877–886.

118. Agarwal, S.; Wendorff, J.H.; Greiner, A. Use of electrospinning technique for biomedical applications. Polymer 2008, 49, 5603–5621.

119. Stitzel, J.; Liu, J.; Lee, S.J.; Komura, M.; Berry, J.; Soker, S.; Lim, G.; Van Dyke, M.; Czerw, R.; Yoo, J.J.; Atala, A. Controlled fabrication of a biological vascular substitute. Biomaterials 2006, 27, 1088–1094.

120. Buttafoco, L.; Kolkman, N.G.; Engbers-Buijtenhuijs, P.; Poot, A.A.; Dijkstra, P.J.; Vermes, I.; Feijen, J. Electrospinning of collagen and elastin for tissue engineering applications. Biomaterials 2006, 27, 724–734.

121. Wong, C.; Liu, X.; Xu, Z.; Lin, T.; Wang, X. Elastin and collagen enhances electrospun aligned polyurethane as scaffolds for vascular graft. J. Mater. Sci. Mater. Med. 2013, 24, 1865–1874.

122. McClure, M.J.; Simpson, D.G.; Bowlin, G.L. Tri-layered vascular grafts composed of polycaprolactone, elastin, collagen, and silk: Optimization of graft properties. J. Mech. Behav. Biomed. Mater. 2012, 10, 48–61.

123. McClure, M.J.; Sell, S.A.; Simpson, D.G.; Walpoth, B.H.; Bowlin, G.L. A three-layered electrospun matrix to mimic native arterial architecture using polycaprolactone, elastin, and collagen: A preliminary study. Acta Biomater. 2010, 6, 2422–2433.

124. Soletti, L.; Hong, Y.; Guan, J.; Stankus, J.J.; El-Kurdi, M.S.; Wagner, W.R.; Vorp, D.A. A bilayered elastomeric scaffold for tissue engineering of small diameter vascular grafts. Acta Biomater. 2010, 6, 110–122.

125. Bonani, W.; Maniglio, D.; Motta, A.; Tan, W.; Migliaresi, C. Biohybrid nanofiber constructs with anisotropic biomechanical properties. J. Biomed. Mater. Res. B 2011, 96, 276–286.

126. Anderson, J.S.; Price, T.M.; Hanson, S.R.; Harker, L.A. In vitro endothelialization of small-caliber vascular grafts. Surgery 1987, 101, 577–586.

127. Ao, P.Y.; Hawthorne, W.J.; Vicaretti, M.; Fletcher, J.P. Development of intimal hyperplasia in six different vascular prostheses. Eur. J. Vasc. Endovasc. Surg. 2000, 20, 241–249.

128. Devine, C.; McCollum, C. Heparin-bonded Dacron or polytetrafluorethylene for femoropopliteal bypass: five-year results of a prospective randomized multicenter clinical trial. J. Vasc. Surg. 2004, 40, 924–931.

129. Feugier, P.; Black, R.A.; Hunt, J.A.; How, T.V. Attachment, morphology and adherence of human endothelial cells to vascular prosthesis materials under the action of shear stress. Biomaterials 2005, 26, 1457–1466.

130. Williamson, M.R.; Black, R.; Kielty, C. PCL-PU composite vascular scaffold production for vascular tissue engineering: attachment, proliferation and bioactivity of human vascular endothelial cells. Biomaterials 2006, 27, 3608–3616.

131. Bergmeister, H.; Grasl, C.; Walter, I.; Plasenzotti, R.; Stoiber, M.; Schreiber, C.; Losert, U.; Weigel, G.; Schima, H. Electrospun small-

diameter polyurethane vascular grafts: ingrowth and differentiation of vascular-specific host cells. Artif. Organs 2012, 36, 54–61.

132. Soffer, L.; Wang, X.; Zhang, X.; Kluge, J.; Dorfmann, L.; Kaplan, D.L.; Leisk, G. Silk-based electrospun tubular scaffolds for tissue-engineered vascular grafts. J. Biomater. Sci. Polym. Ed. 2008, 19, 653–664.

133. Pektok, E.; Nottelet, B.; Tille, J.-C.; Gurny, R.; Kalangos, A.; Moeller, M.; Walpoth, B.H. Degradation and healing characteristics of small-diameter poly(ε-Caprolactone) vascular grafts in the rat systemic arterial circulation. Circulation 2008, 118, 2563–2570.

134. Zhu, Y.; Cao, Y.; Pan, J.; Liu, Y. Macro-alignment of electrospun fibers for vascular tissue engineering. J. Biomed. Mater. Res. B 2010, 92, 508–516.

135. Ju, Y.M.; Choi, J.S.; Atala, A.; Yoo, J.J.; Lee, S.J. Bilayered scaffold for engineering cellularized blood vessels. Biomaterials 2010, 31, 4313–4321.

136. Dutoya, S.; Verna, A.; Lefèbvre, F.; Rabaud, M. Elastin-derived protein coating onto poly(ethylene terephthalate). Technical, microstructural and biological studies. Biomaterials 2000, 21, 1521–1529.

137. Sell, S.A.; McClure, M.J.; Barnes, C.P.; Knapp, D.C.; Walpoth, B.H.; Simpson, D.G.; Bowlin, G.L. Electrospun polydioxanone-elastin blends: potential for bioresorbable vascular grafts. Biomed Mater. 2006, 1, 72–80.

138. Wolfe, P.S.; Madurantakam, P.; Garg, K.; Sell, S.A.; Beckman, M.J.; Bowlin, G.L. Evaluation of thrombogenic potential of electrospun bioresorbable vascular graft materials: Acute monocyte tissue factor expression. J. Biomed. Mater. Res. A 2010, 92, 1321–1328.

139. Liu, Y.; Xiang, K.; Chen, H.; Li, Y.; Hu, Q. Composite vascular repair grafts via micro-imprinting and electrospinning. AIP Adv. 2015, 5.

CHAPTER 10

A dynamic approach to assess international competitiveness of Vietnam's garment and textile industry

Huong Thanh Vu* and Lam Cat Pham

Faculty of International Business and Economics, University of Economics and Business, Vietnam National University – Hanoi, 144 Xuan Thuy, Cau Giay, Hanoi, Vietnam

ABSTRACT

Garment and textile (G&T) industry has been playing as a driving force for the socio-economic development of Vietnam. With the international integration process and rising challenges from the global market, there is a need to examine international competitiveness of Vietnam's G&T industry to find out what Vietnam should focus on to enhance its position in the global market place. This paper, by using the Generalized Double Diamond Model (GDDM), analyzed international competitiveness of Vietnam's G&T industry and compared it with China. The results showed that Vietnam was less competitive than China in all four attributes of the GDDM. The lowest competitiveness of Vietnam in comparison with China was Related and Supporting industries, followed by Factor Conditions. Therefore, the paper argued that although Vietnam should improve all of the four attributes in the long term, Vietnam must put a high priority on developing Related and Supporting Industries and then enhance Factor Conditions while maintaining its strengths over China in terms of G&T export growths and favorable business context.

KEYWORDS

Garment and textile; Vietnam; China; International competitiveness; Generalized Double Diamond Model

1. BACKGROUND

In Vietnam, garment and textile industry (G&T) has been a key exporting industry and contributed considerably to the social and economic development. In 2014, G&T was the second biggest exporting industry and contributed 11.51 % to the total GDP of Vietnam (General Departnment of Vietnam Customs 2015). In the international market, with the share of nearly 3 % in the world G&T exports in 2013, Vietnam has become the 7th biggest G&T exporter after China, the EU, India, Turkey, Bangladesh and the US (ITC 2015). Together with the increasing integration in the global marketplace, Vietnam's G&T industry is facing with fiercer competition from other competitors. In order for the G&T industry of Vietnam to compete successfully and move up in the international G&T market, the understanding of its international competitiveness is of great importance.

International competitiveness is viewed as a strategic phenomenon inherent in the fields of international marketing, international business and international management, and refers to the attributes that make organizations more competitive than others in the global market. Organizations can be understood broadly as a region, a nation, an industry or perhaps a strategic group. Therefore, the term international competitiveness is a multi-level phenomenon working in the global market (Hult 2012). Porter (1990) argued that the competitiveness is created, not inherited and claimed that the source of competitiveness is the competitive advantage, which is created and sustained through a highly localized process. Porter (1990) described the competitive advantage as four main attributes that allow an organization to outperform its competitors. Those attributes, individually and as a system, constitute the Diamond Model of national advantage, which serves as a playing field that each nation establishes and operates for its industries. The competitiveness will increase when each of these attributes is improved. Based on Porter (1990) and D'Cruz and Rugman (1993), the Generalized Double Diamond Model (GDDM) introduced by Moon et al. (1995, 1998) viewed international competitiveness as a broader term to incorporate multinational activities and government in the Diamond Model. At the sector level, D'Cruz (1992) argued that international competitiveness of an industry is the collective ability of firms in that sector to compete internationally. According to Momaya (1998), international competitiveness at industry level is often considered as the results of strategies and actions of firms operating in that sector. Competitiveness is also represented by the relative productivity and its ability to create value added. It allows an industry to maintain and improve position in the global market and can only be assessed by comparing with the same industry in another country (Depperu and Cerrato 2005). With all the ideas above, international competitiveness of an industry in short can be understood as its ability to compete internationally and can be measured through different attributes in comparison with the same industry in other nations.

Even though Vietnam's G&T industry is a common topic for researchers in Vietnam, the previous literature on its international competitiveness is limited. The past studies examining international competitiveness of Vietnam's G&T adopted three main approaches namely the value chain, strengths, weaknesses, opportunities and threats (SWOT) and the Diamond Model of Michael Porter. Vitas (2006) used SWOT and value chain analysis to assess Vietnam's G&T international competitiveness through a great deal of indicators, for example cost, production time, customs procedures, policies and supporting industries in comparison with some main competitors such as China, India, the US, Bangladesh and Thailand. A mixed picture was pointed out in the paper and eventually it was not clear of whether Vietnam's international competitiveness in G&T industry was higher or lower compared to that of other competitors. Also using the value chain approach, Truong et al. (2010), Dang and Dinh (2011), and Luong (2012) argued that Vietnam's G&T international competitive was still low because of its participation in the lowest end of global G&T value chain. Nguyen (2012) using SWOT and the Diamond Model, and IPP and CIEM (2013) adopting the value chain and the Diamond Model drew out the same conclusions that Vietnam was at low international competitiveness mainly because of its dependence on the outside raw material, weak supporting industries and low productivity. Asian Foundation and CIEM (2012) also found out that Vietnam's G&T international competitiveness was modest and affected by tariff, customs, financial policies, labor, technology, materials input, market, and products quality. The previous studies recommended that in order to improve the international competitiveness of Vietnam's G&T industry, Vietnam should increase the localization rate, develop Vietnam's brand name, and increase the added value to move up in the value chain.

The nature of international competitiveness of an industry as stated above is its ability to compete internationally and must be assessed by comparing with the same industry in other countries. One common point of all previous papers relating to international competitiveness of Vietnam's G&T is that although they were informative to describe the current development of this industry, they failed to measure or quantify international competitiveness of Vietnam's G&T aggregately in indexes and therefore it is hard to position Vietnam's international competiveness in the global marketplace. In addition, the past studies proposed a wide range of measures for Vietnam's G&T industry but the priority measures were unclear. To fill this gap, the paper concentrates on analyzing and measuring international competitiveness of Vietnam's G&T industry by using the GDDM. With this methodology and framework, the contribution of this paper is twofold. Firstly, the paper develops a specific framework for assessing international competitiveness of G&T industry. Secondly, this is the pioneering study adopting the GDDM to examine international competitiveness of an industry in Vietnam.

In this paper, the GDDM, with the analysis of both domestic and international attributes, helps answer how internationally competitive Vietnam's G&T is quantitatively compared to the benchmark country, China, and draw out better implications for Vietnam to improve international competitiveness of G&T industry. The paper is structured as below. After the introduction, the second part introduces the framework used while the third part explains the methodology and selection of proxies for the GDDM. In the next part, the paper presents the main results on international competitiveness of Vietnam's G&T and the final part points out some conclusions and implications, which are essential for Vietnam's G&T industry to enhance its international competitiveness in the future.

2. ANALYTICAL FRAMEWORK

Among the models that are adopted to explain nation or industry competitiveness, the widely used one is Michael Porter's Diamond Model introduced firstly in Porter (1990). According to this model, four main attributes that underlie conditions or platform for determination of the national competitive advantage are "Factor conditions", "Demand conditions", "Related and Supporting Industries", and "Firm Strategy, Structure and Rivalry". Porter (1990) also proposed government policies and chance as exogenous shocks, which supported the whole system of national competitiveness with four above-mentioned attributes.

In spite of being influential and widely used, Michael Porter's Diamond Model of competitiveness still has some severe limitations. Firstly, the Diamond Model leaves out the multinational activities such as inbound and outbound foreign direct investment (FDI) and is mainly fixed on a large home base country (Cartwright 1993; Cho 1994; D'Cruz and Rugman 1993; Dunning 2003; Moon et al. 1998; Williams and Morgan 2010). Secondly, the Diamond Model is successful at explaining international competitiveness of big countries like the US and Japan, but not appropriate to analyze the competitiveness of advanced, smaller and open countries like Canada (D'Cruz and Rugman 1993). Therefore, Rugman and Verbeke (1993) proposed an alteration to Porter's Diamond Model, which is called the Double Diamond Model (DDM). This model covers the same four groups of attributes of competitiveness as the Diamond Model but takes into account the activities of multinational enterprises, which have to rely on both home-base and foreign determinants to sustain its competitive advantage, and suggests that managers should build upon both domestic and foreign diamonds to become globally competitive in terms of survival, profitability and growth (Liu and Hsu 2009). However, there are still problems with the DDM. Although this model can be used to explain quite well the cases of countries like Canada and New Zealand, it fails to analyze the competitiveness of all other small open countries such as Korea and Singapore (Son and Kenji 2013). In fact, multinational firms from these small countries have to rely not only on domestic determinants, but also on the resources and markets internationally and especially are likely to link more with global than domestic industrial structure. Therefore, Moon et al. (1998) developed the GDDM, which is suitable for all small open economies (Balcarová 2010; Son and Kenji 2013).

The GDDM consists of two main diamonds (Fig. 1). The inner represents the Domestic Diamond, which is similar to the diamond of Michael Porter. The outside one is the International Diamond, which represents all four attributes in international context. In these two diamonds, chance is included and treated as exogenous variables. Government influence, on the other hand, is included as an important endogenous variable that directly influences all four determinants. The dotted Global Diamond, between the Domestic and International Diamond, represents international competitiveness of an industry as determined by both domestic and international parameters. Difference between the Global Diamond and the Domestic Diamond of Michael Porter is the result of integrating multinational activities into the model.

Compared to the Diamond Model of Porter (1990), this model has three important extensions: (1) incorporates the multinational activities, (2) be able to function the competitiveness paradigm which allows a comparison of size and shape of the Domestic and International diamonds and (3) fits all small open nations in which firms are likely to be concerned more with global than domestic industrial structure (Sardy and Fetscherin 2009; Son and Kenji 2013). With these extensions, this model has been proven to be more generalized and useful in analyzing international competitiveness of different countries at multi-levels, nationally or industrially. For example, Sardy and Fetscherin (2009) analyzed and compared international competitiveness of automotive industry between China, India and South Korea, three countries at different sizes but to be three of ten biggest automotive producers in the global market place. Results from the Global Diamond of these countries pointed out that China's automotive industry was as competitive as South Korea's in terms of Factor Conditions, Demand Conditions and Related and Supporting Industries but was more competitive than India's. Son and Kenji (2013) adopted the GDDM to compare international competitiveness of Korean and Japanese fashion industries based on 32 proxies for four attributes including Factor conditions, Demand conditions, Related and Supporting Industries and Firm Strategy, Structure and Rivalry. They collected secondary data from different sources of Korea and Japan in different years from 2007 to 2010, then calculated international competitiveness indexes of two nations and found out that Korea was less competitive than Japan. While Son and Kenji (2013) applied the GDDM for industrial competiveness comparison, Liu and Hsu (2009) used this model to analyze the overall competitiveness of two open economies including Taiwan and Korea. The results showed that Taiwan was superior to Korea in all attributes except for Demand Condition in the Domestic Diamond. With the same objectives of comparing China and Korea's international competitiveness of fashion industries, Kim et al. (2006) and Son et al. (2007) adopted the GDDM and suggested an entry strategy for the Chinese fashion market. From the previous typical literature, it can be seen that the GDDM can provide a better insight of international competitiveness due to incorporating multinational activities into the Diamond Mode and might be used to compare competiveness of at multi-levels.

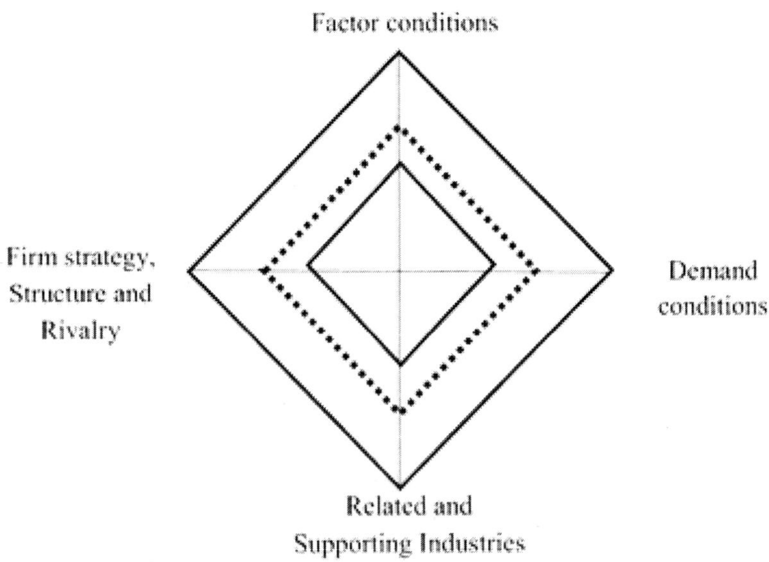

Figure 1 . The Generalized Double Diamond Model. *Source*: Moon et al. (1998)

3. METHODOLOGY AND DATA

The objective of this paper is to assess international competitiveness of Vietnam's G&T industry. In this regard, a methodology that helps quantifying Vietnam's competitiveness and allowing a comparison of this competitiveness level to that of a benchmark country will be required. The GDDM is therefore selected as this model satisfies these two requirements. The model is also proved to be suitable for an open and developing country like Vietnam.

The comparison country chosen in this study was China, which is not only Vietnam's neighboring country sharing the similar cultures and traditions, but also the top G&T exporter in the world. Domestically, the Vietnamese G&T industry competes with China's in providing G&T products to Vietnam's citizens. Internationally, the position that China's G&T industry holds at the moment is the one that Vietnam's G&T industry heads for. Moreover, both China and Vietnam are among top G&T exporters in the international marketplace. Comparing the competitiveness of Vietnam's G&T industry to that of China is necessary for Vietnam to know where it is at the moment in the race with China. Therefore, although China is the second biggest country in the world in terms of GDP, the comparison of two nations at G&T industry level is acceptable as long as cautious analysis is taken when examining proxies that are not at industry level like total population and total GDP.

The key issue is the choice of proxies capturing four attributes that are assessed in the GDDM. The GDDM is developed from the original Diamond Model of Porter, which consists of four groups of attributes namely Factor conditions, Demand conditions, Related and Supporting industries, and Firm Strategy, Structure and Rivalry. In fact, the model has generated over 100 proxies that are used to capture international competitiveness. However, like other previous studies by Rugman and Verbeke (1993), Sardy and Fetscherin (2009), Balcarová (2010), Williams and Morgan (2010), and Son and Kenji (2013), this paper does not cover all proxies but chooses certain proxies which best capture international competitiveness of studied industry. Totally, 27 proxies that describe four attributes, taking into consideration G&T industry-related features, were selected to act as determinants of the model (Table 1).

4. FACTOR CONDITIONS

According to Porter (1990), the domestic Factor Conditions include both basic and advanced factors. Basic factors refer to natural resources, climate conditions, location, unskilled labor, and semiskilled labor that are inherited and require little investment to be utilized in the production process. Advanced factors such as highly skilled workers, highly educated personnel, and Research & Development (R&D), on the other hand, are created and upgraded through reinvestment and innovation.

Table 1 . Variables and proxies of GDDM for Vietnam's G&T industry in comparison with China's

Attributes	Variables	Proxies
Factor Conditions	Domestic	
	Basic factors	Wage of worker in G&T industry (USD/h)
		Number of workers and laborers in G&T industry (million people)
		Labor productivity in G&T industry (shirts/worker/day)
	Advanced factors	R&D expenditure (% of GDP)
	International	
	Advanced factors	Manufacturing inward FDI flows (billion USD)
		Manufacturing outward FDI flows (billion USD)
Demand Conditions	Domestic	
	Size	Total population (million people)
		GDP (billion USD)
		Employment rate (%)
	Sophistication	GDP per capita (USD)
		Household rate of expenditure on G&T out of gross income (%)
		Educational index
	International	
	Size	Total export value of G&T industry (billion USD)
		Average export growth rate of G&T industry (%)
Related and Supporting Industries	Domestic	
	Supporting Industries	Cotton output (1000 t)
		Yarn output (million tons)
	Supporting infrastructures	Rail lines (total route—km)
		Roads, paved (% of total roads)
		ICT index
	International	
	Supporting industries	Cotton exports (1000 t)
		Yarn and fabric exports (billion USD)
	Supporting infrastructures	Container port traffics (TEU: 20 foot equivalent unit)
		Air transport (registered carrier departures worldwide)
Firm Strategy, Structure and Rivalry	Domestic	
	Rivalry	Intensity of local competition
	Business context	World Bank DTF points
	International	
	Rivalry	Market share of the country in G&T global market (%)
	Business context	Average import tariff rate faced by G&T industry (%)

Sources of data are provided in Additional file 1

Source: Developed by the authors based on Rugman and Verbeke (1993), Sardy and Fetscherin (2009), Balcarová (2010), Williams and Morgan (2010), Son and Kenji (2013)

Four proxies were used to assess the domestic Factor Conditions (Table 1). Because G&T is a labor-intensive industry, the paper puts priority to select labor-related proxies including (1) the wages of G&T workers, (2) the number of workers and laborers in G&T industry and (3) the labor productivity in G&T industry to represent the basic factor conditions. Low wages and high number of workers represent cheap labor and labor abundance, implying a possible motive for expansion of the industry and therefore high competitiveness (Brown and Sessions 2001; Pizer 2000;

Sardy and Fetscherin 2009). Some studies also showed that high productivity captures high competitiveness of the industry (Daniel 2000; Han et al. 2015). As variables for advanced factor conditions, (4) R&D expenditures were selected as a proxy for future growth and innovation of the industry.

International factors were assessed based on two aspects: (1) inward and (2) outward FDI (Table 1). While the inward FDI shows the foreign investment in domestic market, the outward FDI represents the outward investment made by domestic firms. Because the data of inward and outward FDI by specific sector were not available in Vietnam, the data of manufacturing inward and outward FDI were used instead. The more inward and outward manufacturing FDI are, the higher competitiveness the manufacturing including G&T is in the international market (Moon and Youn 2010).

5. DEMAND CONDITIONS

Porter (1990) emphasized the role of size and sophistication of Domestic Demand in shaping the competitiveness. While size of home demand forces firms to expand their production to take advantage of economics of scale, sophistication of demand drives firms to continuously change and innovate to meet the high demands in terms of product quality and varieties (Smit 2010).

In this paper, (1) the total population, (2) GDP and (3) the employment rate were used to represent the size of domestic market demand (Table 1). As G&T is a necessity in daily life, the population and GDP are good proxies to represent the domestic demand for G&T products. Meanwhile the employment rate is an additional index to represent the extent to which people can afford satisfy their demand. According to Son and Kenji (2013), the sophistication of clothes buyers is related to fashion buying behaviors such as expenditure on clothes and the frequency of purchasing clothes. Hirschman (1980) and Barnes and McTavish (1983) also argued that consumer sophistication is driven by the fact that the consumers are better informed and educated. Therefore, (4) GDP per capita, (5) household rate of expenditure on G&T products and (6) educational index were used as proxies for sophistication of demand.

The international demand factors were determined through size of the international market, which can be represented by (1) total G&T export value and (2) growth rate of G&T export value (Table 1). The higher export values and growth rates shows the higher and more stable demand for a country's product, implying higher competitiveness. A country normally exports to multiple foreign countries; therefore, there is no appropriate proxy for assessing the sophistication of international demand (Sardy and Fetscherin 2009).

6. RELATED AND SUPPORTING INDUSTRIES

Related and Supporting Industries are essential for ability of an industry to compete in the international market as a industry is more likely to be successful if its supporting industries have a competitive advantage (Sardy and Fetscherin 2009). According to Porter (1990), Related and Supporting Industries attribute refers to the presence or absence in the nation of related industries and suppliers that are internationally competitive. They include the upstream and downstream firms as well as the supporting infrastructure like transportations and communication involved in the value chain. The upstream industries include firms that produce input materials and the downstream ones that are in charge of distributing G&T products to final customers. For Domestic Diamond, this paper used (1) the amount of cotton produced and (2) the amount of yarn produced as proxies to reflect the domestic downstream industries because cotton and yarn are two important inputs of G&T industry. In today's globalization, transportation and communication are essential to promote the competitiveness of an industry (Hult 2012; Son and Kenji 2013; Williams and Morgan 2010). Accordingly, (3) rail lines and (4) the share of paved roads were used as indices for transportation while (5) ICT development index for communication (Table 1).

The international factors were analyzed based on the presence of internationally competitive Related and Supporting Industries. (1) Cotton exports and (2) yarn exports were used to indicate how strong these industries were in the global market through its expansion into related and supporting international industries (Table 1). These two proxies were adopted also because cotton and yard consumption is heavily driving G&T industry. In addition, infrastructure for international transportation are important for improving international competiveness because they facilitate international trade transactions and increase the levels of multinational activities with higher efficiency (Daniel 2000; ITS Global 2008). Therefore, (3) container port traffics and (4) the number of registered air departures worldwide were used to represent the ability to ship goods abroad. While container port traffic measures the flow of containers from land to sea transport modes in 20-foot equivalent units (TEU), the registered air departures provides information on the number of domestic takeoffs and takeoffs abroad of air carriers registered in the country.

7. FIRM STRATEGY, STRUCTURE AND RIVALRY

This attribute reflects the context in which firms are created, organized and managed and the domestic competition environment (Porter 1990). While Porter (1990) focused on the rivalry and considered it as the most critical driver of competitive advantage of a country or an industry, national competitive advantages can be also gained from a good business context (Liu and Hsu 2009).

Competition is the spur that drives firms to look for ways to save costs, increase efficiency and encourage innovation, which are all means of increasing competitive advantages (ITS Global 2008; Mitschke 2008). In this paper, (1) intensity of local competition index[1] was used to show the level of domestic competition while the domestic business environment was represented by (3) DTF (Distance to Frontier). DTF is a measurement developed by the World Bank in Doing Business Report to show the distance of each economy to the "frontier," which represents the best performance observed on each of the indicators[2] across all economies. Value 0 shows the lowest performance and 100 reveals the highest performance. Therefore, DTF is a good proxy for domestic business context of one country in comparison with others.

The international factors were also analyzed in two aspects namely rivalry and business context. (1) Market share of the country in G&T global market was used for rivalry proxy and (2) the average import tariff rate representing international business context (Table 1). The higher market share but lower average tariff rate implies higher competitiveness.

To measure 27 selected above-mentioned proxies, data used in this paper were the secondary data derived from various sources of Vietnam, China, and the international organizations such as the World Bank, United Nation Development Program (UNDP), International Trade Center, International Telecommunication Union and the United States Department of Agriculture. The latest available data for both Vietnam and China were used for each proxy. One point worth commenting is that the paper aims at assessing international competitiveness of Vietnam's G&T industry through comparing it with China's. Therefore, data of all proxies are not necessary to be collected for 1 year but data for a proxy must be in 1 year for both nations. This is also data selection approach used in previous literature by Moon et al. (1998), Liu and Hsu (2009), Sardy and Fetscherin (2009), Balcarová (2010), Williams and Morgan (2010), Son and Kenji (2013).

After data for the above-mentioned proxies were collected, they were then translated into scores to quantify international competitiveness of G&T industry in Vietnam and the benchmark nation, China. The method of score translation used in this study was similar to the one used in Rugman and Verbeke (1993) and Sardy and Fetscherin (2009). Firstly, the value of 100 was set for a benchmark country in all proxies, and based on that, the value of each proxy in another country would be calculated accordingly. In this paper, the value of 100 was set for all proxies of China, the country of comparison and the competitiveness index of Vietnam's G&T industry was then calculated accordingly. If a proxy where the higher value indicated more competitiveness, data of China were taken as the base or denominator in calculations. For instance, in the case of labor productivity in G&T industry, a Vietnam's G&T worker can produce seven shirts/day while that number of Chinese worker is 16 (Table 2). Because the higher productivity indicates more competitiveness, the competitiveness value of Vietnam for this proxy is 43.75 % (7/16) (Additional file 2). In contrast, if a proxy where the higher value indicated less competitiveness, data of China were be taken as the numerator. For instance, wage of a worker in G&T industry of Vietnam is 0.74 compared to 2.65 of China (Table 2). Then, the competitiveness index of Vietnam for this proxy is 358.11 % (2.65/0.74) (Additional file 2). Secondly, for each country, the index values of Domestic Diamond and International Diamond were calculated by using the simple average of all selected proxies and variables because they were considered equally important in determining international competitiveness. Thirdly, the global indexes were calculated by taking the simple average of the Domestic and International indexes, representing international competitiveness of Vietnam G&T industry.

8. RESULTS AND DISCUSSION

From descriptive data for Vietnam and China's G&T industries shown in Table 2, the paper calculated and analyzed competitiveness indexes of two countries. Results of the Domestic Diamond, International Diamond and Global Diamond are presented bellowed.

9. DOMESTIC DIAMOND

Except for Factor Conditions, Vietnam's G&T industry was less competitive than China's in all domestic determinants (Fig. 2). The reason behind the higher measurement of Vietnam's factor conditions was the abundant source of unskilled labor with low costs. Vietnam in 2014 was the 14th most populous country in the world and had a golden population structure with the number of working people being twice as much as the number of dependent people (CIA 2014). However, more than 80 % of labors in Vietnam were untrained (General Statistics Office 2015) and therefore were low paid. According to Werner International, the average wages for labors in Vietnam's G&T industry in 2014 was only 0.74 USD/h, the lowest among Southeast Asian nations and 3.5 times lower than that in China (Table 2). In addition, the fee for labor training was also low in Vietnam due to the fact that most of unskilled labors in Vietnam's G&T industry learnt how to work by doing themselves and receiving advices from more experienced workers. G&T is a labor-intensive industry and thus the labor cost has a substantial share in the total cost (IPP and CIEM 2013), leading to big cost competitiveness for Vietnam's G&T industry. This cost competitiveness to some extent compensated for Vietnam's weaknesses in productivity (Yen 2012), the number of workers in G&T industry (Bui 2014, Kane 2014) and the amount of R&D investment compared to China. As the result, Vietnam surpassed China in Factor Conditions.

The domestic Firm Strategy, Structures and Rivalry of Vietnam were only about 1 % less competitive to China's (Fig. 2). Thanks to improvement in business environment, Vietnam's business context was 102.94 % better than China and the level of local competition in Vietnam was relatively high, equal to 94.44 % of China's (Schawab and Martín 2014) (Additional file 5), implying a fierce competition in Vietnam's domestic market. The first source of this competition came from the international integration policy of Vietnam, in which Vietnam opened market to foreign G&T producers to make investment in Vietnam. The second source was low barrier to enter but significant barrier to exit G&T industry. While the low technological requirements and abundant labor made it easier for newcomers to enter G&T industry in Vietnam, the specialized machines in contrast put the existing firms in a difficult situation to get out of the industry. The other source of competition came from the foreign rivalry in selling G&T products. The foreign G&T firms in Vietnam with higher technology, larger scale and bigger experience often achieved the higher share of G&T exports (Bui 2014). The low diversity of Vietnam's G&T products resulting from the insufficient R&D investment also made the competition among domestic firms more drastic. Totally, China was more competitive than Vietnam in this attribute but the gap between two countries was modest.

The domestic Demand Conditions of Vietnam were far less completive than that of China (Fig. 2) because it reported lower measurements relating to five of six proxies, namely total population, GDP, GDP per capita, rate of expenditure on G&T and educational index (Table 2; Additional file 3). However, even though Domestic Conditions of Vietnam was less competitive, there were signs of possibility for Vietnam to improve its competitiveness in the future. The domestic market in Vietnam was small compared to China with the population of being equal to 6.65 % of China's, but Vietnam was a promising market for the development of G&T industry. In fact, Vietnam was a diverse market with increasing demand for G&T product. While the diversity was shown by 54 ethnic groups with different clothing cultures and preferences, the increasing demand was indicated by the high growth rate of domestic demand. According to the survey of household living standards conducted by the General Statistics Office of Vietnam in 2012, the annual average expenditures on garments, hat, shoes and sandals of a Vietnamese household increased more than 1.5 times from 2008 to 2012. This resulted from the increase in the employment rate and GDP per capita in Vietnam during this period. Beside the size, the sophistication of domestic demand also showed some positive signs. Despite being weak compared to China, the education in Vietnam progressed in the recent years, resulting in an increase in Education Index of Vietnam from 0.49 in 2008 to 0.61 in 2013.

Table 2 . Descriptive data for GDDM of Vietnam and China's G&T industry

Attributes	Variables	Proxies	Vietnam	China
Factor Conditions	Domestic		0.74	2.65
	Basic factors	Wage of G&T worker (USD/h) (2014)		
		Number of workers and laborers in G&T industry (million people) (2013)	2.5	23
		Labor productivity in G&T industry (shirts/worker/day) (2012)	7	16
	Advanced factors	R&D expenditures (% of GDP) (2013)	1.48	2.08
	International			
	Advanced factors	Manufacturing inward FDI flow (billion USD) (2013)	17.14	455.54
		Manufacturing outward FDI flow (billion USD) (2013)	1.96	71.97
Demand Conditions	Domestic			
	Size	Total population (million people) (2014)	90.7	1364.3
		GDP (billion USD) (2014)	186.2	10,360.1
		Employment rate (%) (2013)	76.0	68.0
	Sophistication	GDP per capita (USD) (2014)	2052.3	7593.9
		Household rate of expenditure on G&T out of gross income (%) (2010)	3.19	6.96
		Educational index (2013)	0.51	0.61
	International			
	Size	G&T Export value (billion USD) (2014)	26.18	287.59
		Average G&T export growth rate (%) (2012–2014)	16.17	6.20
Related and Supporting Industries	Domestic			
	Supporting Industries	Cotton output (1000 t) (2014–2015)	1.36	6532
		Yarn output (million ton) (2013)	0.72	32
	Supporting infrastructures	Rail lines (total route—km) (2012)	2347	66,298
		Roads, paved (% of total roads) (2010)	47.6	60.9
		ICT index (2013)	4.09	4.64
	International			
	Supporting industries	Cotton exports (1000 t) (2014)	0.00	16,000
		Yarn and Fabric exports (billion USD) (2013)	3.26	106.90
	Supporting infrastructures	Container port traffics (TEU: 20 foot equivalent unit) (2013)	8,121,019	170,080,330
		Air transport (registered carrier departures worldwide) (2014)	144,630	3,356,756
Firm Strategy, Structure and Rivalry				
	Domestic			
	Rivalry	Intensity of local competition (2013–2014 weighted average)	5.10	5.40
	Business context	World Bank DTF point (2014)	64.42	62.58
	International			
	Rivalry	Market share of the country in G&T global market (%) (2014)	3.16	34.69
	Business context	Average import tariff rate faced by G&T industry (%) (2013)	12.4	12.8

Source: Synthesized and calculated by authors

Calculation of competitiveness index of each proxy was shown in Addition files 2, 3, 4, and 5

Calculation of competitiveness index of the GDDM model was shown in Additional file 6

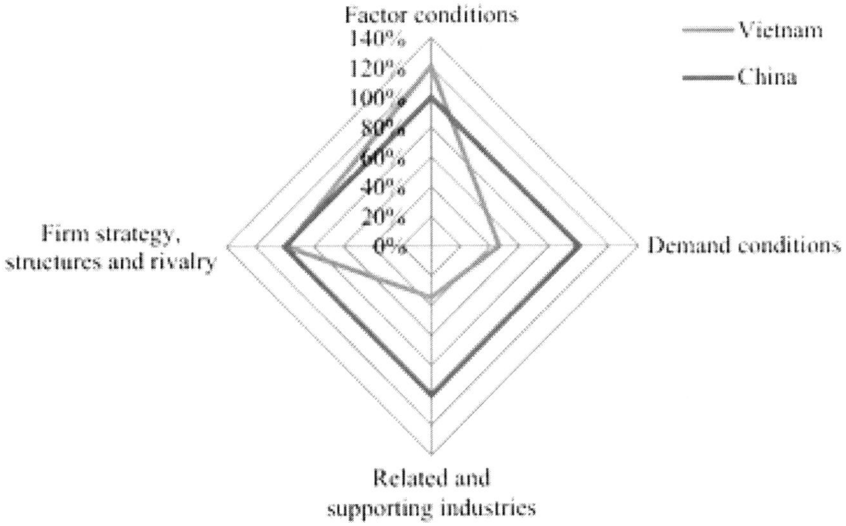

Figure 2 . Domestic Diamond of Vietnam's G&T industry in comparison with China's

China was superior to Vietnam in all five proxies of Related and Supporting Industries. Vietnam's competitive value of this attribute was only equal to 34.42 % of China (Fig. 2). Though the supporting infrastructures such as transportation and communication have recently experienced improvements in Vietnam, problems are still being found in the upstream and downstream industries.

Regarding the upstream industries, both Vietnam's cotton and yarn industries are revealing big weaknesses. Vietnam has consistently confronted the lack of cotton and yarn for G&T production. In Vietnam, only 2.5 thousand hectares was used for cotton cultivation in marketing year (MY) 2014/2015 compared to 4.4 million hectares in China because of unfavorable weather condition and restricted agricultural land. Therefore, cotton output of Vietnam reached 1.36 thousand tons, equivalent to only around 0.02 % of that of China (Additional file 4). The situation is forecasted to be worse because of the decreasing cotton planted area in the years to come and the low productivity of cotton farmers, resulting in the fact that Vietnam's G&T industry will rely more heavily on cotton imports. The domestic cotton production met only around 2 % of total domestic demand and the rest of 98 % must be imported (Vu 2014). Regarding the yarn industry, there has been a paradox. Though the yarn industry experienced some developments when the total output in 2013 rose to 720 thousand ton, equal to 2.1 % of the world total yarn output, only 30 % of the output could be used domestically while the remaining part had to be exported (Vo and Wilder 2015). The reason behind this paradox was the low quality and the lack of diversity in Vietnam's yarn. Vietnam's yarn industry has just focused on low-end products, which cannot satisfy the demand of domestic G&T industry. In comparison with China, yarn output of Vietnam was only equivalent to 2.25 % of that of China. China has always been the world's biggest producer of cotton and yarn due to special status of its agriculture and superiority in resources for G&T industry (Yuan and Xu 2007; Meador and Wu 2014).

While Vietnam's upstream industries were really weak compared to China's, the downstream industries that are related to marketing and distribution have also experienced the same situation. 73 % of Vietnam's G&T exports applied cut–make–trim (CMT) method, by which the Vietnamese G&T firms received orders from their partners abroad, manufactured and then sent back the final products, which would be distributed and sold by foreign partners. Or Vietnam's G&T firms exported products to destinations as instructed by foreign partners. Therefore, marketing and distribution network of Vietnam's G&T enterprises has been underdeveloped and relied largely on foreign distributors (Bui 2014; Vu 2014). Vietnam's G&T enterprises have also participated in the lowest end of global G&T value chain (Truong et al. 2010; Dang and Dinh 2011; Luong 2012). China in contrast gradually moved up in G&T global value chain by shifting to the higher value-added stages rather than CMT. The Chinese G&T companies conducted vertical integration and were able to offer everything from design input to packaging, customs and shipping services. The Chinese companies were also innovative in choosing their business model for their own markets (McNamara 2008).

The weak upstream and downstream industries of Vietnam together with the loose connectivity between them have also resulted in the lack of G&T clusters in Vietnam. In fact, Vietnam had only a handful number of G&T clusters with the biggest one located in the south. This cluster was the result of the co-operation between Ho Chi Minh City, Dong Nai province and Binh Duong province, contributing 56.4 % to total G&T output, 39.4 % to total export value, and 30 % to total labor in G&T industry in 2011 (IPP and CIEM 2013). Vietnam was therefore far behind China with 151 G&T clusters by May 2011 (EUSME Centre 2011). As forming a cluster is vital in improving international competitiveness of an industry (Porter 1990), the lack of G&T clusters has deteriorated international competitiveness of Vietnam's G&T industry considerably in comparison with China. With the above-mentioned reasons, Related and Supporting Industries become the most important and different attribute affecting the competitiveness of Vietnam's G&T in Domestic Diamond.

10. INTERNATIONAL DIAMOND

The International Diamond of Vietnam in G&T industry was much worse than China's. China surpassed Vietnam in all attributes, except for Demand Conditions (Fig. 3).

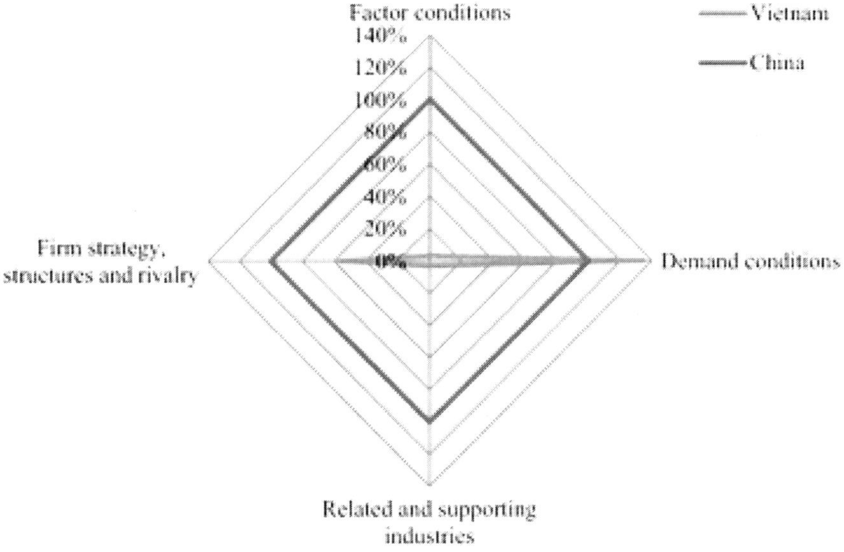

Figure 3 . International Diamond of Vietnam's G&T industry in comparison with China's

Vietnam showed better Demand Conditions than China in the international context (Fig. 3) because G&T exports of Vietnam witnessed a high growth rate of more than 16 % compared to only 6.2 % of China in the period 2012–2014 (Table 2). There were two reasons for this miraculous rate. Firstly, the world economy recovered from the global crisis, leading to higher income and demand for goods and services all over the world. Secondly, Vietnam took well advantage of the EU–Vietnam Partnership and Cooperation Agreement (PCA) signed in 2012 and the EU's debt crisis, when the EU's demand for luxury goods decreased but for necessity goods like food and clothes increased. In addition, in this difficult time, the EU's consumers had a tendency to come back to products with reasonable prices and quality like those made by Vietnam (Vietnam Trade Promotion Agency 2013). In 2012–2014, when the Eurozone was deep in the debt crisis, the growth rate of Vietnam's G&T exports to the EU reached more than 8.4 % while this rate for China was −2.3 %. Besides, Vietnam-Japan Economic Partnership Agreement, the newly signed Vietnam-Korean FTA and the forthcoming

European Union and Vietnam FTA (EVFTA) are also motives for Vietnam to expand exporting G&T products to these key partners. Therefore, though the absolute exports of Vietnam's G&T industry were still low compared to China's, the remarkable growth rates made Vietnam more competitive than China in international Demand Conditions attribute.

Ranking second for Vietnam in the International Diamond was Firm Strategy, Structures and Rivalry attribute (Fig. 3). Vietnam reported lower measurement in international rivalry but exceeded China in business context (Table 2; Additional file 5). With the efforts of Vietnam's government to integrate into the world economy through accessing the World Trade Organization, and signing ten multilateral and bilateral FTAs up to now (Vu and Nguyen 2015), the tariff faced by Vietnam's G&T producers in the global market considerably reduced and was lower than that faced by China. Until now, the average tariff faced by Vietnam in the field of G&T was 12.4 % compared to 12.8 % of China (Table 2). In the near future, when Trans-Pacific Partnership Agreement (TPP) and EVFTA, in which China is so far the outsider, are concluded, the tariffs faced by Vietnam's G&T are likely to reduce substantially because TPP and EVFTA involve the most important G&T partners of Vietnam such as the US, the EU and Japan.

China overwhelmed Vietnam in the international Factor Conditions (Fig. 3) because of its high inward and outward FDI (Zhou and Leung 2015). The manufacturing FDI inflows in Vietnam equaled only 3.76 % and similarly the outward FDI of Vietnam was only 2.72 % as much as China's (Additional file 2).

Similarly, the Related and Supporting Industries of Vietnam reported much lower measurements than China's with all four proxies of being less competitive. Vietnam did not export cotton, and the yarn and fabric exports equaled only about 3 % of China's (Additional file 4). The ability to transport goods to the international market was also weak because the index of container port traffic was equivalent to 4.77 % and the index of the air transport being 4.31 % of China's. This attribute of Vietnam was the weakest in the GDDM model and the most difference between Vietnam and China, and therefore needs comprehensive attention to be improved in the future.

11. GLOBAL DIAMOND

Integrating the Domestic and International Diamond provides the Global Diamond, which shows international competitiveness of Vietnam in G&T industry (Fig. 4).

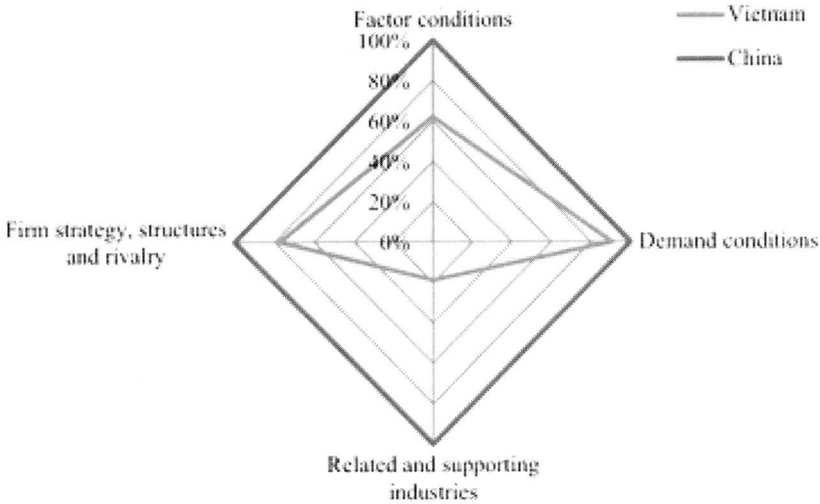

Figure 4 Global Diamond of Vietnam's G&T industry in comparison with China's

International competitiveness of Vietnam in G&T industry was lower than China's in all four GDDM attributes. The biggest gap and also the most important difference between Vietnam and China's G&T competiveness can be seen in Related and Supporting industries, in which Vietnam was around 81 % lower than China (Additional file 6). In contrast, the lowest gap of 9.4 % was reported in terms of Demand Conditions. Factor Conditions and Firm Strategy, Structures and Rivalry of Vietnam were 37.9 and 22.6 % lower respectively. Therefore, in order for Vietnam to improve competitiveness in the global market, the Vietnamese government should put priority on long-term projects and policies to boost the Related and Supporting Industries, and pay more attention to enhance Factor Conditions for G&T industry.

12. CONCLUSIONS

By using the GDDM, this paper analyzed international competitiveness of Vietnam's G&T industry through comparing with China's. The results showed that out of 27 proxies selected, Vietnam was superior to China in terms of only five proxies namely wages, employment rate, average G&T export growth, DTF point and average import tariff rate. As a result, the Domestic Diamond of Vietnam's G&T industry showed lower measurements than China in all attributes, except for Factor Conditions. With the International Diamond, Vietnam was only more competitive in

Demand Conditions. Totally, the Global Diamond showed that Vietnam's G&T industry was far less internationally competitive than China's in all four attributes.

The biggest gap between Vietnam and China's international competitiveness in G&T industry was realized in Related and Supporting industries, in which Vietnam's competitive index was more than 81 % lower than China's. The low competitiveness of Vietnam originated from consistent problems of inadequate downstream and upstream industries including the insufficiency of cotton and yarn production; foreign-reliant G&T marketing and distribution; and weak transportation system for traded goods. Besides, the shortage of G&T cluster in Vietnam deteriorated the competitiveness of this industry internationally.

The second biggest gap was reported in Factor Conditions, in which Vietnam's lower competitiveness came from the low labor productivity, limited R&D expenditure and especially far lowers inward and outward FDI in the manufacturing sector of Vietnam. Vietnam's G&T industry development has so far relied heavily on low-cost labor that was not sustainable for future growth.

The third biggest gap between Vietnam and China in Global Diamond was Firm Strategy, Structures and Rivalry. Vietnam was better than China in business context with higher DTP point and lower average import tariff as a result of the dynamic international integration but China exceeded Vietnam in terms of intensity of local competition and global market share. Totally, Vietnam's international competitiveness index was 38 % lower than that of China.

In the final attribute, Demand Conditions, Vietnam's international competeveness index was approximate to China's and only 9.4 % lower. This was due to Vietnam achieved a high and stable growth rate of G&T exports in the recent years and relatively high educational index. However, other criteria such as GDP per capita and rate of expenditure on G&T in Vietnam were lower than in China.

With all the above results, in order to improve international competitiveness of G&T industry, it is of great importance for Vietnam to enhance all of the four attributes. However, this paper argues that firstly Vietnam should put a high priority on promoting the Related and Supporting Industries. Develop cotton production, diversify and improve quality of yarn, support firms to build up its own distribution channel, set up G&T clusters and upgrade transportation for shipping goods abroad are urgent measures to enhance G&T Related and Supporting industries in Vietnam. Next is the task to improve Factor Condition by investing more in training G&T workers to enhance productivity and educating high skilled labors to improve R&D activities in the industry. Finally, Vietnam should put effort to maintain its current strengths over China in terms of G&T export growths and favorable business context through taking advantages of its existing FTAs and the promising EVFTA and TPP.

This paper has contributed to the existing literature by using the GDDM approach to analyze international competitiveness of Vietnam's G&T industry. However, it still has limitation and can be improved in the future. In fact, due to limited data as well as differences in statistical system and methods between China and Vietnam, some data were not available for both nations. Therefore, some proxies for the whole manufacturing sector were used instead of those specifically for G&T industry including inward FDI, outward FDI and R&D expenditure. This shortcoming will be resolved in the future research when the data problem is addressed. Given Vietnam and China's statistical system and methods are in status-quo, data limitation can be partly solved in future research through conducting survey or in-deep interview of Vietnam's G&T enterprises and Vietnam Textile and Apparel Association to get G&T data on productivity, R&D expenditure and FDI.

Moreover, the future research can take into consideration some more G&T-related indicators such as design power in clothing and textiles (Factor Conditions); demanding needs of consumers, and preference for local brands in oversea market (Demand Conditions); G&T cluster, and educational facility related to G&T industry (Related and Supporting Industries); and the rate of added value in G&T industry (Firm Strategy, Structure and Rivalry) to provide better results on Vietnam's competitiveness of G&T industry.

Finally, this paper chooses China, which is a developing but much bigger than Vietnam in terms of GDP, population and area, as the benchmark country to compare with Vietnam. In order to provide multi-dimensional analysis of Vietnam's G&T international competitiveness, identify more precisely Vietnam's G&T international competitiveness gap with its main competitors and more importantly recommend the progress measures for Vietnam to improve its international competitiveness, the future research can take into consideration of some other comparing nations with more similar size such as Bangladesh, Turkey and Cambodia.

13. FOOTNOTES

1Intensity of local competition in an index used in the Global Competitiveness Report of World Economic Forum to measure the level of competition in the domestic market. 1 = not intense at all; 7: extremely intense.

2Ten indicators are reported in Dong Business Report, namely starting a business, dealing with construction permits, getting electricity, registering property, getting credit, protecting minority investors, paying taxes, trading across borders, enforcing contracts and resolving insolvency.

AUTHORS' CONTRIBUTIONS

PCL has proposed the general methodological approaches, contributed to data collection and drafted the manuscript. VTH has made a significant contribution to the methodological approach, analysis framework, data collection and analysis, and finalized the manuscript. All authors read and approved the final manuscript.

COMPETING INTERESTS

The authors declare that they have no competing interests.

REFERENCES

1. Asian Foundation, CIEM (2012) The competitiveness of export enterprises in three sectors: garment and textiles, fisheries and electronics in Vietnam. The Asian Foundation, Center Institute for Economic Management, Hanoi
2. Balcarová P (2010) The generalized double diamond model—application for Czech Republic and Slovakia. In: 2nd international PhD student conference in new economic challenges, University of Masaryk University, Brno, 20–21 January
3. Barnes JG, McTavish R (1983) Segmenting industrial markets by buyer sophistication. Eur J Mark 17(6):16–33
4. Brown S, Sessions J (2001) International competition and the labor market prospects of union and nonunion workers. J Labor Res 22(3):669–677
5. Bui VT (2014) Garment and textiles report. FPT Securities, Hanoi
6. Cartwright WR (1993) Multiple linked "diamonds" and the international competitiveness of export-dependent industries: the New Zealand experience. Manag Int Rev 33(2):55–71

7. Cho DS (1994) A dynamic approach to international competitiveness: the case of Korea. Asia Pacific Bus Rev 1(1):1743–1792

8. CIA (2014) Vietnam in the World Factbook. Central Intelligence Agency, Washington D.C. http://www.cia.gov/library/publications/the-world-factbook/geos/vm.html. (Accessed 10 Apr 2015)

9. Dang TTN, Dinh CK (2011) Policy research brief: Vietnam's garment and textiles value chain Ho Chi Minh. Fullbright, Vietnam

10. Daniel C (2000) The question of international competitiveness. Int Adv Econ Res 6(3):417–426

11. D'Cruz JR (1992) New compacts for Canadian competitiveness. Kodak Canada, Toronto

12. D'Cruz JR, Rugman AM (1993) Developing international competitiveness: the five partners model. Bus Q 58:60–72

13. Depperu D, Cerrato D (2005) Analyzing international competitiveness at the firm level: concepts and measures Working Paper No. 32. University Cattolica del Sacro Cuore, Piacenza

14. Dunning JH (2003) The role of foreign direct investment in upgrading China's competitiveness. J Int Bus Econ 4(4):1–13

15. EUSME Centre (2011) Report: textile and apparel market in China. EUSME Centre, Beijing

16. General Departnment of Vietnam Customs (2015) Statistical yearbook of customs on Vietnam's imports and exports in 2014 Hanoi. Financial Publishing House, Hanoi

17. General Statistics Office (2015) Statistical handbook of Vietnam: 2014. Statistical Publising House, Hanoi

18. Han J, Park H, Ock J (2015) An international competitiveness evaluation model in the global construction industry. J Civ Eng 19(3):465–477

19. Hirschman EC (1980) Innovativeness, novelty seeking, and consumer creativity. J Consum Res 7(3):283–295

20. Hult GTM (2012) A focus on international competitiveness. J Acad Sci 2012(40):195–201

21. IPP, CIEM (2013) Assessing the competitiveness of garment and textiles cluster in Ho Chi Minh city and other surrounding areas Ho Chi Minh. Institute of Public Policy, Center Institute for Economic Management, Ho Chi Minh city

22. ITC (2015) Trade Map, International Trade Center, Geneva. http://www.trademap.org/Index.aspx?AspxAutoDetectCookieSupport=1. (Accessed 12 June 2015)

23. ITS Global (2008) Improving international competitiveness in Australian business. Business Councils of Australia, Melbourne

24. Kane G (2014) Facts on China's garment industry. Clean Clothes Campaign, Amsterdam

25. Kim M-J, Kwak D-R, Lee Y (2006) Analysis of international competitiveness of apparel industry in Korea and China based on the Generalized Double Diamond Model. J Korean Soc Cloth Text 30(9/10):1354–1365

26. Liu DY, Hsu HF (2009) An international comparison of empirical generalized double diamond model approaches to Taiwan and Korea. Compet Rev Int Bus J 19(3):160–174

27. Luong TL (2012) The global garment and textiles value chain and participation of Vietnam's garment and textiles industry. Dissertation, University of Economcis and Business, Vientam National University - Hanoi

28. McNamara K (2008) Garment industry: the role of information and communication technologies (ICTs) in exploing the value chain. The World Bank, Washington, DC

29. Meador MMM, Wu X (2014) China—peoples republic of cotton and products annual. United States Department of Agriculture, Washington, DC

30. Mitschke A (2008) The influence of national competition policy on the international competitiveness of nations: a contribution to the debate on international competition rules. Springer, Berlin

31. Momaya K (1998) Evaluating international competitiveness at the industry level. Vikalpa 23:39–46

32. Moon HC, Youn KJ (2010) Comparing the competitiveness of Korea and Singapore after ten years of Asian economics. Rev Bus Hist 25(1):75–91

33. Moon HC, Rugman AM, Verbeke A (1995) The generalized double diamond approach to international competitiveness. In: Rugman AM, Den JV, Verbeke A (eds) Beyond the diamond—research in global strategic management, vol 5. Emerald Group, Bingley

34. Moon HC, Rugman AM, Verbeke A (1998) A generalized double diamond approach to the global competitiveness of Korea and Singapore. Int Bus Rev 7(2):135–150

35. Nguyen BV (2012) The competitiveness of Vietnam's garment and textiles exports to the EU in the context of WTO integration. Dissertation, University of Economics and Business, Vietnam National University - Hanoi

36. Pizer SD (2000) Does international competition undermine wage differentials and increase inequality? J Int Econ 52(2):259–282

37. Porter ME (1990) The competitive advantage of nations. Free Press, New York

38. Rugman AM, Verbeke A (1993) Foreign subsidiaries and multinational strategic management: an extension and correction of Porter's single diamond framework. Manag Int Rev 33:71–84

39. Sardy M, Fetscherin M (2009) A double diamond comparison of the automotive industry of China, India, and South Korea. Compet Forum 7(1):6–16

40. Schawab K, Martín XS (2014) The global competitiveness report: 2014–2015. World Economic Forum, Geneva

41. Smit AJ (2010) The competitive advantage of nations: is Porter's diamond framework: a new theory that explains the international competitiveness of countries? South Afr Bus Rev 14(1):105–130

42. Son MY, Kenji Y (2013) A comparative analysis on the competitiveness of Korean and Japanese fashion industry by applying Generalized Double Diamond Model. Asia Mark J 15(1):57–81

43. Son MY, Kim YJ, Ji H-K (2007) Utilizing of the Chinese fashion market for globalization of Korean fashion industry: focused on the competitiveness of Korean and Chinese fashion industry by applying the Double Diamond Model. J Korean Soc Cloth Text 31(4):507–518

44. Truong HT, Nguyen TBT, Nguyen TL (2010) The value chain approach to industrial upgrading of Vietnam textile industry. J Sci Technol 2(37):158–168

45. Tuyet Y (2012) For the sustainable development of garment and textiles industry. J Ind 1:40–41

46. Vietnam Trade Promotion Agency (2013) Export promotion report: 2012–2013—trade promotion at local level Hanoi. Vietnam Trade Promotion Agency, Hanoi

47. Vitas (2006) Vietnam's garment and textiles export strategy 2006–2010. Vietnam Textile and Apparel Association, Hanoi

48. Vo TK, Wilder DA (2015) Vietnam: cotton and products annual—commodity report global agricultural information network. United States Department of Agriculture, Washington, DC

49. Vu NA (2014) Sectoral report of VietinbankSc: garment and textiles industry of Vietnam. Vietnam VietinBankSc, Hanoi

50. Vu TH, Nguyen TMP (2015) The European-Vietnam Free Trade Agreement: a perspective of opportunities and challenges. Communist Rev 104(8):117–120

51. Williams DA, Morgan B (2010) Benchmarking for international competitiveness: lessons for public policy. In: Proceedings of the 7th quest for global competitive conference, Puerto Rico

52. Yen T (2012) For the sustainable development of garment and textiles industry. J Ind 1:40–41

53. Yuan T, Xu F (2007) China's textile industry international competitive advantage and policy suggestion. J Wash Inst China Stud 2(1):75–84

54. Zhou L, Leung D (2015) China's overseas investment, explained in 10 graphics. World Resources Institute, Washington D.C. http://www.wri.org/blog/2015/01/china's-overseas-investments-explained-10-graphics. (Accessed 18 Feb 2015)

CHAPTER 11

Textile Materials for the Design of Wearable Antennas: A Survey

Rita Salvado [1,*], Caroline Loss [1], Ricardo Gonçalves [2] and Pedro Pinho [2]

[1] Unidade de Investigação Materiais Têxteis e Papeleiros, Universidade da Beira Interior, 6201-001 Covilhã, Portugal

[2] Instituto de Telecomunicações, Campus Universitário de Santiago, 3810-135 Aveiro, Portugal

ABSTRACT

In the broad context of Wireless Body Sensor Networks for healthcare and pervasive applications, the design of wearable antennas offers the possibility of ubiquitous monitoring, communication and energy harvesting and storage. Specific requirements for wearable antennas are a planar structure and flexible construction materials. Several properties of the materials influence the behaviour of the antenna. For instance, the bandwidth and the efficiency of a planar microstrip antenna are mainly determined by the permittivity and the thickness of the substrate. The use of textiles in wearable antennas requires the characterization of their properties. Specific electrical conductive textiles are available on the market and have been successfully used. Ordinary textile fabrics have been used as substrates. However, little information can be found on the electromagnetic properties of regular textiles. Therefore this paper is mainly focused on the analysis of the dielectric properties of normal fabrics. In general, textiles present a very low dielectric constant that reduces the surface wave losses and increases the impedance bandwidth of the antenna. However, textile materials are constantly exchanging water molecules with the surroundings, which affects their electromagnetic properties. In addition, textile fabrics are porous, anisotropic and compressible materials whose thickness and density might change with low pressures. Therefore it is important to know how these characteristics influence the behaviour of the antenna in order to minimize unwanted effects. This paper presents a survey of the key points for the design and development of textile antennas, from the choice of the textile materials to the framing of the antenna. An analysis of the textile materials that have been used is also presented.

KEYWORDS

wearable antenna; flexible antenna; textile materials; dielectric properties

1. INTRODUCTION

Body worn systems endowed with sensing, processing, actuation, communication and energy harvesting and storage abilities are emerging as a solution to the challenges of ubiquitous monitoring of people in applications such as healthcare, lifestyle, protection and safety [1]. Accordingly, the new generation of clothing will be able to sense, communicate data and harvest energy in a nonintrusive way [1,2]. The wearable antenna is thus the bond that integrates cloth into the communication system, making electronic devices less obtrusive. To achieve good results, wearable antennas have to be thin, lightweight, low maintenance, robust, inexpensive and easily integrated in radio frequency (RF) circuits [3–6]. Thus, planar antennas are the preferred type of antenna as, despite the fact their maximum attainable bandwidth-efficiency is significantly lower than the theoretical limit for electrically small antennas, they allow an excellent integration of the antenna with the RF circuits, feeding lines and matching circuits on a standard multilayer board material [4]. Therefore, they might be integrated in cloth in a minimally intrusive way [2,5]. In particular, the microstrip patch antennas are good candidates for body-worn applications, as they mainly radiate perpendicularly to the planar structure and also their ground plane efficiently shields the body tissues [5,7]. Specific requirements for the design of wearable antennas are thus: planar structure; flexible conductive materials in the patch and ground plane; and flexible dielectric materials [3,5,8]. The characteristics of the materials are crucial for the behaviour of the antenna. For instance, the permittivity and the thickness of the dielectric substrate mainly determine the bandwidth and the efficiency performance of the planar antenna [4]. Also, the conductivity of the ground plane and of the patch is an important factor in the efficiency of the antenna and must be the highest possible.

Textile materials, being universally used and easily available, are possible materials to design wearable antennas for in- and on-Body Area Networks (BAN). The characterization of their electric and electromagnetic properties is essential for the design of the antenna [8]. Electrical properties of conductive textiles have been accurately characterized using the Transmission Line Method [9], Cavity Method [10] and MoM-segment Method [11], and their surface resistivity are often given by specialized producers. Therefore, specific conductive textiles, sometimes designated electrotextiles, that are commercially available have been successfully used in antennas [5,12–16]. Ordinary textile fabrics have been used as dielectric substrates. However, little information is found on the electromagnetic properties of regular commercialized textiles. Therefore, this paper is focused on the analysis of textile materials, mainly dielectric fabrics, that have been used in antennas [2,17–22].

The following section presents an overview of the influence of some features of the textile materials, dielectrics and conductive ones, in the behaviour of the antennas. It also gives some guidelines for the choice of materials for the design of textile antennas. Later, Section 3 reviews the textile materials that have been used to develop different wearable antennas, focusing on the regular fabrics used as dielectrics and listing their relative permittivity and loss tangent values. Furthermore, Section 4 presents an overview of guidelines for the construction of planar and wearable antennas and the techniques used to assemble the various materials. Finally, Section 5 presents the main conclusions of this work.

2. IMPORTANT FEATURES OF TEXTILE MATERIALS IN THE DESIGN OF WEARABLE ANTENNAS

Fabrics are planar fibrous materials which properties are mainly determined by the properties of the component fibres and the structure of the yarns and/or of the fabric. They are porous materials, in which the density of the fibres, air volume and size of the pores determine general behaviour, for instance, air permeability and thermal insulation. Accordingly, fabrics are flexible and compressible materials which thickness and density might change with low pressures. Moreover, the main orientation of the fibres and/or yarns introduces an intrinsic planar anisotropy of general properties. Plus, fibres are constantly exchanging water molecules with the surroundings, which affects their morphology and properties. All these features are somehow difficult to control in real applications of textiles, therefore it is important to know how they may influence the behaviour of the antenna in order to minimize any unwanted effects. Moreover, some characteristics of the textile materials known to influence the performance of the antenna and referred in the literature are reviewed in this section.

2.1. The Dielectric Constant (Relative Permittivity) of the Fabrics

The constitutive parameter of dielectrics is the permittivity, ε, that is a complex value parameter. It is usually expressed as a relative value ε_r: $\varepsilon = \varepsilon_0\varepsilon_r = \varepsilon_0(\varepsilon'_r - j\varepsilon''_r)$, where ε_0 is the permittivity of vacuum, which is 8.854×10^{-12} F/m [17]. In general, the dielectric properties depend on the frequency, temperature, and surface roughness [17], and also on the moisture content, purity and homogeneity of the material [18]. The real part of the relative permittivity, ε'_r, is called the dielectric constant, but one must note that it is not constant in frequency. The ratio of the imaginary part to real part is called the loss tangent, $\tan\delta = \varepsilon''_r / \varepsilon'_r$.

Some researchers have studied and reviewed the dielectric properties of textiles [19–21]. As textile materials are anisotropic materials, their characterization also depends on the electric field orientation. This anisotropy is fully described with a permittivity tensor, although in most practical applications like the ones surveyed in this paper, a specific component of this tensor is enough to characterize the behaviour of the textile material for a specific application. Thus, the relative permittivity describes the behaviour of the material tested under a specific electric field orientation and frequency.

The dielectric behaviour of textile materials depends on the properties of the constituent fibres and polymers [19], and on the fibre packing density in the fibrous material [18,20]. However, textile fabrics are rough, porous and heterogeneous, having air in between the fibres, making their characterization difficult [23]. In addition, the ability of the fibres to absorb moisture must also be considered in the characterization of the dielectric behaviour of textiles, as will be explained further in Section 2.4. Thus, the accurate measurement of dielectric characteristics of textiles is challenging and different experimental techniques have been used, such as the Cavity Perturbation Method [4,10], MoM-segment method [9], Resonance Method [21], Free Space Method [24] and Transmission Line Method [25–27]. Among these techniques, the simplest ones and thus very promising ones are the techniques based on the measurement of the behaviour of transmission lines [25–27].

In general, textiles present a very low dielectric constant as they are very porous materials and the presence of air approaches the relative permittivity to one. As an example, Table 1 shows the dielectric properties of normal textile fabrics, possible dielectric substrates, that were obtained with a waveguide cavity method, under 2.6 GHz [10].

Table 1. Dielectric Properties of normal fabrics tested in [10].

Nonconductive Fabric	ε'_r	**tanδ**
Cordura®	1.90	0.0098
Cotton	1.60	0.0400
100% Polyester	1.90	0.0045
Quartzel® Fabric	1.95	0.0004
Cordura/Lycra®	1.50	0.0093

The low dielectric constant reduces the surface wave losses which are tied to guided wave propagation within the substrates. Therefore, lowering the dielectric constant increases spatial waves and hence increases the impedance bandwidth of the antenna, allowing the development of antennas with acceptable efficiency and high gain [3,28–30]. Again, one should note that the relative permittivity value changes with the moisture content of the substrate affecting the bandwidth of the antenna [2,29].

2.2. Thickness of the Dielectric Fabrics

The bandwidth and efficiency performance of a planar microstrip antenna is mainly determined by the substrate dielectric constant and its thickness [4,31]. As referred before, the changes in permittivity may change the antenna bandwidth, but lowering the substrate permittivity can also increase the resonance frequency of the antenna.

As textile materials present a quite narrow range of permittivity values, it is therefore their thickness, which values may present much larger variations, that will mainly determine the bandwidth as well as the input impedance of the antenna and so its resonance frequency [28]. The thickness of the dielectric material is thus crucial in the design of antennas [2,31]. For a fixed relative permittivity, the substrate thickness may be chosen to maximize the bandwidth of the planar antenna. However, this value may not optimize the antenna efficiency. Therefore, the choice of the thickness of the dielectric material is a compromise between efficiency and bandwidth of the antenna [4,31].

The influence of the thickness on the bandwidth (BW) of the antenna may be explained by Equation (1), where Q is the antenna quality factor:

$$BW \sim 1/Q \tag{1}$$

The Q factor is influenced by the space wave (Q_{rad}) losses, the conduction ohmic (Q_c) losses, the surface waves (Q_{sw}) and dielectric (Q_d) losses as shown in Equation (2) [32]:

$$\frac{1}{Q_t} = \frac{1}{Q_{rad}} + \frac{1}{Q_c} + \frac{1}{Q_d} + \frac{1}{Q_{SW}} \tag{2}$$

For thin substrates($h \ll \lambda_0$) the quality factor associated with radiation (Q_{rad}) is usually the dominant factor and is inversely proportional to the height of the substrate [32]. Therefore, increasing the height of the substrate lowers the Q factor (Q_t). As the Q-factor decreases with an increased aperture between the patch and the ground planes of the antenna, a thicker substrate allows a larger antenna bandwidth [5].

Moreover, the thickness of the substrate also influences the geometric sizing of the antenna. This means that a thick substrate with low relative permittivity (value between 1 and 2) results in a large patch and a thin substrate with the same dielectric constant results in a smaller patch [5].

There are commercially available fabrics with a very diverse range of thickness values. Plus, nominal thickness values are given in any technical data sheet, allowing a careful choice of the material based on the required thickness. Moreover, accurate values of the thickness of fabrics under specified pressure are easily obtained by simple standard methods, such as ISO 5084:1996 and ASTM D374-99(2004) or with a Digimatic Indicator [31]. Therefore, the thickness of the fabrics is a feature that may guide the search for suitable textile dielectrics.

2.3. The Electrical Surface Resistivity of the Conductive Fabrics

Fabrics are planar materials and therefore their electrical behaviour may be quantified by the surface resistance and characterized by the surface resistivity. The surface resistance, which unit is (Ω), is the ratio of a DC voltage to the current flowing between electrodes of specific configuration that are in contact with the same face of the material under test [33]. The surface resistivity is the ratio of the DC voltage drop per unit length to the surface current per unit width. Surface resistivity is thus a property of the material, not depending on the configuration of the electrodes used for the measurement. It is usually expressed in Ohm/square (Ω/sq; Ω/□ [34].

Despite the existence of several standard methods, i.e.,

-AATCC Test Method 76-2011: Electrical Surface Resistivity of Fabrics

-ASTM D4496: 2004 Standard Test Method for D C Resistance or Conductance of Moderately Conductive Materials

-ISO 10965:2011: Textile floor coverings—Determination of electrical resistance

-ISO 21178:2005: Light conveyor belts—Determination of electrical resistance

to measure the surface resistance and resistivity of textile fabrics, they are dedicated to moderately conductive materials and are still aim of analysis [35]. An accurate characterization of highly conductive fabrics demands other techniques, such as for instance the ones based on transmission lines and waveguide cavities [9–11,36].

For the antenna design, the relevant parameter is the conductivity of the fabric, σ, which unit is Siemens per meter (S/m). It is related to the surface resistivity, ρ_s, by Equation (3), where t is the thickness of the fabric:

$$\sigma = 1/(\rho_s \cdot t) \tag{3}$$

Besides the dielectric constant of the substrate, the choice of the conductive fabric for the patch and the ground planes is also very important to assure a good performance of the antenna. In general these fabrics must have a very low electrical surface resistance in order to minimize the electric losses and thus increase the antenna efficiency. Despite the fact that the surface resistance value should be constant over the area of the antenna [2], the fabric may present some heterogeneities, such as for instance some discontinuities in the electric current. If these discontinuities are parallel to the surface current they will not interfere with the electromagnetic fields [37], but if discontinuities impede the flow of the electrical current, the fabric resistance will increase [2]. The structure of the fabric should thus be considered, seeking to determine the density, the continuity and the alignment of the conductive components, which may be fibres, filaments or surface covers. In [2] the studied fabrics having conductive fibres performed better than the coated fabrics, because of discontinuities that increased the surface resistance of the coated fabrics. In [10] higher conductive thread density in the woven fabric results in higher effective conductivity.

Wovens and knits with electric surface resistivity below 1 Ω/□ are commercially available (e.g., Less EMF Inc. at http://www.lessemf.com or Shieldex Trading at http://www.shielextrading.net) and have been successfully used [2,24,26]. However, knits may present higher anisotropy than wovens, showing different electric surface resistance along the longitudinal and the transversal directions and this anisotropy may increase with the knit deformation. Indeed, in fabrics composed of conductive threads, conductive paths exist in all directions through the conductive threads and/or inter-contact points across threads [10]. According to Locher et al. [2], for the studied knit material that is composed of silver plated polyamide fibres, deformations under 8% of elongation along the direction of the wales slightly change the surface resistivity. However, when elongating the knitted fabric along the direction of the courses the electric surface resistivity results stable up to 3% of elongation but then increases and at 8% of elongation it reaches triple the initial value.

The influence of the structure of the fabrics on the surface resistivity may be better understood looking at Figure 1 that shows schemes of regular patterns of fabrics: the jersey knit and the satin 5 weave patterns. If the conductive threads in the weave pattern are along the intended direction for current flow, woven patterns are much more efficient in terms of electrical conduction than knit patterns, because the conductive paths in woven are better aligned with the current direction, which minimizes the conductive losses, as concluded in [10].

Moreover, fabrics present differences between their faces that should be consider when positioning the conductive fabric in a planar assemblage. For instance, in Figure 1 the right and back sides of the satin 5 woven are clearly different. In [10], a satin 5 woven composed of conductive yarns (represented by white rods in Figure 1) and non-conductive yarns (represented by black rods) was tested in a microstrip resonator measurement to study the effect of the differences among faces on the electrical properties of planar structures, as wearable antennas. It was proved that keeping the conductive face against the dielectric substrate is preferred for lowering the electrical losses.

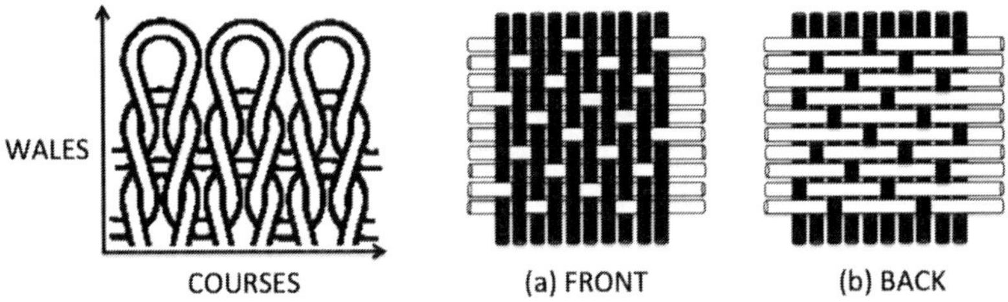

Figure 1. Schemes of: (left) Jersey knit; (middle and right) satin 5 woven; (a) front and (b) back.

2.4. The Moisture Content of the Fabrics

Textiles always establish a dynamic equilibrium with the temperature and humidity of the air surroundings they are in contact with, as the fibres are constantly exchanging water molecules with the air. However, the amount of water that a material takes until reaching this equilibrium depends on the type of material. The extent to which a material is sensitive to moisture is described by its regain, which is defined, by the ratio of the mass of absorbed water in specimen to the mass of dry specimen, expressed as a percentage, [20]. In [20], page 169 shows the relation between regain and relative humidity of the air (RH), for various textile fibres, compiling studies made by several authors. Indeed, for the same RH conditions, there are textile fibres with largely different moisture contents. For instance, at 65% RH, wool fibre might present a regain of 14.5%, cotton might present a regain of 7.5% and polyester fibre might present a regain of 0.2% [20].

In general, the moisture absorption changes the properties of the fabrics. Because of this, textile metrology is performed always in a conditioned environment (20 °C and 65% RH). Also, for commercial transactions there are national legislations setting nominal values of moisture content that are values close to the regain obtained at 65% relative humidity.

Water has a dielectric constant of $\varepsilon'_r = 78$ at 2.45 GHz and 25 °C [31]. Although this value depends on the salinity, temperature and frequency [38], water has a much higher and more stable dielectric constant than textile fabrics, whose dielectric constant is generally in the range $\varepsilon'_r = 1 - 2$ because of their high porosity, as previously presented. Therefore, when water is absorbed by the textile fibres or is trapped into the fabric structure, it changes the electromagnetic properties of the fabric, increasing its dielectric constant and loss [2,5,19,20,31]. Charts presenting the relationship between the RH of the air or the moisture content of various fibres and their dielectric properties can be found in [20]. Additionally, several authors have been correlating the electromagnetic properties of fabrics to their absorption properties [31,38].

Likewise, the absorbed water by or trapped into the textile components of the antenna dramatically changes the behaviour of the antenna. The higher permittivity of the water drives the performance of the antenna, reducing its resonance frequency [2,6] and bandwidth [9,29,31].

Beyond these effects, when textile fibres absorb water they swell transversely and axially, causing tightening of the fabrics [20,39]. This tightness affects the dimensional stability of the fabrics and therefore affects the dimensional stability of the antenna, influencing its behaviour [2]. The swelling of the fibres also contributes to the change of the dielectric properties, as swollen fibres decrease the porosity of the fabric [39]. Indeed, as seen above, the low relative permittivity values of the fabrics are mainly determined by the presence of air in them. In [20] (page 227), experimental results from several authors are compiled, showing that values of swelling in water vary the same way as the values of the regain of the fibres. For instance, when cotton is in water, its cross section area might swell 40%, whilst polyamide only swells 3.2% [20].

Moreover, water absorption by the fibres is an exothermic reaction and the released heat of wetting is the greatest for the highly absorbing fibres. This affects the temperature and thus the electromagnetic properties of the material. More details and data quantifying these effects can be found in [20].

Therefore, climatic changes altering the relative humidity of the air or environmental water, ice and snow conditions will influence the textile antenna performance [31,40]. In the same way, when the antenna is used close to the skin, the hydrophobicity and the regain of the fabric are very important features [28,41] as the fabric will absorb moisture from the skin.

In general, the characteristics of antennas based on textile materials with small moisture absorption values (regain less than 3%) are more stable [31]. Therefore, materials with low regain are preferable for use as substrates and the same conclusion applies to the textile conductive components of the antenna.

2.5. Mechanical Deformations of the Dielectric and Conductive Fabrics

A curvature on a human body consists of a superposition of bends in arbitrary directions. Because of their excellent flexibility and elasticity textile materials adapt well to these surfaces. However, when the textile fabric adapts to the surface topology it bends and deforms, causing changes to its electromagnetic properties and thus influencing the antenna performance [2,29,42]. Indeed, the bending and the elongation of the dielectric fabric influences its permittivity and its thickness, which affects the resonance frequency of the antenna and especially the bandwidth, as previously explained [28].

Moreover, the variation of the geometric dimensions of the textile components of the antenna due to their elongation and/or compression diminishes the geometric precision of the shape of the antenna affecting its behaviour, which may change its resonance frequency [28].

In addition, from the point of view of manufacturing antennas, the elasticity of the fabrics is an inconvenience as it makes difficult the precise definition and cut of the shape of the components and also makes difficult the superposition of the several materials without folds. Wovens and nonwovens, being more stable fabrics than knits, allow higher geometrical accuracies of the frame of the antenna. In general, the accuracy depends on the thickness of the component yarns or fibres. For instance, conductive woven may allow an accuracy of about ±0.15 mm [2]. For these reasons knits are not stable enough for the dielectric of an antenna. However, as further presented in Section 4, they might be suitably stabilized if assembled with a more rigid textile material, such as a fabric with high tensile strength.

3. TEXTILE MATERIALS USED IN WEARABLE ANTENNAS

Wearable antennas are a recent research subject, although one of the first proposals on the subject appeared in 2001 [3], when Salonen et al. [22], presented a Planar Inverted F Antenna (PIFA) for dual-band operation, built on a flexible unspecified substrate. It was intended to be placed in the sleeve of clothing and operate at GSM (900 MHz) and Bluetooth (2.4 GHz) frequency, although the lower band was not achieved, the antenna still showed good performance, even for human-body presence, around the 2.4 GHz band. Later in 2003, they presented an antenna built on a textile substrate intended for WLAN applications [43], where results are claimed to be acceptable.

Furthermore, in 2004 Salonen et al. [28] proposed a GPS antenna with circular polarization, in which they have experimented with five different synthetic fabric materials as dielectric substrates. The conductive parts were made of copper tape. The dielectric synthetic materials sued were: (1) Vellux®, which is a 5 mm thick fabric covered on both surfaces with thin layers of plastic foam; (2) synthetic felt, which is a 4 mm thick nonwoven in which fibres are looser on the surface than in the centre; (3) Delinova 200®, which is a strong fabric made of polyamide Cordura® fibres laminated with Gore-Tex membrane, weighing about 370 g/m^2 and having a thickness of 0.5 mm; (4) fleece, which is a very soft polyester fabric with 4 mm thickness, commonly used in sportswear; (5) upholstery fabric, which is composed of three fabric layers bound together resulting in a thin (1.1 mm) fabric of polyester and acrylic that has firmness. The relative permittivity, ε'_r, of the five fabrics was measured by a cavity perturbation method, at 1.575 GHz, and the values ranged between 1.1 and 1.7. Among the studied fabrics, the one made of high tenacity polyamide fibres (Cordura®) was pointed out as the more interesting fabric for the development of a flexible antenna, because of its constant thickness and its high resistance. These properties allow more stable geometric dimensions of the antenna.

More recent demonstrations on wearable antennas for Personal Area Networks (PANs) to operate in the 2.45 GHz Industrial, Scientific and Medical (ISM) band and for GPS applications are presented in [5,12]. In these examples, antennas for wearable protective clothing intended for professional use under rough conditions are presented and their behaviour in various practical scenarios is discussed. High performance aramid fabric that can withstand high temperatures is used as substrate while conductive textiles, like Shieldit® and FlecTron®, are used for the antenna patch and ground plane. These antennas have shown acceptable performance, even in a real environment with human-body presence and when subjected to bending and deformations.

Locher et al. [2] have built four purely textile wearable patch antennas for Bluetooth applications. They have used three electrical conductive fabrics: (1) a nickel-plated woven fabric (with plating thickness about 250 nm applied on the fabric surface); (2) a silver-plated knitted fabric; (3) a silver-copper-nickel-plated woven fabric. Fabric (3) is the one preferred for building textile antennas with geometric precision, as it is woven and not knitted and its electric surface resistance was more homogeneous than the one of fabric (1). For the dielectric substrate, they used two types of fabrics: (1) woollen felt of 1,050 g/m^2 with a thickness of 3.5 mm and (2) polyamide spacer fabric, of 530 g/m^2, with a thickness of 6 mm. The felt was dimensionally more stable and harder to bend, whereas the spacer fabric was lighter and more elastic due to its knitting-based structure. The dielectric properties were measured by a transmission line method, at a frequency of 2.4 GHz, obtaining as results for the felt: permittivity $\varepsilon'_r = 1.45$ and loss tangent $\tan\delta = 0.02$, and for the spacer fabric: permittivity $\varepsilon'_r = 1.14$ and the loss tangent was negligible. The four different antennas produced have shown good performance and could satisfy the Bluetooth specifications, even when subjected to bending effects. However the antennas lose their circular polarization when subject to bending.

The same year, Tronquo et al. [13] presented rectangular-ring textile antennas for body area networks (BAN) that are circularly polarized, covering a bandwidth of more than 190 MHz. For the conductive antenna patch and ground plane they used the conductive fabric named Flectron®, that is a thin copper plated fabric with low surface resistivity, lesser then 0.1 Ω/□ For the dielectric substrate they used a fleece fabric of 2.56 mm thickness. Its dielectric properties were measured testing antennas and they obtained a relative permittivity of $\varepsilon'_r = 1.25$.

In 2007, Zhu and Langley [18] developed a dual-band coplanar patch antenna integrating electromagnetic bandgap material (EGB), to operate at the 2.45 and 5.8 GHz wireless bands. The conductive parts were made of Zelt® fabric while the dielectric substrate was a thin felt, with 1,1 mm thickness, and with relative permittivity $\varepsilon'_r = 1.30$ and $\tan\delta = 0.02$.

Matthews and Pettitt presented in [44] three types of antennas which are integrated into clothing, a broadband wire dipole, a bowtie and a spiral antenna, operating in frequencies from 100 MHz to 1 GHz. They have tested different materials (textiles and others), different frames and manufacturing techniques. Among the tested conductive materials there are conductive ribbon, conductive paint and ink, conductive nylon fabric (that is also adhesive on the back face), phosphor bronze mesh fabric (also adhesive on the back face), conductive thread, liquid crystal polymer (LCP) and copper coated fabric. The phosphor bronze mesh, LCP and copper coated fabric have the advantage that the antennas can be directly soldered to. In some antennas, a conducting epoxy was used to bond materials, but this shows some lack of robustness. In terms of radio frequency (RF) performance of the designed antennas, the spiral antenna, in which the spiral is broidered with conductive thread, performed worse than any other antenna and was clearly loss. Overall, based on RF performance, the most attractive materials to design wearable antennas were the textile fabrics: the conductive nylon and the copper coated fabrics.

In [16] the stability and efficiency of wearable and washable antennas are discussed for textile antennas in which the conductive parts were screen-printed with conductive ink. These antennas have shown acceptable performance. The combination of screen-printing with a breathable thermoplastic polyurethane (TPU) coating assured the performance was maintained even after several wash cycles. An embroidered technique was used in [45,46] to sew conductive fibres into polymer and fabric substrates, and it was proved that by increasing the density of the embroidering stitching, the conductivity of the conductive section is increased and also the accuracy of fabricated prototypes, which allowed better conformance with the simulations. Dipole, spiral and microstrip patch antennas were fabricated with this technique and presented very good RF performance when compared to the corresponding rigid copper structures. Moreover, these antennas when made on flexible polymer substrates conform well to curved surfaces and thus maintain their performance. Table 2 summarizes the main characteristics of the textile materials that have been used to develop wearable antennas, focusing mainly in the dielectric materials.

Table 2. Comparison of the textile materials used to design wearable antennas.

Reference	Application	Dielectric material				Conductive material	Performance
		Material	h (mm)	ε_r	tan δ		
[22]	GSM (900 MHz) and Bluetooth (2.4 GHz)	Unspecified material	0.236	3.29	0.0004	-	Acceptable
[43]	WLAN (2.4 GHz)	Fleece fabric	3	1.04	-	Knitted copper fabric	Acceptable
[28]	GPS (1.5 GHz)	*Cordura®*	0.5	Between 1.1 and 1.7	-	Copper tape	Good
[5,12]	ISM (2.4 GHz) and GPS (1.5 GHz)	Fleece fabric	2.56	1.25	-	*Flectron*	Acceptable to Good
[13]	ISM (900 MHz)	Polyurethane protective Foam	11	1.16	0.01	*Flectron*	Acceptable
[14,15]	WLAN (2.4 GHz and 5.8 GHz)	Felt	1.1	1.30	0.02	*Zelt*	Acceptable
[16]	ISM (2.4 GHz)	Cotton/Polyester	2.808	1.6	0.02	*Flectron*/Conductive ink	Acceptable
[6,46]	Not specific (2–2.4 GHz)	Polydimethylsiloxane (PDMS)	-	3.0–13	0.02	Embroidered conductive fibres	Good
[2]	Bluetooth (2.4 GHz)	Polyamide spacer fabric	6	1.14	Negligible	Silver-copper-nickel plated woven fabric	Good
[2]	Bluetooth (2.4 GHz)	Woollen felt	3.5	1.45	0.02	Silver-copper-nickel plated woven fabric	Good

The performance of the textile antennas presented earlier can be improved with integrated solutions, for instance, if diversity techniques such as Multiple-Input Multiple Output (MIMO) are considered [47], or, as shown in [48], with the introduction of a low-noise amplifier (LNA) in a wearable garment which was used to achieve an active integrated antenna, increasing the sensitivity and the gain of the overall system.

Declercq et al. [13] showed another integrated solution consisting of an aperture-coupled antenna on a textile and foam substrate, with a flexible solar cell, for tracking and monitoring solutions. Instead of integrating a LNA to increase the wearable antenna performance, Zhu and Langley [11,12] developed a dual-band coplanar patch antenna, to operate in the 2.45

and 5.8 GHz wireless bands, in which they integrated an electromagnetic band gap (EBG) to reduce body-presence effects and increase antenna gain. As shown in Table 2, the conductive parts were made of Zelt® fabric while the dielectric substrate was a thin felt with relative permittivity $\varepsilon'_r = 1.30$ and $\tan\delta = 0.02$. They proved that the introduction of the 3 × 3 array EBG with the coplanar patch could reduce the radiation towards the body by 10 dB, while increasing the antenna gain in 3 dB.

4. CONSTRUCTION OF WEARABLE ANTENNAS

After choosing the textile materials to design an antenna, their assemblage in the antenna is also crucial and specific, as they are very deformable materials. Thus, the conformation of the conductive patch with the dielectric substrate is critical [2]. Many authors have been improving the manufacturing processes [2,5,24,27] to construct textile antennas and some guidelines can be summarized as follows:

(1)The geometrical dimensions of the patch should remain stable while connecting to the dielectric substrate as the mechanical stabilization of both materials is essential to preserve the desired antenna characteristics [2,31]. In the microstrip patch antennas developed by Hertleer et al. [31] an alteration of no more than 0.5 mm on the length or the width of the patch influenced the performance of the antenna by causing a slight shift of the antenna characteristics. For this reason, woven fabrics, being more stable, are preferred to make the patches. However, the antenna geometrical stability can be achieved if at least one component is less deformable. For example, bonding using an adhesive sheet with a deformable patch, such as a conductive knit, with a less deformable substrate, such a woven dielectric, results in a stable frame [2,28].

(2)The techniques used to connect the various layers must not affect the electrical properties of the patch, such as its surface resistivity, nor the properties of the substrate. Connections using adhesive sheets or conductive fabrics with a thermal adhesive face have shown good results [2,5,29,31]. Indeed, the adhesive remains at the interface of the materials and therefore the surface resistance of the patch and the relative permittivity of the substrate are not significantly changed. However, in [29] the authors show that the adhesive layer introduces extra losses in the substrate. This process of attachment of the superposed layers is very simple to perform by a simple ironing operation. However, attention should be made to the ironing process, in special if the patch is made of a fabric with metallic components. Indeed, the oxidation of the metallic component, due to the hot moistening of the fabric, may increase the surface resistance of the fabric and so decrease the efficiency of the textile antenna [27].

Connection with seams is an alternative technique [7,31] but it presents some difficulties. Firstly, the seam must be plane, without wrinkling, what might be difficult to achieve with deformable materials. Secondly, the stitch passes through all materials: the patch, then the substrate and further the ground plane of the antenna, which may cause electrical shorts between them. In Locher et al. [2] report that the sewing needle has pulled conductive fibres from the patch through the substrate, shorting the patch with the ground plane.

Another technique is connecting with liquid adhesives [31]. However, it is difficult to apply a thin layer of glue. This difficulty introduces heterogeneity and in the zones where there are accumulations the glue may play the role of insulator between the conductive yarns of the patch. Furthermore, these adhesives are usually stiff and brittle, and so they cannot be applied in an area-wide manner on textiles as they will interfere in their flexibility [2]. In order to obtain a uniform thickness of the attachment of the several layers, Tronquo et al. [29,31] perform an additional stitch, in addition to the glue.

In [5] a smooth fabric was added to both faces of the substrate to optimise the attachment of the conductive components.

(3)The positioning of the textile components must consider the differences between right and back faces, in terms of roughness and of density of conductive elements [7,10]. In [10], a satin 5 woven was tested in a microstrip resonator, placing it in two positions: (1) with the right face against the dielectric substrate and (2) with the back face, the conductive one, against the dielectric substrate. It was observed that when the conductive face is placed on the top of the substrate and so underneath the nonconductive yarns of the nonconductive face, most of the electrical field is contained in the substrate. Thus, the dielectric loss in the nonconductive yarns is minimized.

(4)The core of the antenna may be obtained by stacking low-loss fabrics [7], adjusting this way the desired thickness of the substrate. However this introduces heterogeneities in the substrate due to the extra layers of air between the fabrics, influencing its dielectric properties.

(5)Finally, the connections at the antenna terminals may also be critical as in wearable and flexible antennas these connections have to be mechanically robust. In general, textile fabrics cannot be directly soldered to (an exception is Flectron® that already showed good resistance to soldering [5]). Therefore conductive epoxy has been used, but some concerns remain as this connection is not very resistant [44].

5. CONCLUSIONS

The developed wearable antennas are mainly planar ones, specifically microstrip patch antennas, because they mainly radiate perpendicularly to the planar structure and also their ground plane efficiently shields the human body. The bandwidth and efficiency performance of a planar microstrip antenna is mainly determined by the substrate dielectric constant and its thickness.

In general, textiles present a very low dielectric constant, between 1 and 2, as they are very porous materials and the presence of air approaches the relative permittivity to one. The low dielectric constant reduces the surface wave losses that are tied to guided wave propagation within the substrates. Therefore, lowering the dielectric constant increases spatial waves and hence increases the impedance bandwidth of the antenna. However, lowering the substrate permittivity can also increase the resonance frequency of the antenna, allowing the development of antennas with acceptable efficiency and high gain.

In addition, the ability of the fibres to absorb moisture must also be considered in the characterization of the dielectric behaviour of textiles. Water has a much higher and more stable dielectric constant than textile fabrics. Therefore, when water is absorbed by the textile fibres or is trapped in the fabric structure, it changes the electromagnetic properties of the fabric, increasing its dielectric constant and loss tangent. Likewise, the absorbed water by or trapped into the textile substrate reduces the resonance frequency and bandwidth of the antenna. In general, the characteristics of antennas based on textile materials with small moisture absorption (regain less than 3%) are more stable. Therefore, materials with such low regain values are preferable for use as substrates and as conductive components of the antenna.

The thickness of the dielectric material is also crucial in the design of antennas. For a fixed relative permittivity, the substrate thickness may be chosen to maximize the bandwidth of the planar antenna. However, this value may not optimize the antenna efficiency. Therefore, the choice of the thickness of the dielectric material is a compromise between efficiency and bandwidth of the antenna. Moreover, the thickness of the substrate also influences the geometric sizing of the antenna.

The conductive fabrics for the patch and the ground planes must have a very low electrical surface resistivity in order to minimize the electric losses and so increase the antenna efficiency. There are several conductive textile fabrics, and also yarns, available on the market that has been successfully used in planar antennas. Coated fabrics might perform worse than the fabrics having conductive fibres, because of discontinuities that may increase the surface resistivity. Wovens and nonwovens, being more stable fabrics than knits, allow higher geometrical accuracies of the frame of the antenna and thus may be preferred to make the patch. In general, the sizing accuracy of the patch made of woven fabric depends on the thickness of the component yarns.

After choosing the textile materials, their assemblage may also be critical, as the elongation and bending causes mechanical deformations that interfere with the antenna behaviour. The presence of at least one textile material presenting high tensile strength, high bending rigidity and stable geometry stabilizes the frame of the antenna. When connecting the various layers making up the antenna, the positioning of the textile fabrics must consider differences between right and back faces in terms of roughness and of density of conductive elements, in order to minimize losses.

Finally, despite the fact the performance of the textile antennas can be improved with integrated solutions and despite the major progresses already achieved in the development of wearable antennas, further study and better characterization of ordinary and conductive textile materials may still offer relevant improvements to their design and to the optimization of their behaviour.

ACKNOWLEDGMENTS

The authors acknowledge the Portuguese FCT/MCTES for financing the project PTDC/EEA-TEL/122681/2010-PROENERGY-WSN-Prototypes for Efficient Energy Self-Sustainable Wireless Sensor Networks and the European Union (FP7-PEOPLE-2009-IAPP) for financing the project No. 251373 INSYSM (Intelligent Systems for Structures Strengthening and Monitoring).

REFERENCES

1. Bonfiglio, A.; De Rossi, D. Wearable Monitoring Systems, 1st ed.; Springer: New York, NY, USA, 2011; p. 100.
2. Locher, I.; Klemm, M.; Kirstein, T.; Tröster, G. Design and Characterization of Purely Textile Patch Antennas. IEEE Trans. Adv. Pack. **2006**, 29, 777–788.
3. Grupta, B.; Sankaralingam, S.; Dhar, S. Development of Wearable and Implantable Antennas in the Last Decade: A Review. Proceedings of Mediterranean Microwave Symposium (MMS), Guzelyurt, Turkey, 25–27 August 2010; pp. 251–267.

4. Brebels, S.; Ryckaert, J.; Boris, C.; Donnay, S.; De Raedt, W.; Beyne, E.; Mertens, R.P. SOP Integration and Codesign of Antennas. IEEE Trans. Adv. Pack. **2004**, 27, 341–351.
5. Hertleer, C.; Rogier, H.; Member, S.; Vallozzi, L.; Langenhove, L.V. A Textile Antenna for Off-Body Communication Integrated into Protective Clothing for Firefighters. IEEE Trans. Adv. Pack. **2009**, 57, 919–925.
6. Zhang, L.; Wang, Z.; Psychoudakis, D.; Volakis, J.L. Flexible Textile Antennas for Body-Worn Communication. Proceedings of IEEE International Workshop on Antenna Technology, Tucson, ZA, USA, 5–7 March 2012; pp. 205–208.
7. Kaija, T.; Lilja, J.; Salonen, P. Exposing Textile Antennas for Hash Environment. Proceedings of 2010 Military Communications Conference, San Jose, CA, USA, 31 October–3 November 2010; pp. 737–742.
8. Liu, N.; Lu, Y.; Qiu, S.; Li, P. Electromagnetic Properties of Electro-Textile for Wearable Antennas Applications. Front. Electr. Electron. Eng. China **2011**, 6, 553–566.
9. Shaw, R.; Long, B.; Werner, D.; Gavrin, A. The Characterization of Conductive Textile Materials Intended for Radio Frequency Application. IEEE Trans. Anten. Propag. **2007**, 49, 28–40.
10. Ouyang, Y.; Chappell, W.J. High Frequency Properties of Electrotextiles for Wearable Antenna Applications. IEEE trans. Anten. Propag. **2008**, 56, 381–389.
11. Cottet, D.; Gryzb, J.; Kistein, T.; Tröster, G. Electrical Characterization of Textile Transmission Lines. IEEE Trans. Adv. Pack. **2003**, 26, 182–190.
12. Hertleer, C.; Rogier, H.; Vallozzi, L.; Declercq, F. A Textile Antennas Based on High-Performance Fabrics. Proceedings of 2nd European Conference on Antennas and Propagation, Edinburgh, UK, 11– 16 November 2007; pp. 1–5.
13. Declercq, F.; Georgiadis, A.; Rogier, H. Wearable Aperture-Coupled Shorted Solar Patch Antenna for Remote Tracking and Monitoring Applications. Proceedings of 5th European Conference on Antennas and Propagation, Rome, Italy, 11–15 April 2011; pp. 2992–2996.
14. Zhu, S.; Langley, R. Dual-Band Wearable Antennas over EBG Substrate. Electr. Lett. **2007**, 43, 141–142.
15. Zhu, S.; Langley, R. Dual-Band Wearable Textile Antennas over EGB Substrate. IEEE Trans. Anten. Propag. **2009**, 57, 926–935.
16. Scarpello, M.L.; Kazani, I.; Hertleer, C.; Rogier, H.; Ginste, D.V. Stability and Efficiency of Screen-Printed Wearable and Washable Antennas. IEEE Anten. Wireless Propag. Lett. **2012**, 11, 838–841.
17. Baker-Jarvis, J.; Janezic, M.D.; DeGroot, D.C. High-Frequency Dielectric Measurements. IEEE Trans. Instrum. Meas. **2010**, 13, 24–31.
18. Brandão, D.P.L. Tecnologia da Electrecidade: Materiais Usados em Electrotecnia; Fundação Calouste Gulbenkian: Lisbon, Portugal, 1983.

19. Bal, K.; Kothari, V.K. Measurement of Dielectric Properties of Textile Materials and Their Applications. Indian J. Fibre Text. **2009**, 34, 191–199.
20. Morton, W.E.; Hearle, W.S. Physical Properties of Textile Fibres, 4th ed.; Woodhead Publishing: Cambridge, UK, 2008.
21. Sankaralingam, S.; Bhaskar, G. Determination of Dielectric Constant of Fabric Materials and Their Use as Substrates for Design and Development of Antennas for Wearable Applications. IEEE Trans. Instrum. Meas. **2010**, 59, 3122–3130.
22. Salonen, P.; Keskilammi, M.; Rantanen, J.; Sydanheimo, L. A Novel Bluetooth Antenna on Flexible Substrate for Smart Clothing. Proceedings of Conference of IEEE International Conference on Systems, Man, and Cybernetics, Tucson, AZ, USA, 7– 10 October 2001; pp. 789–794.
23. Hasar, U.C. A New Microwave Method for Electrical Characterization of Low-Loss Materials. IEEE Microw. Wirel. Compon. Lett. **2009**, 19, 801–803.
24. Harmer, S.W.; Rezgui, N.; Bowring, N.; Luklinska, Z.; Ren, G. Determination of the Complex Permittivity of Textiles and Leather in the 14–40 GHz, Millimetre-Wave Band Using a Free-Wave Transmittance Only Method. IET Microw. Anten. Propag. **2008**, 2, 606–614.
25. Yue, H.; Virga, K.L.; Prince, J.L. Dielectric Constant and Loss Tangent Measurement Using a Stripline Fixture. IEEE Trans. Compon. Pack. A **1998**, 21, 441–446.
26. Moretti, A. Estudo do Brim Santista Visando Aplicações em Antenas têxteis. M.S. Thesis, Universidade Estadual de Campinas, Campinas, Brazil, 2011.
27. Declercq, F.; Rogier, H.; Hertleer, C. Permittivity and Loss Tangent Characterization for Garment Antennas Based on a New Matrix-Pencil Two-Line Method. IEEE Trans. Anten. Propag. **2008**, 56, 2548–2554.
28. Salonen, P.; Rahmat-samii, Y.; Schafhth, M.; Kivikoski, M. Effect of Textile Materials on Wearable Antenna Performance: A Case Study of GPS Antenna. Proceedings of IEEE Antennas and Propagation Society International Symposium, Monterey, CA, USA, 20– 25 June 2004; pp. 459–462.
29. Tronquo, A.; Rogier, H.; Hertleer, C.; Langenhove, L.V. Applying Textile Materials for the Design of Antennas for Wireless Body Area Networks. Proceedings of EuCap2006: First European Conference on Antennas and Propagation, Nice, France, 6– 10 November 2006.
30. Hertleer, C.; Tronquo, A.; Rogier, H.; Langenhove, L.V. The Use of Textile Materials to Design Wearable Microstrip Patch Antennas. Text. Res. J. **2008**, 78, 651–658.
31. Hertleer, C.; Laere, A.V., Rogier, H.; Langenhove, L.V. Influence of Relative Humidity on Textile Antenna Performance. Text. Res. J. **2009**, 80, 177–183.

32. Balanis, C.A. Antenna Theory: Analysis and Design, 3rd ed.; Wiley Interscience: Hoboken, NJ, USA, 2005.
33. Webster, H.G. The Measurement, Instrumentation, and Sensors Handbook; Springer: Berlin, Germany, 1999.
34. Maryniak, W.A.; Uehara, T.; Noras, M.A. Surface Resistivity and Surface Resistance Measurements—Using a Concentric Ring Probe Technique. Application Note; Trek, Inc.: Medina, NY, USA, 2003; pp. 1–4.
35. Kacprzyk, R. Measurements of the Volume and Surface Resistance of Textile Materials. Fibres Text. East. Eur. **2011**, 19, 47–49.
36. Lilja, J.; Salonen, P.; Maagt, P.D.; Zell, N.K. Characterization of Conductive Textile Materials for Softwear Antenna Characterization of Conductive Textiles. Antennas and Propagation Society International Symposium, Charleston, SC, USA, 1– 5 June 2009.
37. Salonen, P.; Rahmat-Samii, Y.; Hurme, H.; Kivikoski, M. Effect of Conductive Material on Wearable Antenna Performance: A Case Study of WLAN Antennas. 20– 25 June 2004; pp. 455–458.
38. Wang, X.; Xu, W.; Wenbin, L. Study on the Electrical Resistance of Textiles under Wet Conditions. Text. Res. J. **2009**, 79, 753–760.
39. Canteri, A.; Avancini, F.; Neves, J.D.; Crespim, L. Resistência a Penetração/Absorção de Água; Universidade Estadual de Maringá: Goioerê, Brazil, 2000; pp. 1–40.
40. Kaija, T.; Lilja, J.; Salonen, P. Exposing Textile Antennas for Harsh Environment. Proceedings of 2010 Military Communications Conference, San Jose, CA, USA, 31 October–3 November 2010; pp. 737–742.
41. Volakis, J.L.; Gullu, K. Novel Materials for RF Devices. Proceedings of IEEE Antennas and Propagation Society International Symposium, Honolulu, HI, USA, 9– 15 June 2007; pp. 1701–1704.
42. Bayram, Y.; Zhou, Y.; Shim, B.S.; Xu, S.; Zhu, J.; Kotov, N.A.; Volakis, J.L. E-Textile Conductors and Polymer Composites for Conformal Lightweight Antennas. IEEE Trans. Anten. Propag. **2010**, 58, 2732–2736.
43. Salonen, P.; Hurme, H. A Novel Fabric WLAN Antenna for Wearable Applications. Proceedings of IEEE International Symposium of Antennas and Propagation Society, 22– 27 June 2003; pp. 100–103.
44. Matthews, J.C.G.; Pettitt, G. Development of Flexible, Wearable Antennas. Proceedings of EuCAP 2009: 3rd European Conference on Antennas and Propagation, Berlin, Germany, 23– 27 March 2009; pp. 273–277.
45. Wang, Z.; Zhang, L.; Bayram, Y.; Volakis, J.L. Embroidered Conductive Fibers on Polymer Composite for Conformal Antennas. IEEE Trans. Anten. Propag. **2012**, 60, 4141–4147.
46. Volakis, J.L.; Zhang, L.; Wang, Z.; Bayram, Y. Embroidered Flexible RF Electronics. Proceedings of IEEE International Workshop on Antenna Technology, Tucson, AZ, USA, 5– 7 March 2012.

47. Vallozzi, L.; Torre, V.P.; Hertleer, C.; Rogier, H.; Moeneclaey, M.; Verhaevert, J. Wireless Communications for Firefighters Using Dual-Polarized Textile Antennas Integrated in Their Garment. IEEE Trans. Anten. Propag. **2010**, 58, 1357–1368.
48. Dierck, A.; Declercq, F.; Rogier, H. Review of Active Textile Antenna Co-Design and Optimization Strategies. Proceedings of IEEE International Conference on RFID-Technologies and Applications, Sitges, Spain, 15– 16 September 2011; pp. 194–201.

CHAPTER 12

Medical Smart Textiles Based on Fiber Optic Technology: An Overview

Carlo Massaroni, Paola Saccomandi and Emiliano Schena *

Center for Integrated Research, Università campus Bio-Medico, Alvaro del Portillo, 21, Rome 00128, Italy

ABSTRACT

The growing interest in the development of smart textiles for medical applications is driven by the aim to increase the mobility of patients who need a continuous monitoring of such physiological parameters. At the same time, the use of fiber optic sensors (FOSs) is gaining large acceptance as an alternative to traditional electrical and mechanical sensors for the monitoring of thermal and mechanical parameters. The potential impact of FOSs is related to their good metrological properties, their small size and their flexibility, as well as to their immunity from electromagnetic field. Their main advantage is the possibility to use textile based on fiber optic in a magnetic resonance imaging environment, where standard electronic sensors cannot be employed. This last feature makes FOSs suitable for monitoring biological parameters (e.g., respiratory and heartbeat monitoring) during magnetic resonance procedures. Research interest in combining FOSs and textiles into a single structure to develop wearable sensors is rapidly growing. In this review we provide an overview of the state-of-the-art of textiles, which use FOSs for monitoring of mechanical parameters of physiological interest. In particular we briefly describe the working principle of FOSs employed in this field and their relevant advantages and disadvantages. Also reviewed are their applications for the monitoring of mechanical parameters of physiological interest.

KEYWORDS

smart textiles; fiber optic sensors; fiber Bragg grating sensors; respiratory monitoring; macrobending sensors; hetero-core fiber optics; Magnetic Resonance Imaging; MR-compatibility

1. INTRODUCTION

The growing interest in smart textiles for medical applications is driven by the aim to increase the mobility of patients who need a continuous monitoring of physiological parameters [1,2,3,4]. Smart textiles are able to interact with the environment; therefore they embed one or more sensors to monitor various mechanical, thermal and chemical parameters (e.g., strain, temperature, displacement, oxygen blood saturation) [5,6,7]. During the last decades the use of fiber optic-based sensors (FOSs) has been gaining acceptance in a large number of applications in the fields of civil engineering, the automotive industry and medicine [8,9,10], among others. These sensors allow the measurement of physical and chemical parameters employing a large number of working principles and configurations [4]. FOSs can be divided into intrinsic sensors, where the fiber optic constitutes the sensing element, and extrinsic ones, where the fiber optic is only used as a medium to transport light. There are a number of reviews and books that focus on the description of FOSs [11,12,13,14,15], hence it is not possible to describe all the applications and working principles of FOSs in one journal article.

FOSs have good metrological properties (i.e., low zero drift and sensitivity drift, good accuracy and good sensitivity and large bandwidth), they offer the possibility to implement distributed sensors and they are immune to electromagnetic interferences. These features make FOSs an emerging solution for the monitoring of physiological parameters and more generally for applications in medicine [9]. In fact, thanks to the mentioned valuable characteristics, FOSs can compete with other traditional electrical and mechanical sensors, and in many fields of application the superiority of their performances has been demonstrated. Lastly, the possibility to develop Magnetic Resonance (MR)-compatible sensors further motivates the growing interest in this technology.

Fiber optic technology is particularly attractive for application in smart textiles because it allows both sensing and signal transmission. Moreover, polymer optical fibers (POFs) are cheap, lightweight, flexible and robust, and they are able to measure high strain values without damage.

During the last decade, several groups of research have focused their efforts on obtaining a substantial development in the integration of smart textiles and fiber optic technology [16,17] as reported in a recent review [18].

In this paper, we aim at describing the use of smart textiles in medicine employing FOSs; in particular we focus on their use for monitoring physiological parameters by measuring mechanical variables and on the metrological properties and performances of the devices reported in literature. Also reviewed are the working principles of FOSs most frequently used in smart textiles for the mentioned applications. In Section 2 the working principles of the FOSs used in smart textiles and their applications in medicine are described. In Section 3 the applications in physiological monitoring of smart textiles based on FOSs along with their main advantages and drawbacks are reviewed.

2. WORKING PRINCIPLE OF FIBER OPTIC SENSORS USED IN SMART TEXTILES

FOSs can be designed using a large number of working principles. In this section the ones used to develop smart textiles for measuring mechanical variables in physiological monitoring are reviewed. In particular we focus on sensors based on the fiber Bragg grating technology and on intensity-based FOSs. The next subsections describe the working principle of these sensors with a brief description of their application in medicine.

2.1. Fiber Bragg Grating Sensors

Fiber Bragg grating (FBG) sensors consist of a periodic perturbation of the refractive index along the fiber core length obtained by exposure of the core to an intense optical interference pattern. Hill reported the first fabrication of FBGs in the late 1970s [19], but a major breakthrough occurred a decade later, when Meltz et al. improved the technique for their fabrication [20]. This last study can be considered the milestone, which enabled the development of FBG sensors for a large number of applications.

Basically, an FBG can be considered as a short segment of a fiber optic (usually FBG longer than 3 mm–6 mm are employed, although for particular application smaller FBG are required), which reflects a narrow range of wavelengths and transmits all others. A schematic representation of the working principle of an FBG sensor and of its response to strain is shown in Figure 1.

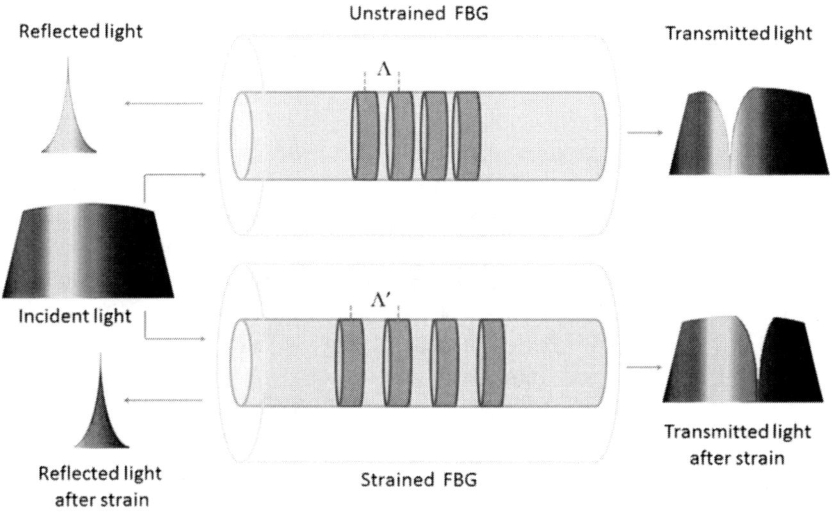

Figure 1. Schematic of the working principle of Fiber Bragg grating (FBG) sensors, and its response to strain.

The wavelength of the input light that is back-reflected (λ_B) is sensitive to temperature and strain. In fact, λ_B can be expressed by the following equation, which is the first-order Bragg condition of the grating:

$$\lambda_B = 2 \cdot \Lambda \cdot \eta_{eff} \tag{1}$$

where Λ is the spatial period of the grating and η_{eff} is the effective refractive index of the fiber core. Both the terms are sensitive to strain and temperature, as a consequence, the use of a proper configuration allows estimating strain and temperature or both, by monitoring changes of λ_B. The dependence of λ_B can be described by the fractional changes of Λ and η_{eff}:

$$\frac{\Delta\lambda_B}{\lambda_B} = \frac{\Delta\Lambda}{\Lambda} + \frac{\Delta\eta_{eff}}{\eta_{eff}} \tag{2}$$

Actually, due to the different sensitivity of Λ and η_{eff} to temperature changes and strain, the FBGs sensors used to estimate strain are based on the λ_B shift due to Λ changes, the FBG sensors used to monitor temperature are based on the λ_B shift due to η_{eff} changes. Equation (1) shows an important advantage of the FBG: their output is not affected by fluctuation of source intensity.

FBG sensors are employed in a large number of industrial fields to monitor different physical variables, including, among others, temperature, pressure, flow and vibrations. Moreover, FBGs have been largely used to monitor physiological parameters and more in general in medical fields, such as stroke volume, blood pressure and heartbeats [21,22], in microsurgery [23,24], foot pressure in diabetic patients [25], in biomechanical studies [26,27], in the monitoring of temperature during thermal ablation of cancer [28,29,30,31], in respiratory monitoring system [32,33], and in tactile sensing [34,35].

FBGs sensors are considered to hold great potential for application in medicine due to their good metrological properties. Moreover their characteristics of biocompatibility, non-toxicity and chemical inertness, as well as their small size and flexibility make them particularly attractive for invasive measurements during in vivo trials [36]. Lastly, they are suitable for application in environments with high electromagnetic noise, thanks to their immunity to electromagnetic interferences and their intrinsic MR-compatibility [37].

2.2. Intensity Modulated FOSs

Intensity-modulated FOSs modulate light intensity, measured by a secondary element (e.g., a photodiode), in response to an environmental effect. A simple configuration of this kind of sensor is shown in Figure 2A. Two optical fibers are held in close proximity to each other; the light is injected into one of the optical fibers; as the light expands into a cone of light, its intensity, emitted by the first fiber and conveyed into the other one, depends on the distance (d) between the two fiber tips. Therefore, the light intensity can be considered an indirect measurement of the distance between the two fibers and of other physical variables influencing this distance. A similar configuration can be designed either by using a single fiber and a mirror (Figure 2B) or by using more than one fiber to obtain a differential configuration. The differential solution allows neglecting the influence of the light source intensity on sensor output. Another configuration to develop intensity-modulated FOSs is underpinned by the phenomenon that the light is lost from an optical fiber when it is bent. In particular a bent radius causes a leakage of the light traveling within the core of the fiber into the cladding with a resulting intensity modulation of light propagating through an optical fiber (Figure 2C). Macrobending sensors based on hetero-core fibers have been proposed to measure several physical properties [38,39].

Intensity-modulated FOSs have been used in different medical fields starting from the late 1960s, when Lekholm and Lindstrom proposed a sensor for intravascular pressure monitoring [40]. A similar sensor was proposed to monitor the intracranial pressure [41], and commercially available intracranial pressure intensity-modulated FOSs are produced by Camino Laboratories Inc. Their performances are largely investigated in clinical settings [42]. Moreover these sensors are used for pressure and temperature monitoring [43].

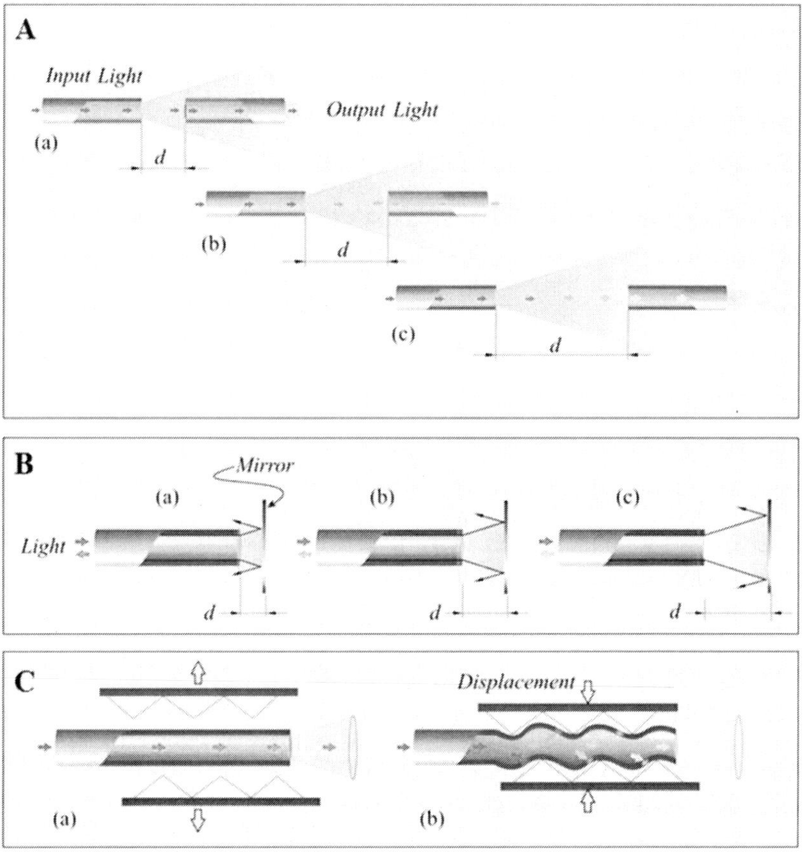

Figure 2. (**A**) Schematic of the working principle of an intensity modulated sensor using two fiber optic; (**B**) schematic of the working principle of an intensity modulated sensor using a fiber optic and a mirror; (**C**) schematic of the light lost from the fiber core caused by bending (adapted from [10]).

Macrobending FOSs find applications in medicine mainly in the monitoring of respiratory movements [44,45,46]. These sensors are largely used in smart textiles, therefore their medical applications will be described in more detail in the following section. A particular approach based on the bending of the fiber is also used to develop flow sensors for mechanical ventilation [47,48].

3. SMART TEXTILES BASED ON FIBER OPTIC SENSORS: MEDICAL APPLICATIONS

Smart technical textiles are by definition textiles that can interact with their environment. Their ability to sense physical and chemical parameters can be accessed by embedding several kinds of sensors. The use of FOSs on textile is particularly attractive in some medical fields because of the possibility to use this technology during MR procedures, the cost reduction of key optical components and the improvement of the component quality, as well as the good metrological properties of these sensors. In particular, polymer optical fibers (POFs) match well with the requirements for application in smart textiles, being cost effective, lightweight and robust; moreover POFs are able to measure high strain values of several ten percent without fiber damage [49].

Significant developments of the integration of FOSs in textiles were driven by several groups of research, and in particular by the group involved in the European project OFSETH (optical fiber sensors embedded into technical textile for healthcare). OFSETH aimed at investigating the possibility to use FOSs embedded into textiles to monitor several physiological parameters.

In the following two subsections the application of smart textiles based on FBG sensors and on intensity modulated FOSs are described.

3.1. Smart Textiles Based on FBG Sensors: Medical Applications

FBG is one of the most frequently employed technologies to design smart textiles based on fiber optics, thanks to their good sensitivity to strain. This characteristic allows developing several configurations based on the measurement of the strain experienced by the FBG sensor to monitor different parameters of physiological interest.

In particular this solution has been employed to monitor respiratory movements. The use of FBG to monitor respiratory movements and breathing rate has been demonstrated in the past [32,50,51], but only during the last decade have they been embedded in smart textiles. A number of studies regarding this topic have been proposed by the groups involved in the OFSETH project. A smart textile embedding two different FOSs (i.e., FBG and macrobending) for respiratory monitoring has been developed. FOSs and the use of MR-compatible connectors allow the use of the proposed smart textiles on anesthetized patients during MR procedures. Indeed, these sensors are free from metallic or electrical conductive wires (when using a custom made MR-compatible connector); in addition, they are remotely interrogated via an optical fiber cable allowing the location of the monitoring unit outside of the MR field [52,53], as schematically reported in Figure 3.

The high FBG sensitivity to strain (i.e., $\approx$1.2 pm·$\mu\varepsilon^{-1}$) allows discriminating small strains; therefore they have been used to monitor thoracic movements that are smaller than abdominal ones. Two different methods to embed the FBG sensors within textiles (stitching and crochet) have been proposed. The calibration of the system, which embeds the FBG by stitching (see Figure 4A), has been performed by stretching the textiles in steps of 0.4% up to 40%. During the calibration, the FBG experiences strains up to 0.8%. The system shows good linearity from about 8.5% to 35%–40% of textile stretching, with a sensitivity of 0.35 nm/% and an accuracy better than 0.1% of elongation. Only preliminary experiments on the textiles developed with the crochet method have been performed. The integration of this sensor with a different FOS has been assessed on ten healthy volunteers [54]. Trials to evaluate the long-term properties and the stability of the respiratory sensor by an ad hoc developed simulator have been performed as well [55]. In particular the mechanical robustness of the sensor has been tested with more than 90,000 cycles in 129 hours with a simulated breathing rate between 10 breaths per minute and 12 breaths per minute [56]. A further valuable characteristic of the proposed system (see Figure 4B) is that it enables the continuous measurement of respiratory movement providing free access to all vital organs for medical staff actions [57].

A recent interesting study proposed a simple wearable system based on a single FBG sensor, which allows the simultaneous detection of both heart rate and respiratory cycles [58]. The main innovation is related to the structure in which the FBG sensor is embedded, which is [9] a PVC laminate resulting in a strain-sensitive foil, manufactured by an industrial spread-coating process. This integrated solution shows a sensitivity of 0.8 pm·$\mu\varepsilon^{-1}$. The authors performed trials on healthy volunteers, using two filters for breathing rate monitoring (band-pass filter tuned in the range 0.1–0.4 Hz) and cardiac frequency monitoring (band-pass in the range 0.5–1.3 Hz). The same group of research developed a system for breathing rate and cardiac frequency monitoring able to work in a wide range of temperature [59]. The authors selected the polychloroethanediyl as carrier material due to its excellent performance/cost and to the good sensitivity to strain. The system showed a linear response for elongation ranging from 0.6% to 1.6%. In this range the FBG output experienced an increase of about 8 nm; therefore the sensitivity was about 8 nm/%. They also investigated the sensors' output changes with changes in temperature and found an increase of Bragg wavelength of about 1.5 nm for a temperature increase of 140 °C (a thermal sensitivity of about 10.7×10^{-3} nm/°C).

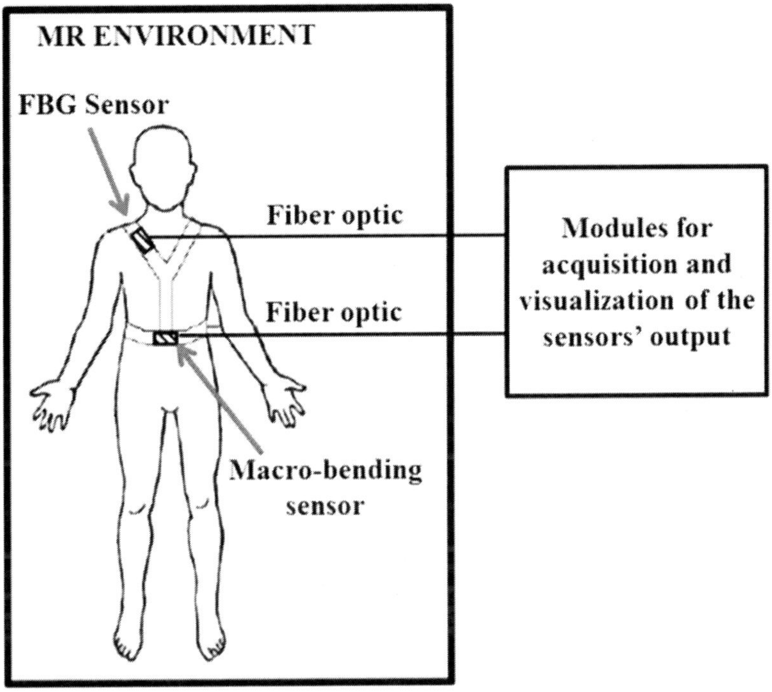

Figure 3. Schematic representation of the monitoring systems proposed by and developed in the optical fiber sensors embedded into technical textile for healthcare (OFSETH) project.

Other simple solutions have been proposed to monitor respiration and heart activity. In particular, Dziuda and coauthors developed a system which consisted of a Polymethyl methacrylate (PMMA) board with the size of 220 × 95 mm² and a thickness of 1.5 mm. An FBG sensor was attached with epoxy adhesive [60]. They experimentally assessed the error of the system in the measurement of breaths per minute and heartbeats per minute during magnetic resonance imaging examinations, showing promising results (about one breath per minute and about three heartbeats per minute). The same group of researchers reported a system based on two FBG sensors positioned orthogonally to each other [61]. This group validated their system on three patients during MRI procedures by comparing their results with the ones obtained by an MRI-compatible portable module [62]. The results were promising (a relative error lower than 8% can be considered satisfactory considering that the system is intended for monitoring rather than diagnosis). Recently, they proposed a system for heart rate monitoring. It is MR-compatible and has been tested on seven volunteers showing a root mean square error of less than six beats per minute [63].

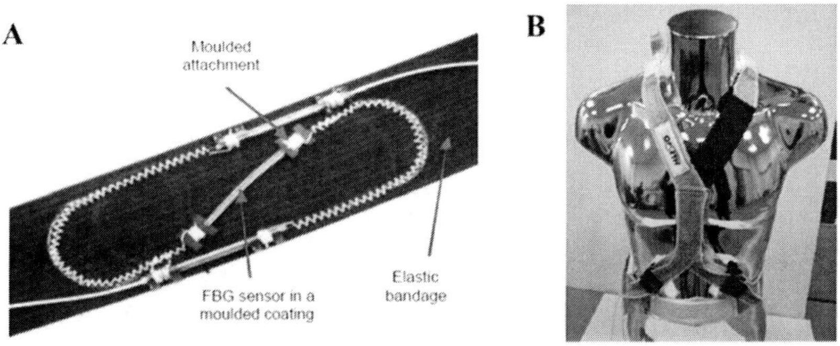

Figure 4. (**A**) Design of the FBG sensor developed in OFSETH project (adapted from [57]); (**B**) MRI-compatible sensing harness which embeds the fiber optic sensors for respiratory monitoring (adapted from [57]).

Allsop and coauthors developed a wearable system for respiratory function monitoring [64]. Their system employs an array of 40 in-line FBG sensors that produce 20 curvature sensors at different locations, each sensor consisting of two FBGs. They carried out experiments to measure the absolute volumetric changes of the human torso and estimated an error of 6% on the average volume.

Lastly, Li and coworkers developed a wearable sensor in intelligent clothing based on FBG for human body temperature monitoring [65]. They partly embedded an FBG in a polymer filled strip to improve the sensitivity of the measuring system. This way they obtained a temperature sensitivity of 150 $pm \cdot °C^{-1}$, almost 15 times higher than that of the bare FBG. To measure human body temperature, they distributed five FBGs in five places (i.e., left chest, right chest, left armpit, right armpit and at center of the upper back). Moreover, they developed a model to estimate the body temperature by the data of the five FBGs. With this method they found an accuracy of ±0.1 °C.

The main characteristics of the abovementioned smart textiles and their applications are reported in Table 1.

Table 1. Smart textiles and wearable systems based on fiber optic sensors: working principle, medical application and metrological properties.

Reference	Working Principle	Medical Application	Metrological Properties and Other Features
[52,53,56]	Silica FBGs	Respiratory monitoring during MRI procedures	Non-invasive; MR compatible; good linearity in a wide range of strains with sensitivity = 0.35 nm/%; accuracy better than 0.1% of elongation
[58,59]	Silica FBGs	Cardiac and Respiratory monitoring during MRI procedures	Non-invasive; MR compatible; sensitivity of 0.8 nm/$\mu\varepsilon^{-1}$
[59]	Silica FBGs	Cardiac and Respiratory monitoring	Non-invasive; Sensitivity of 8 nm/%; good linear trend; thermal sensitivity ≈ 10.7×10^{-3} nm/°C
[60,62]	Silica FBGs	Cardiac and Respiratory monitoring during MRI procedures	Non-invasive; MR compatible; Simple design; Good accuracy in terms of breathing rate (±1 bpm) and heartbeat per minute (±3 bpm); relative error in patients during MRI procedures <8%
[61]	Silica FBGs	Cardiac and Respiratory monitoring during MRI	Non-invasive; Simple design; Flat frequency response in the range of interest (0.5 Hz up to 20 Hz); maximum relative error of 12%
[63]	Silica FBG	Heart rate monitoring	Non-invasive; MR compatible; Root mean square error lower than 6 beats per minute
[64]	Silica FBGs	Respiratory function monitoring	Non-invasive; 6% of error on the average volume
[65]	Bare FBG	Body temperature monitoring	Non-invasive; Sensitivity of 150 pm/°C in the range of interest (from 33 °C to 42 °C); accuracy 0.1 °C
[52,53,56]	Macro-bending/OTDR technique	Respiratory monitoring	Non-invasive; MR compatible; Good sensitivity stability after 172800 cycles (variations < 10%); low cost component for their interrogation
[66]	Intensity modulated	Respiratory monitoring	Non-invasive; MR compatible; low cost component
[67,68]	Intensity modulated	Respiratory monitoring	Non-invasive; low cost component
[69,70]	Macrobending hetero-core fiber optic	Respiratory monitoring	Non-invasive; low cost component; good agreement with the breathing rate measured by a commercial device
[71]	microbending	Respiratory monitoring during MRI procedures	Non-invasive; MR compatible; Accuracy better than ±2 breaths per minute
[72]	microbending	Respiratory rate and heart rate	Non-invasive; MR compatible; Accuracy better than ±2 breaths or beats per minute for respiratory monitoring heart rate
[73]	microbending	Heartbeat and respiratory monitoring	Non-invasive; low cost component; good agreement with the heart beat measured by a commercial device

3.2. Smart Textiles Based on Intensity-Modulated FOSs

FOSs based on intensity modulation and in particular macrobending sensors are employed to develop smart textiles for monitoring of physiological parameters.

Textiles based on these kinds of FOSs were proposed by the groups of research involved in the OFSETH project. In particular, they developed smart textiles based on macrobending FOS for the monitoring of respiratory rate on patients during MRI procedures [52,53]. The sensors are MR-compatible and the monitoring devices are located out of the magnetic resonance environment (see Figure 3). A standard single mode fiber was embedded within a textile as shown in Figure 5A. This macrobending sensor, due to the lower sensitivity than the FBG sensors, was used to monitor abdominal movements, because these movements are larger than thoracic ones (Figure 5B). The authors found large oscillations in the sensor's output during the stretching of the textile, which could cause a wrong computation of the breathing rate. As a consequence, they designed a sensor with more periods of the U-shape on the textile bandage (Figure 5B). This solution allowed to substantially increase the sensitivity (e.g., for a textile elongation of 38%, the sensor response increased from less than 3 dB with a single-loop design to more than 28 dB with a 10 loops-design). Lastly they developed a sensor based on the optical time-domain reflectometry (OTDR) in polymer optical fiber (POF). Basically, the macrobending entails a change of the backscattering in POF that can be sensed by the technique of OTDR [56]. This solution allows measuring strain in different locations of a single fiber (distributed measurement). A respiratory simulator was employed to test the robustness of both the macrobending sensor and the sensor based on the OTDR technique. Both these sensors show low variations after cycles at 10 breaths per minute and elongation up to 3% and 5% for the OTDR sensor and for the macrobending one, respectively. These sensors were also tested on healthy volunteers.

An interesting solution to develop an MR-compatible sensor wearable sensor for respiratory monitoring based on intensity modulation was presented in [66]. It consists of a PMMA tube, a mirror, a spring and a plastic optical fiber. Abdominal movement causes a variation of the distance between the mirror and the distal end of the plastic optical fiber that is related to the intensity of reflected light coupled to the fiber (as shown in the schematic representation in Figure 2B). The authors also tested the sensor during an MR procedure, and they did not find any negative effects related to patient safety and image quality.

Krehel and coauthors developed a textile for respiratory monitoring based on FOS previously described [67,68]. Basically, the working principle of the sensor employed can be explained as follows: the fiber optic geometry changes when a force (or pressure) is applied on the fiber; these geometry changes affect the wave guiding properties and hence induce light loss in the optical fiber. They characterized the sensor in [67], showing a range of measurement for force applied on 3 cm of fiber length up to 40 N, with a discrimination threshold of 0.05 N. They also performed the feasibility assessment of the wearable system for breathing rate monitoring [68]. The trials were performed at two breathing rates and at different positions of the sensing textile on the human body. Lastly, the comparison with a commercial device for respiratory measurements was performed using the Bland Altman analysis. The results showed that a large part of the differences between the measurements obtained by the FOS textile and the commercial device was concentrated in the range ±3 min^{-1}, and the limits of agreement were about ±6 min^{-1}.

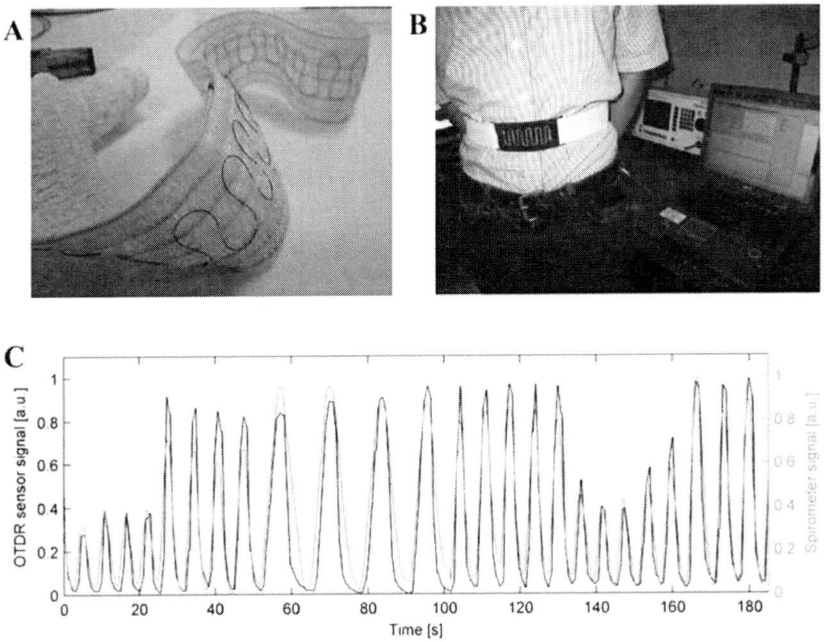

Figure 5. (**A**) Textiles based on macrobending FOS embedding a standard single mode fiber (adapted from [57]); (**B**) Macrobending sensors for monitoring of abdominal movements (adapted from [57]); (**C**) Output of the OTDR sensors during the monitoring of abdominal movement (adapted from [57]).

Alemdar and coauthors developed a smart textile based on macrobending hetero-core fiber optic for respiratory movement analysis [69,70]. They embedded within a textile, a periodic macrobending hetero-core fiber optic that formed a periodic "U". They experimentally assessed the increase of sensitivity to strain with the number of loops (ranging from one to seven loops) and the influence of the loop length on sensitivity. Then, the most sensitive configuration was tested to monitor the abdominal movement on a healthy volunteer.

Lau et al. designed a simple microbend fiber optic sensor for respiratory monitoring during MRI procedures [71]. They assessed the sensor feasibility during MRI procedures on twenty healthy volunteers, showing an accuracy of ±2 breaths per minute in the measurement of frequency rate. They also developed a system to monitor both respiratory rate and heart rate with a similar approach [72]. They tested the system on 11 volunteers during an MRI procedure, showing an accuracy of two breaths per minute for respiratory rate monitoring, and two beats per minute for heart rate monitoring.

Recently, a smart textile based on the use of a periodic fiber optic microbend sensor has been proposed for heartbeat and respiratory monitoring [73]. The authors integrated a section of multimode optical fiber sandwiched between parallel strips acting as a microbender onto an elastic substrate. Basically, the system monitors respiration and heartbeat by detecting the vibration caused by these actions. The microbending results in light loss, which is detected by a photodetector. After a set of in vitro experiments, they tested the textile on healthy volunteers. The measured heartbeat was compared to the one measured by a commercial device, the measured respiratory rate with the number of cycles manually counted in one minute.

The main characteristics of the abovementioned smart textiles and their applications are reported in Table 1.

4. DISCUSSION

Smart textiles are used in a number of industrial fields such as civil engineering, transport and medicine, amongst others. Regarding the application in medicine and healthcare, they allow the continuous monitoring of physiological parameters of great importance (e.g., respiratory activity and heartbeat).

The monitoring of these parameters can be performed by several approaches, and different devices are commercially available. The main advantage to integrate FOSs into smart textiles is the possibility to develop MR-compatible systems. The number of MR scanners, the request for high field devices and for MR procedures in several medical branches (e.g., cardiology, surgery, orthopedics and neurology) are increasing worldwide. This trend reflects the growing need for MR-compatible systems able to monitor physical parameters inside the scanner, in order to provide real-time feedback about the status of the patient.

As a consequence, smart textiles based on the use of FOSs can be considered a potential new market niche in the field of healthcare monitoring. In this scenario is important to have a small size and lightweight system, and more in general a system allowing a normal range of motion and preserve the patient's comfort. As a consequence, intensity-based fiber optic sensors seem to be indicated more than FBG ones that require the use of the interrogator. Moreover, also in homecare application the privacy and confidentiality of the patient must be respected. Solutions, such as enclosed rooms without traffic or others present and data transmission over secure lines can help to fulfill this requirement. Lastly, the problems of washability and, in some cases, sterilizability of FOS-based smart textiles are not well addressed. This feature must be considered because it can be a cost-effective option and can avoid the use of these systems for mono-patient or disposable applications. Therefore, future studies should address the robustness of these textiles against stress and washing cycles or their sterilizability The advantage is that FOSs are intrinsically safer than conventional sensors as they do not have any electrical connection with the patient and may be used at medium-term for monitoring different physiological parameters in order to replace standard sensors during MRI procedures. In this scenario it is also important to take into account the issue related to the connection of fibers. As reported by Kinet et al. [74] different solutions can be employed (e.g., commercially available standard connectors for optical fibers and ad hoc designed connectors). For the applications in the MRI environment, it is crucial to employ ad hoc designed, MR-compatible connectors.

Moreover, they can answer to the increasing demand for wearable systems for continuous monitoring of physiological parameters. A comfortable and wearable measuring system may significantly improve the quality of life of patients.

The main concerns related to the extensive use of smart textiles based on FOSs and their spread in commercially available products are related to the necessity to improve their stability due to light coupling (in particular in homecare long-term applications), and the abovementioned issue related to the connection of fibers (in particular in application during MRI procedures). The continuous improvement of the quality of the key optical components and their cost reduction, as well as the good metrological properties of these sensors motivates the increasing interest in this topic.

A recent review published by Quandt et al. in 2014 [18], described body-monitoring systems based on optical fibers and their application in the field of physiological monitoring. In this review, we devoted our attention especially to smart textiles based on FOSs for monitoring mechanical parameters of physiological interest. Also briefly reviewed are the working principles of the FOSs most used in these applications. Lastly, we focused on the metrological properties and on the performances of the systems reported in literature. In particular the most employed FOSs are based on FBG technology and on the modulation of intensity (in particular on macrobending). They are largely used for heartbeat and respiratory monitoring. These two kinds of sensors show complementary advantages and disadvantages in terms of sensitivity to movement and costs. FBG sensors are able to measure small strains thanks to their high sensitivity; on the other hand, in order to have good performances, interrogation must take place through an expensive device. The interrogation of intensity based FOSs is very simple and involves low cost and compact components, and their integration into textile fabrics may be very straightforward; on the other hand they are less sensitive than FBG sensors and are usually employed to monitor large movements (e.g., abdominal movements).

The use of smart textiles based on FOS in respiratory and heartbeat monitoring is just in its beginning stages, but their good metrological performances and the possibility to monitor patients in real time during MRI procedures motivates to continue research effort devoted to this topic.

AUTHOR CONTRIBUTIONS

Carlo Massaroni, Paola Saccomandi and Emiliano Schena contributed equally to this manuscript; Carlo Massaroni, Paola Saccomandi and Emiliano Schena wrote the article.

CONFLICTS OF INTEREST

The authors declare no conflict of interest.

REFERENCES

1. Lymberis, A.; Olsson, S. Intelligent Biomedical Clothing for Personal Health and Disease Management: State of the Art and Future Vision. Telemed. J. E-Health 2003, 9, 379–386.

2. Lobodzinski, S.S.; Laks, M.M. New devices for very long-term ECG monitoring. Cardiol. J. 2012, 19, 210–214.

3. Lobodzinski, S.S. ECG patch monitors for assessment of cardiac rhythm abnormalities. Prog. Cardiovas. Dis. 2013, 56, 224–229.

4. Inan, O.T.; Etemadi, M.; Wiard, R.M.; Giovangrandi, L.; Kovacs, G.T. Robust ballistocardiogram acquisition for home monitoring. Physiol. Meas. 2009, 30, 169–185.

5. Zhang, X.; Tao, X. Smart textiles: Passive smart. Text. Asia 2001, 45–49.

6. Zhang, X.; Tao, X. Smart textiles: Active smart. Text. Asia 2001, 49–52.

7. Krehel, M.; Wolf, M.; Boesel, L.F.; Rossi, R.M.; Bona, G.L.; Scherer, L.G. Development of a luminous textile for reflective pulse oximetry measurements. Biomed. Opt. Express 2014, 5, 2537–2547.

8. Li, H.N.; Li, D.S.; Song, G.B. Recent applications of fiber optic sensors to health monitoring in civil engineering. Eng. Struct. 2014, 26, 1647–1657.

9. Da Silva, A.F.; Gonçalves, A.F.; de Almeida Ferreira, L.A.; Araújo, F.M.M.; Mendes, P.M.; Correia, J.H. PVC smart sensing foil for advanced strain measurements. IEEE Sens. J. 2010, 10, 1149–1155.

10. Silvestri, S.; Schena, E. Optical-fiber measurement systems for medical applications. In Optelectronics: Devices and Applications; Predeep, P., Ed.; InTech: Rijeka, Croatia, 2011.

11. Lee, B. Review of the present status of optical fiber sensors. Opt. Fiber Technol. 2003, 9, 57–79.

12. Marazuela, M.; Moreno-Bondi, M. Fiber-optic biosensors—An overview. Anal. Bioanal. Chem. 2002, 372, 664–682.

13. Grattan, K.T.V.; Meggitt, B.T. Optical Fiber Sensor Technology, Vol. 2—Devices and Technology; Chapman & Hall: London, USA, 1998.

14. Udd, E. Fiber Optic Sensors, 1st ed.; John Wiley & Sons, Inc.: Hoboken, New Jersey, 2006.

15. Wang, X.D.; Wolfbeis, O.S. Fiber-optic chemical sensors and biosensors (2008–2012). Anal. Chem. 2012, 85, 487–508.

16. Krebber, K.; Liehr, S.; Witt, J. Smart technical textiles based on fibre optic sensors. In Proceedings of the 22nd International Conference on Optical Fiber Sensors, Beijing, China, 14 October 2012.

17. Ghosh, S.; Amidei, C.; Furrow, K. Development of a sensor-embedded flexible textile structure for apparel or large area applications. Indian J. Fibre Text. 2005, 30, 42–48.

18. Quandt, B.M.; Scherer, L.J.; Boesel, L.F.; Wolf, M.; Bona, G.L.; Rossi, R.M. Body-monitoring and health supervision by means of optical fiber-based sensing systems in medical textiles. Adv. Healthc. Mater. 2015, 4, 330–355.

19. Hill, K.O.; Fujii, Y.; Johnson, D.C.; Kawasaki, B.S. Photosensitivity in optical fiber waveguides: Application to reflection filter fabrication. Appl. Phys. Lett. 1978, 32, 647–649.

20. Meltz, G.; Morey, W.W.; Glenn, W.H. Formation of Bragg gratings in optical fibers by a transverse holographic method. Opt. Lett. 1989, 14, 823–825.

21. Rao, Y.J.; Webb, D.J.; Jackson, D.A.; Zhang, L.; Bennion, I. Optical in-fiber bragg grating sensor systems for medical applications. J. Biomed. Opt. 1998, 3, 38–44.

22. Gurkan, D.; Starodubov, D.; Yuan, X. Monitoring of the heartbeat sounds using an optical fiber bragg grating sensor. In Proceedings of the 4th IEEE Conference Sensors 2005, Irvine, CA, USA, 30 October–3 November 2005.

23. Haines, D.E. Determinants of lesion size during radiofrequency catheter ablation: The role of electrode-tissue contact pressure and duration of energy delivery. J. Cariovasc. Electrophysiol. 1991, 2, 509–515.

24. Yokoyama, K.; Nakagawa, H.; Shah, D.C.; Lambert, H.; Leo, G.; Aeby, N.; Ikeda, A.; Pitha, J.V.; Sharma, T.; Lazzara, R.; et al. Novel contact force sensor incorporated in irrigated radiofrequency ablation catheter predicts lesion size and incidence of steam pop and thrombus. Circ. Arrhythm. Electrophysiol. 2008, 1, 354–362.

25. Hao, J.Z.; Tan, K.M.; Tjin, S.C.; Liaw, C.Y.; Roy Chaudhuri, P.; Cuo, X.; Lu, Cu. Design of a foot-pressure monitoring transducer for diabetic patients based on FBG sensors. In Proceedings of the LEOS, the 16th Annual Meeting of the IEEE, Tucson, AZ, USA, 27–30 October 2003; pp. 23–24.

26. Al-Fakih, E.; Abu Osman, N.A.; Mahamd Adikan, F.R. The use of fiber Bragg grating sensors in biomechanics and rehabilitation applications: The state-of-the-art and ongoing research topics. Sensors 2012, 12, 12890–12926.

27. Tjin, S.C.; Tan, Y.K.; Yow, M.; Lam, Y.Z.; Hao, J. Recording compliance of dental splint use in obstructive sleep apnoea patients by force and temperature modeling. Med. Biol. Eng. Comput. 2001, 39, 182–184.

28. Saccomandi, P.; Schena, E.; Caponero, M.A.; di Matteo, F.M.; Martino, M.; Pandolfi, M.; Silvestri, S. Theoretical analysis and experimental evaluation of laser-induced interstitial thermotherapy in ex vivo porcine pancreas. IEEE Trans. Biomed. Eng. 2012, 59, 2958–2964.

29. Di Matteo, F.; Martino, M.; Rea, R.; Pandolfi, M.; Panzera, F.; Stigliano, E.; Schena, E.; Saccomandi, P.; Silvestri, S.; Pacella, C.M.; et al. US-guided application of Nd: YAG laser in porcine pancreatic tissue: An ex vivo study and numerical simulation. Gastrointest. Endosc. 2013, 78, 750–755.

30. Saccomandi, P.; Schena, E.; Giurazza, F.; del Vescovo, R.; Caponero, M.A.; Mortato, L.; Panzera, F.; Cazzato, R.L.; Grasso, F.R.; di Matteo, F.M.; et al. Temperature monitoring and lesion volume estimation during double-applicator laser-induced thermotherapy in ex vivo swine pancreas: A preliminary study. Laser Med. Sci. 2014, 29, 607–614.

31. Saccomandi, P.; Schena, E.; di Matteo, F.M.; Pandolfi, M.; Martino, M.; Rea, R.; Silvestri, S. Laser Interstitial Thermotherapy for pancreatic tumor ablation: Theoretical model and experimental validation. In Proceedings of the 2011 Annual International Conference of the IEEE on Engineering in Medicine and Biology Society, EMBC, Boston, MA, USA, 30 August–3 September 2011; pp. 5585–5588.

32. Wehrle, G.; Nohama, P.; Kalinowski, H.J.; Torres, P.I.; Valente, L.C.G. A fibre optic Bragg grating strain sensor for monitoring ventilatory movements. Meas. Sci. Technol. 2001, 12, 805.

33. Allsop, T.; Miller, M.; Bennion, I.; Carroll, K.; Lloyd, G.; Webb, D.J. Application of long-period-grating sensors to respiratory plethysmography. J. Biomed. Opt. 2007, 12.

34. Cowie, B.M.; Webb, D.J.; Tam, B.; Slack, P.; Brett, P.N. Fibre Bragg grating sensors for distributive tactile sensing. Meas. Sci. Technol. 2007, 18.

35. Saccomandi, P.; Schena, E.; Oddo, C.M.; Zollo, L.; Silvestri, S.; Guglielmelli, E. Microfabricated Tactile Sensors for Biomedical Applications: A Review. Biosensors 2014, 4, 422–448.

36. Mishra, V.; Singh, N.; Tiwari, U.; Kapur, P. Fiber grating sensors in medicine: Current and emerging applications. Sens. Actuat A Phys. 2011, 167, 279–290.

37. Taffoni, F.; Formica, D.; Saccomandi, P.; Pino, G.D.; Schena, E. Optical fiber-based MR-compatible sensors for medical applications: An overview. Sensors 2013, 13, 14105–14120.

38. Watanabe, K.; Tajima, K.; Kubota, Y. Macrobending characteristics of a hetero-core splice fiber optic sensor for displacement and liquid detection. IEICE Trans. Electron. 2000, 83, 309–314.

39. Nishiyama, M.; Sasaki, H.; Watanabe, K. A deformation sensitive pad-structure embedded with hetero-core optic fiber sensors. Sens. Actuat A Phys. 2007, 136, 205–211.

40. Lekholm, A.; Lindstrom, L. Optoelectronic transducer for intravascular measurements of pressure variations. Med. Biol. Eng. 1969, 7, 333–335.

41. Wald, A.; Post, K.; Ransohoff, J.; Hass, W.; Epstein, F. A new technique for monitoring epidural intracranial pressure. Med. Instrum. 1977, 11, 352–354. [PubMed]

42. Piper, I.; Barnes, A.; Smith, D.; Dunn, L. The Camino intracranial pressure sensor: Is it optimal technology? An internal audit with a review of current intracranial pressure monitoring technologies. Neurosurgery 2001, 49, 1158–1165. [PubMed]

43. Wolthuis, R.A.; Mitchell, G.L.; Hartl, J.C.; Saaski, E. Development of a dual function sensor system for measuring pressure and temperature at the tip of a single optical fiber. IEEE Trans. Biomed. Eng. 1993, 40, 298–302.

44. Babchenko, A.; Khanokh, B.; Shomer, Y.; Nitzan, M. Fiber optic sensor for the measurement of respiratory chest circumference changes. J. Biomed. Opt. 1999, 4, 224–229.

45. Maletras, F.X.; Augousti, A.T.; Mason, J. Construction and calibration of a new design of Fiber Optic Respiratory Plethysmograph (FORP). In Proceedings of the SPIE 4444, Optomechanical Design and Engineering 2001, San Diego, CA, USA, 29 July 2001; p. 285.

46. Nishyama, M.; Miyamoto, M.; Watanabe, K. Respiration and body movement analysis during sleep in bed using hetero-core fiber optic pressure sensors without constraint to human activity. J. Biomed. Opt. 2011, 16.

47. Schena, E.; Saccomandi, P.; Silvestri, S. A high sensitivity fiber optic macro-bend based gas flow rate transducer for low flow rates: Theory, working principle, and static calibration. Rev. Sci. Instrum. 2013, 84.

48. Saccomandi, P.; Schena, E.; Silvestri, S. A novel target-type low pressure drop bidirectional optoelectronic air flow sensor for infant artificial ventilation: Measurement principle and static calibration. Rev. Sci. Instrum. 2011, 82.

49. Husdi, I.R.; Nakamura, K.; Ueha, S. Sensing characteristics of plastic optical fibres measured by optical time-domain refelectometry. Meas. Sci. Technol. 2004, 15, 1553–1559.

50. Augousti, A.T.; Maletras, F.X.; Mason, J. Improved fibre optic respiratory monitoring using a figure-of-eight coil. Physiol. Meas. 2005, 26, 585–590.

51. Augousti, A.T.; Raza, A.; Graves, M. Design and characterisation of a fibre optic respiratory plethysmograph (FORP). In Proceedings of the SPIE 1996, Biomedical Sensing, Imaging, and Tracking Technologies I, San Jose, CA, USA, 27 January 1996; pp. 250–257.

52. Narbonneau, F.; Kinet, D.; Paquet, B.; Depré, A.; de Jonckheere, J.; Logier, R.; Zinke, J.; Witt, J.; Krebber, K. Smart textile embedding optical fibre sensors for healthcare monitoring during MRI. Adv. Sci. Tech. 2008, 60, 134–143.

53. Grillet, A.; Kinet, D.; Witt, J.; Schukar, M.; Krebber, K.; Pirotte, F.; Depré, A. Optical fiber sensors embedded into medical textiles for healthcare monitoring. IEEE Sens. J. 2008, 8, 1215–1222.

54. De Jonckheere, J.; Jeanne, M.; Narbonneau, F.; Witt, J.; Paquet, B.; Kinet, D.; Kreber, K.; Logier, R. OFSETH: A breathing motions monitoring system for patients under MRI. In Proceedings of the Annual International Conference of the IEEE Engineering in Medicine and Biology Society, Minneapolis, USA, 2-6 September 2009; pp. 1016–1019.

55. Narbonneau, F.; D'Angelo, L.T.; Witt, J.; Paquet, B.; Kinet, D.; Kreber, K.; Logier, R. FBG-based smart textiles for continuous monitoring of respiratory movements for healthcare applications. In Proceedings of the 12th IEEE International Conference on e-Health Networking Applications and Services (Healthcom), Lyon, France, 1–3 July 2010; pp. 277–282.

56. Witt, J.; Narbonneau, F.; Schukar, M.; Krebber, K.; de Jonckheere, J.; Jeanne, M.; Kinet, D.; Paquet, B.; Depré, A.; D'Angelo, L.T.; et al. Medical textiles with embedded fiber optic sensors for monitoring of respiratory movement. IEEE Sens. J. 2012, 12, 246–254.

57. Krebber, K. Smart Technical Textiles Based on Fiber Optic Sensors. In Current Developments in Optical Fiber Technology; InTech: Rijeka, Croatia, 2013; pp. 319–344.

58. Silva, A.F.; Carmo, J.P.; Mendes, P.M.; Correia, J.H. Simultaneous cardiac and respiratory frequency measurement based on a single fiber Bragg grating sensor. Meas. Sci. Tech. 2011, 22.

59. Carmo, J.P.; da Silva, A.M. F.; Rocha, R.P.; Correia, J.H. Application of fiber Bragg gratings to wearable garments. IEEE Sens. J. 2012, 12, 261–266.

60. Dziuda, Ĺ.; Skibniewski, F.W.; Krej, M.; Baran, P.M. Fiber Bragg grating-based sensor for monitoring respiration and heart activity during magnetic resonance imaging examinations. J. Biomed. Opt. 2013, 18.

61. Dziuda, Ĺ.; Skibniewski, F.W.; Krej, M.; Lewandowski, J. Monitoring respiration and cardiac activity using fiber Bragg grating-based sensor. IEEE Trans. Biomed. Eng. 2012, 59, 1934–1942.

62. Dziuda, L.; Krej, M.; Skibniewski, F.W. Fiber Bragg grating strain sensor incorporated to monitor patient vital signs during MRI. IEEE Sens. J. 2013, 13, 4986–4991.

63. Krej, M.; Dziuda, L.; Skibniewski, F. A method of detecting heartbeat locations in the ballistocardiographic signal from the fiber-optic vital signs sensor. IEEE J. Biomed. Health Inform. 2015.

64. Allsop, T.; Bhamber, R.; Lloyd, G.; Miller, M.R.; Dixon, A.; Webb, D.; Castañón, J.D. A.; Bennion, I. Respiratory function monitoring using a real-time three-dimensional fiber-optic shaping sensing scheme based upon fiber Bragg gratings. J. Biomed. Opt. 2012, 17.

65. Li, H.; Yang, H.; Li, E.; Liu, Z.; Wei, K. Wearable sensors in intelligent clothing for measuring human body temperature based on optical fiber Bragg grating. Opt. Express 2012, 20, 11740–11752.

66. Yoo, W.J.; Jang, K.W.; Seo, J.K.; Heo, J.Y.; Moon, J.S.; Park, J.Y.; Lee, B.S. Development of respiration sensors using plastic optical fiber for respiratory monitoring inside MRI system. J. Opt. Soc. Korea 2010, 14, 235–239.

67. Krehel, M.; Rossi, R.M.; Bona, G.L.; Scherer, L.J. Characterization of flexible copolymer optical fibers for force sensing applications. Sensors 2013, 13, 11956–11968.

68. Krehel, M.; Schmid, M.; Rossi, R.M.; Boesel, L.F.; Bona, G.L.; Scherer, L.J. An optical fibre-based sensor for respiratory monitoring. Sensors 2014, 14, 13088–13101.

69. Alemdar, K.; Likoglu, S.; Fidanboylu, K.; Toker, O. A novel periodic macrobending hetero-core fiber optic sensor embedded in textiles. In Proceedings of the SPIE 9062, Smart Sensor Phenomena, Technology, Networks, and Systems Integration, San Diego, CA, USA, 8 March 2014.

70. Alemdar, K.; Likoglu, S.; Fidanboylu, K.; Toker, O. A novel periodic macrobending hetero-core fiber optic sensor embedded in textiles. In Proceedings of the 8th International Conference on Electrical and Electronics Engineering (ELECO), Bursa, Turkey, 28–30 November 2013; pp. 467–471.

71. Lau, D.; Chen, Z.; Teo, J.T.; Ng, S.H.; Rumpel, H.; Lian, Y.; Yang, H.; Kei, P.L. Intensity-Modulated Microbend Fiber Optic Sensor for Respiratory Monitoring and Gating During MRI. IEEE Trans. Biomed. Eng. 2013, 60, 2655–2662.

72. Chen, Z.; Lau, D.; Teo, J.T.; Ng, S.H.; Yang, X.; Kei, P.L. Simultaneous measurement of breathing rate and heart rate using a microbend multimode fiber optic sensor. J. Biomed. Opt. 2014, 19.

73. Yang, X.; Chen, Z.; Chia, S.; Lam, H.; Ng, S.; Teo, J.; Wu, R. Textile Fiber Optic Micro Bend Sensor Used for Heartbeat and Respiration Monitoring. IEEE Sens. J. 2015, 15, 757–761.

74. Kinet, D.; Mégret, P.; Goossen, K.W.; Qiu, L.; Heider, D.; Caucheteur, C. Fiber Bragg Grating Sensors toward Structural Health Monitoring in Composite Materials: Challenges and Solutions. Sensors 2014, 14, 7394–7419.

Index